DYNAMIC
PATTERNS
IN
COMPLEX
SYSTEMS

Proceedings of a Conference, sponsored by Florida Atlantic University, held in honor of Hermann Haken on the occasion of his 60th birthday.

DYNAMIC PATTERNS IN COMPLEX SYSTEMS

Edited by

J.A.S. Kelso
A.J. Mandell
M.F. Shlesinger

World Scientific
Singapore • New Jersey • Hong Kong

Published by

World Scientific Publishing Co Pte Ltd
5 Toh Tuck Link, Singapore 596224
USA office: 27 Warren Street, Suite 401-402, Hackensack, NJ 07601
UK office: 57 Shelton Street, Covent Garden, London WC2H 9HE

Library of Congress Cataloging-in-Publication Data
Dynamic patterns in complex systems
edited by J.A.S. Kelso, A.J. Mandell, and M.F. Shlesinger,
p. cm.
Proceedings of a conference, sponsored by Florida Atlantic University, held in honor of Hermann Haken on the occasion of his 60th birthday, 21–23 October, 1987, Fort Lauderdale, USA.
ISBN-13 9789971504854 -- ISBN-10 9971504855
1. Pattern perception -- Congresses. 2. Chaotic behavior in systems -- Congresses. 3. Neurobiology -- Congresses. I. Kelso, J .A. Scott. II. Mandell, Arnold J. III. Shlesinger, Michael F. IV. Haken, H. V. Florida Atlantic University.
Q327 .D96 1988
006,4--dc19 88-5568

British Library Cataloguing-in-Publication Data
A catalogue record for this book is available from the British Library.

Acknowledgements

Major funding for this conference was provided by:

National Institute of Mental Health
Office of Naval Research (Physics Division)

We wish to thank Drs. S. H. Koslow and R. Schoenfeld of the Neurosciences Research Branch (NIMH) for their support and encouragement–indeed, for making this cross-disciplinary exchange possible. We would like also to express our gratitude for additional financial assistance from:

Berlex Laboratories, Inc.
Hoechst Roussel Pharmaceuticals, Inc.
Sandoz Research Institute
The Upjohn Company

Our thanks go out to those individuals who presented posters at the conference and who, by doing so, enriched the intellectual atmosphere considerably. The abstracts of their presentations are provided in the present volume so that interested readers may write for further information. Finally, we would like to thank Kay Lioy for her multiple contributions toward the smooth running of the conference itself, and for her assistance in preparing the present volume.

J.A.S. Kelso

A. J. Mandell

M. F. Shlesinger

Table of Contents

Introductory Remarks: Dynamic Patterns

J.A.S. Kelso
Center for Complex Systems
Florida Atlantic University
Boca Raton, Florida 33431 U.S.A.

This conference is the second, of which we hope will be many, to deal with the rapidly emerging field of pattern formation in complex systems. The first, held at NIH in April, 1986 under the title *Perspectives in Biological Dynamics and Theoretical Medicine* (Koslow, S.H., Mandell, A.J. & Shlesinger, M.F. (Eds.) *Annals of the New York Academy of Sciences,* 504, 1987) introduced practicing biologists, medical researchers and clinicians to the fascinating area of nonlinear dynamical systems, including chaos and its manifold implications and applications.

There is growing conviction in science that the processes of pattern formation and self-organization in complex physical, chemical and biochemical systems develop according to a few typical scenarios which are quite independent of the specific system studied. Of course, such convictions or suspicions make it all the more important that the behavior of each specific system be studied in detail, so that the extent to which general laws apply can be evaluated. That is one of the reasons why–at this conference–we have sought to blend theory (including computational models) with experimental data from a variety of different disciplines. Thus, not only are some of the top theorists in attendance here, but eminent experimentalists as well.

This conference was to be held in honor of our colleague, Professor H. Haken, Director of the Institute of Theoretical Physics and Synergetics, University of Stuttgart and Co-Director, with myself, of the Center for Complex Systems at Florida Atlantic University, who celebrates his 60th birthday this year. Every physicist knows of Haken's fundamental contributions to the theory of the laser. Many–scientists and laymen alike–have also heard of *synergetics*, an interdisciplinary approach–founded by Haken nearly 20 years ago– to the study of pattern formation in open, multicomponent systems. It is with great sadness that Professor Haken is unable to attend due to the recent loss of his wife. At a meeting on *Lasers and Synergetics* (*Springer Proceedings in Physics, 19,* 1987) held in Stuttgart in July, 1987, Frau Haken was present though, by all reports, not well. No one, however, was quite aware of just how sick she was. In a letter that I received last week, Hermann thanks those who have come to honor him and wishes our conference much success. This conference follows the interdisciplinary spirit of those that Hermann Haken has run at Schloss Elmau in the Bavarian Alps for some years. I wish he was here to participate in this one on U.S. soil, but it is not to be. We are delighted, nevertheless, that Professor Haken was able to provide us with a paper for inclusion in this volume.

In a few brief comments, I wish to explain something about the subject matter of this conference–*Dynamic Patterns in Complex Systems*–the title of which my co-organizers, Arnold Mandell and Michael Shlesinger, gave me the leeway to select. And perhaps express some expectations of what we may hope to achieve by gathering together on problems of complexity. First–**Patterns.**

The world around us–if we exaggerate the details–can appear extraordinarily complicated. Faced with complexity, an orthodox tendency is to reduce or decompose the system to its smallest elements and study their properties. The knowledge gained about such units may be enough to understand the behavior of the whole system. A concern that's been expressed in many circles is that the more we focus on the material details of each individual system, the farther we remove ourselves from principles of organization–how the material stuff is put together to form patterns. The issue, however, is not really matter versus process, but to find an adequate level of description that will allow us to extract the essential aspects of the system under study. In this conference, the emphasis is less on the 'things' themselves e.g., fluid molecules, particles, chemical species, heart cells, neurons, organisms, computer components, but more on understanding the (dynamic) *behavioral patterns* that are produced by the cooperation among the individual components. We seek to discover the rules or laws through which such patterns are realized. An emphasis on pattern means continual abstraction, winnowing out unnecessary detail in order to isolate the essential features of the pattern and their dynamics (stability and change).

When we find the rules of pattern formation, there seems little doubt that we can use them, for example in forecasting, diagnosing and treating diseases, in designing machines (e.g., for pattern recognition) and so on. The technological and health related applications that follow from a deeper understanding of pattern formation in natural and artificial systems, are, in my view, immense. But the basic scientific intuition is that there's only a small amount of information that's important in the world out there, and that we, as scientists have to find out what it is. This is a different kind of reductionism-to a small set of universal laws for pattern formation. Such laws lie, perhaps, in the balance between stability and fluctuations or maybe, as some will argue here-on the borderline between ordered and chaotic behavior.

Second, the patterns of interest are *dynamic patterns evolving in time* that arise from interactions among a system's many components in an entirely self-organized fashion. Patterns arise spontaneously when, for example, environmental conditions are changed. There is no ordering influence from the outside, and no homunculus sitting inside. Examples are manifold e.g., Bénard convection, concentration patterns in chemical reactions, coherent lightfield of the laser, the vortices and eddies that occur on the road to turbulence in fluids, morphogenesis, and so forth.

Granted nature is complicated, why bother using the redundant term 'complex'? We use two meanings of the word 'complex' (see Figure 1): The first, obvious meaning is that most systems of interest have very many elements. Hence the term *system complexity*. Emphasizing pattern formation is equivalent to bridging the gap between what one element does e.g., a single water molecule, a single neuron, a single heart cell and what many of them can do when they cooperate together in a collective manner. One of the reasons nonlinear dynamics is so exciting is that methods have been found to: a) reduce system complexity (i.e., many degrees of freedom) to only a few degrees of freedom (what Otto Rössler once referred to as a 'miracle'); and b) the consequent low-dimensional dynamics, in turn, have been found to display *complex behavioral patterns*–hence our second meaning of the word 'complex'–including multistability, multiple patterns, flexibility, deterministic chaos.

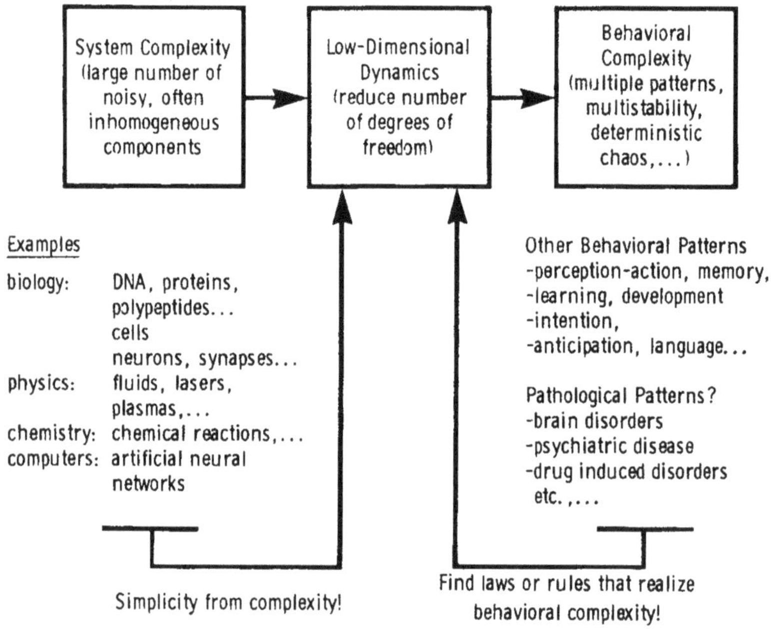

Fig. 1. Features of complex systems and strategies for understanding them.

Understanding the compression of system complexity into something simpler and manageable is a daunting enough scientific task. From our point of view, however, the mapping of complex patterns of behavior onto lower- dimensional dynamic laws (the *reverse* step in Figure 1) may be even more crucial as we consider those 'typically biological' features of physical systems–their ability to perceive, coordinate actions, memorize, learn, anticipate–including intentional and linguistic aspects. In addition, many aberrant behavioral patterns demand understanding. Understanding such behavior will involve rather more than a mere application of ideas from nonequilibrium physics. Even so, it's worth keeping in mind the fact that even simple dynamical systems can display creative and generative behavior. Forgive me, if I plug our own work in this regard, which shows that a wide variety of temporal patterns of human movement behavior can be accounted for by a simple (though nonlinear and stochastic) law, that in turn can be derived from a cooperative coupling between the components. Shaw's dripping faucet is often cited as another simple, yet eternally creative, and even aberrant dynamical system.

Finally, what can an interdisciplinary conference on *Dynamic Patterns in Complex Systems* hope to achieve? First, there is a pragmatic goal–that of sharing and exposing a variety of disciplines to the tools and techniques of nonlinear dynamics, many of these methods developed by the participants here. Such new measurement techniques and ways of thinking, may aid in identifying the key aspects of dynamic patterns in complex systems and their dynamics (stability, change). A second, global aim is to enhance our *understanding* of pattern, especially in living systems. It is not always appreciated that understanding depends on the language used. Here we emphasize the abstract level of the (collective) variables that characterize patterns, at every level of description. Third and relatedly, we need a *language* upon which understanding should be founded. A language that emphasizes dynamical processes at all scales of observation, regardless of the particular system studied. A language through which scientists from different disciplines may communicate. It is my hope that through the foresight of our sponsors, such a language may eventually emerge from meetings like this one.

THE EPISTEMOLOGY OF COMPLEXITY

Robert Rosen
Department of Physiology and Biophysics
Dalhousie University
Halifax, Nova Scotia, Canada

ABSTRACT

A system is called simple if it can be completely syntacticized. A formal system is completely syntactic if its inferential rules can be formalized in the usual sense, and hence generated by a properly programmed mathematical machine (e.g. a Turing machine). A natural or material system is completely syntactic, and hence simple, if its causal structure is such that all of its models are simulable or formalizable in this sense. A system which is not simple is called complex. We discuss a few ramifications of the possibility that there exist complex systems, especially in the material realm.

I. INTRODUCTION

In recent years, the concept of complexity has become elevated to the status of an explanatory principle. Thus, we find it widely said that a given system possesses some favorite attribute (e.g. can learn, evolve, adapt, reproduce, ...) because it is complex. To characterize a system as complex, then, is supposed to convey some important information about it. Conversely, a system which is not complex (i.e., is simple) can do none of these things, simply by virtue of that fact alone. There has thus grown up a highly seductive circle of ideas, which has shifted attention away from individual system attributes to that supreme attribute which simultaneously encompasses and explains them: complexity.

The employment of complexity as an explanatory principle seems to have become fashionable with the work of von Neumann (e.g. 1951). He seemed to believe that all systems could be ranked along a well-ordered scale of complexity, i.e., that complexity was a kind of taxonomic index of systems, which could itself be somehow measured or computed. In addition, he suggested that there was a critical value, or threshold, of complexity, below which systems behaved in their conventional mechanical modes, and below which their complexity could only decrease with time. Above the threshold, however, new realms of behavior were manifested, and such systems could in fact grow more complex with time; they could self-organize, learn, reproduce and evolve. As far as I am aware, von Neumann himself was never very specific about what complexity actually meant, but his accolytes have seemed to favor the view that complexity is synonymous with complication; the more elementary units of which a system is composed, and the more elaborate the modes of interconnection between them, the more complex the system must be.

Clearly, such notions of complexity are of central interest to biology, if for no other reason than that the characteristics of complexity which we noted above are also essentially those of an organism. Indeed, the "threshold of complexity" serves to separate material systems into two subclasses, very much like a class of "inorganic systems" and a class of "organisms". Making a system somehow cross this threshold is thus the creation of life. Clearly, such ideas were very much in the air in the 1940's and 1950's.

I myself began to think about complexity somewhat later, around 1967 or 1968. From the outset, I doubted that one could usefully equate complexity with mere complication. In particular, I did not think of complexity as an observable, as something which could be measured; or if it could, certainly not in terms of something like a mere number. It seemed to me that complication is an artifact of how a system is observed; we can make a system appear more complicated or less complicated by a mere change of variables or coordinate transformation. Complexity, on the other hand, should be something inherent; something invariant, and hence independent of a choice of coordinates. I suggested instead (cf. Rosen, 1969, 1977) that complexity in a material system should rather be measured by the class of inequivalent models which could be made of the system. This turned out to be a fundamentally different way of looking at complexity; one which ties complexity to something about the system rather than presupposing some subjectively preferred way of looking at the system.

It has turned out that several lines of argument, initially quite unrelated, have important implications for this view of complexity. For one thing, complexity in this sense inherently involves formal concepts; it depends on relationships between mathematical objects (namely, those which can be models of a given material system). For another thing, if there are material systems which are complex in this sense, then there is something fundamentally inadequate about the ideas of atomism and reductionism which have dominated modern science for the past three centuries. For these ideas imply, among other things, that every material system has a kind of overarching universal model, from which all others can be constructed by purely formal means, and which is in fact of a specific mathematical type.

In short, it has turned out that systems which are complex in our sense require a whole new epistemology; one which, I believe, will have some fundamental and radical repercussions for the way we view both the material world and the mathematical language in which we describe that world. It is the purpose of the present brief note to sketch some of the features of this epistemology, and to draw attention to a few of its ramifications.

II. NATURAL SYSTEMS AND FORMAL SYSTEMS

Before turning to our discussion of the adjective "complex", we must first say something about the noun it modifies, namely, "system".

As we have argued at great length elsewhere (e.g. Rosen, 1985a), the general theory of systems encompasses two distinct realms. On the one hand, there is the theory of natural systems; the theory of phenomena. Roughly speaking, natural systems are what populate the external world; they are what we can measure and perceive, and comprise the subject-matter of the (natural) sciences and technologies. On the other hand, there is a parallel theory of the internal world, which comprises the theory of formal systems. The world of formal systems is a world of symbols, language and ideas. We shall be especially concerned with those formal systems which may, in the broadest sense, be called mathematics.

Although these two broad realms of system theory are obviously quite different in their subject-matter, there are many profound parallels between them. For instance, in each realm, we find a fundamental notion of entailment. In the world of natural systems, entailment means causality; causality is precisely the name we give to the regularity of the succession of events in the formal world. Without causal entailment, this succession of events would be entirely capricious or whimsical; science, and perhaps even sanity, would be impossible as we now know them, and would be replaced by unentailed historical chronicles. On the other hand, in the realm of formal systems, entailment means inference or implication. Implication plays therefore an analogous role in the formal world to that played by causality in the material; it replaces whimsy by tangible order and structure.

We may note that, in both the natural and the formal worlds, entailment serves to generate chronicles. In general, it is our hope that any chronicle, however historical it may initially appear, in fact arises from some underlying entailment; which serves precisely to explain it. But this is another story.

We can in fact establish much more than a mere analogy between causal entailment in the natural world and inferential entailment in the formal one. Namely, it is the fundamental assertion of Natural Law (cf. Rosen, 1985a) that we can establish exact congruences between causal order in the natural world, and inferential order in appropriate formal systems of propositions about that world. Such a congruence between a natural system and a formal one establishes a fundamental relation between them, which we have called a modelling relation. Modelling relations will provide a basic cornerstone for our treatment of complexity (cf. Rosen, op cit).

Briefly, let us suppose we are given a natural system N, i.e., a circumscribed class of events or phenomena in the external world. Thus, N might be a particle or a set of particles or a machine or an organism or a population of any or all of the above. As we have noted, if the succession of events in N is not entirely arbitrary, then it is governed by relations or constraints we term causal. On the other hand, propositions about N, or more precisely about what happens in N, belong to some formal system (call it F), in which relations between propositions are governed by an inferential structure in F. A posited association between an event in N, and a proposition about that event in F, serves to encode the event into the proposition; conversely, association between a proposition in F and an event in N serves to decode the proposition into the event. We thus have a situation which can be schematized as shown in Figure 1.

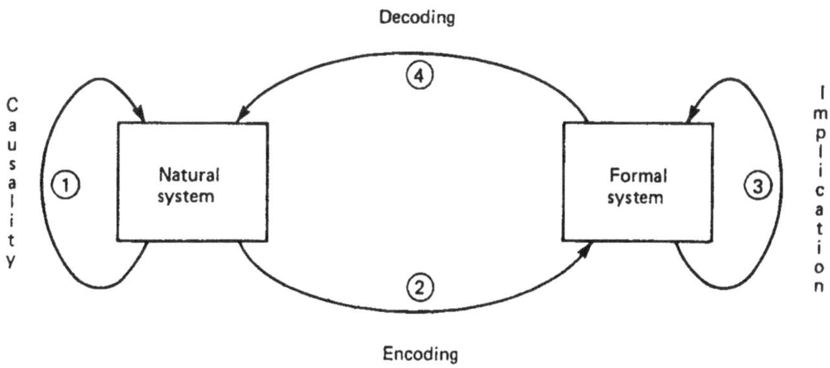

Fig. 1

We shall say that a modelling relation exists between N and F if the diagram of Figure 1 commutes, in the precise sense that

$$\text{arrow } 1 \quad = \quad \text{arrows } 2 + 3 + 4.$$

That is, we always get the same answer, whether we simply sit and observe the unfolding sequence of events in the natural system N, or whether we (a) initially encode events of N into F, (b) utilize the inferential structure of F to derive new propositions (theorems) within F, and (c) decode these theorems back into events in N (i.e., turn our theorems in F into predictions about N). The commutativity of Figure 1 is what we mean by the establishment of a congruence between a causal structure in the natural world and an inferential structure in some appropriate formalism.

It should be noted explicitly that the encoding and decoding arrows in Figure 1, which play the fundamental role in relating the natural and formal worlds, themselves belong to neither world. In particular, they are not constrained either by causality or by inference alone; they are unentailed in an important sense. This fact turns out to have some profound epistemological consequences, but as they are somewhat aside from our present point, we shall not pursue this further here.

There are many important things which flow from the simple diagram of Figure 1. However, we shall confine ourselves to one aspect; namely, given a natural system N, what kind of formalisms F can sit in the right-hand box of that Figure? In other words, what kinds of formal systems F (via suitable encoding and decoding arrows) can be models of N (or, stated another way, can be realized by N)? Above all, what is the internal structure of the class of all such models F? As hinted earlier, this will provide the habitat for our treatment of complexity. However, before turning directly to this aspect, it will be helpful to develop first some fuller understanding of causal relations in natural systems, and of inferential relations in formal ones. This will be taken up in the following two sections. We will then return to a discussion of the main topic: complexity.

III. INFERENCE IN FORMAL SYSTEMS: SYNTAX AND SEMANTICS

The charm of logical inference in formal systems is that its inferential structure passes truth hereditarily from premise (i.e., something in the domain of an inferential rule) to conclusion (i.e., something in the range of the rule).

In logical systems, "truth" can be of (at least) two different types. First, a proposition may be true by virtue of its relation to some referent external to the formalism; a referent which is somehow thereby encoded into the formalism. Or, a proposition may be true on internal grounds alone, true in effect by virtue of its form, independent of any external referent whatever. Truths of this second kind may be called syntactic truths; truths of the first kind may accordingly, for want of a better term, be called semantic.

In general, any formalism or language will contain truths of both types. Indeed, the whole purpose of our natural languages is tied up with external referents. Nevertheless, semantic truths are contingent on something outside, and hence something unentailed. Consequently, semantic truths seem somehow less necessary, and more subjective, than purely syntactic truths. Within mathematics itself, especially over the past century, the view has grown up that, in a sense, the more syntax in a formalism, and the less semantics, the "better" that formalism must be. Hence, the best formalisms are those which consist of syntactic truth alone. Such formalisms are, in many ways, the formal equivalent of closed systems; totally free of every external referent, in which every proposition which is true is so by virtue of its form alone.

The great importance of such purely syntactic systems arises from the formalist program within mathematics itself. This program was initially advocated by David Hilbert, as a way of coping with the paradoxes which had arisen in the early days of Set Theory, and which had grown to threaten much of classical mathematics itself. It was Hilbert's position that every formal system could be reformulated in such a way that every one of its truths could be recaptured by purely syntactic means; i.e., it could be formalized.

In more detail: If we think of a proposition as being built out of symbols in themselves meaningless, then syntactic rules should themselves become translated into rules for the ("spatial") manipulation of these symbols. That is, syntactic rules of inference involve only the number and relative positions (configurations) of these symbols. It was the Formalist belief, or hope, that such syntactic rules sufficed for every formal purpose, and thus that mathematics itself could be freed from every external referent. There was, of course, a price to be paid; namely, that mathematics in the Formalist sense becomes only a kind of game, played according to meaningless rules, for manipulating meaningless sequences of symbols. In such a game, the mathematician becomes only a symbol processor, or word processor, an executor of the rules for symbol manipulation. In return, the benefit gained would be freedom from inconsistency or paradox.

It was in this context that the idea of the syntactic engine, or the mathematical machine, became prominent. These machines (e.g. the Turing machines) comprise the ultimate syntactic processors. In the Formalist sense, such machines are the ultimate mathematicians as well.

These ideas are of basic importance to us because, since they pertain to all formal systems, they pertain *a fortiori* to those formal systems F which can be models of a natural system N. It will turn out that, if such models can themselves be formalized (i.e., freed from every external referent) then this imposes some profound limitations on the causal structure of N itself. To express this, however, we need to develop first the idea of simulation in the syntactical context of mathematical machines.

Let us for definiteness fix attention on the syntactical engines called Turing machines. The first thing to notice about any Turing machine is that it is a composite structure; it consists of a tape (software) and a reading head (hardware). All of the inferential structure, all of the entailment in the system, is concentrated in the hardware; there is no entailment at all in the software. Thus, whatever is written on a tape is itself unentailed, in two senses. First, it is unentailed as a sequence; there is no reason for the inscription of one sequence rather than another. Second, even if the sequence is given, there is no entailment between the elements of the sequence; no way of entailing the nth letter of the sequence from its predecessors. Indeed, a word inscribed on a tape constitutes the supreme example of pure unentailed chronicle.

Let us now introduce a little terminology. Suppose that $A = \{a_1, a_2, \ldots, a_n\}$ is a finite set of alphabet symbols. Then the set of all words which can be written with these symbols (i.e., all finite sequences of symbols) fills out a new set $A^{\#}$. This set $A^{\#}$ is itself endowed with a rudimentary algebraic structure; it has an associative multiplication (concatenation) with unit (the empty word e); it is the free monoid generated by the alphabet A. Any such word $w \in A^{\#}$ inscribed on a tape constitutes software. The hardware, the reading head, syntactically associates with appropriate input words w a uniquely defined output word. Stated more precisely, the hardware of a Turing machine induces a (partial) mapping

$$f : A^{\#} \to A^{\#}.$$

If we write

$$v = f(w)$$

wherever meaningful, then f is an inferential rule on $A^{\#}$, which is by definition of a purely syntactic character. Indeed, we may write

$$w \Rightarrow v;$$

or more correctly,

$$f \Rightarrow (w \Rightarrow v).$$

Let us notice that the implication signs $\Rightarrow$ in this last expression are being used in two quite different senses. As we shall see later, the first of them is analogous to saying that the inferential rule f is the "efficient cause" of the effect v; the second is analogous to saying that the given word w is the "material cause" of the effect v.

The next thing to notice is that we can "trick" the inferential rule f, embodied in the hardware of the Turing machine, to give us other inferential rules on $A^{\#}$. For instance, let us pick some fixed word $w_o \in A^{\#}$, and let us define, for each word w which renders the expressions meaningful,

$$g(w) = f(ww_o).$$

Then we can say that the prefix word w_o constitutes a program, enabling the original inferential rule f to induce a new rule g. In a certain sense, the program w_o serves to describe the new rule g to the old rule f. In the process, we create a referent for the words of $A^{\#}$.

It is significant to notice that any inferential rule f trivially induces itself in this fashion, but the program for this purpose is always the empty word. Thus, a Turing machine's description of itself to itself is always trivial. This is itself a significant observation, which we again cannot pause to explore here.

The concept of a program allows us to write down a new kind of implication, namely,

$$w_o \Rightarrow v.$$

The implication sign in this expression involves a relation distinct from either of those we characterized earlier; as we shall see, it is analogous to saying that the program w_o is the "formal cause" of the effect v.

It is also evident that the idea of programming which we have sketched above carries with it the idea of simulation of one machine by another, which is the notion we are after. Specifically, if we had been given a Turing machine which computes the mapping g, then we could simulate that machine on the entirely different one which computes the mapping f, by means of appropriate programming.

The crux of simulation and programming, for our purposes, is that program is software. Software, as we have noted, is itself unentailed. In particular, the hardware which computes f, the inferential structure which it imposes on $A^{\#}$, has no relation to, or interpretation in, the software which it processes.

This in turn means that the simulation relation between two machines is entirely different from the modelling relation embodied in Figure 1. Modelling involves, as we have seen, the establishment of a congruence between two entailment structures. Simulation, on the other hand, never involves such a congruence.

Before proceeding further, it may be helpful to pause at this point and recapitulate. We began, as we recall, with a consideration of syntactic vs. semantic truth. We recounted the Formalist program, which asserts that every formal system could be rewritten in such a way that all of its truths become purely syntactic, i.e., isolated from all referents external to the system. In such a formalization, inferential rules are expressed entirely as symbol processing or word processing. At this point, we introduced the idea of the mathematical machine, in its guise as the ultimate symbol processor. Along with the machine came the notion of program, and hence of simulation. In fact, in a certain sense, simulation is precisely what mathematical machines do.

If the Formalist program were correct, all of mathematics would be expressible in terms of programs for such machines; all mathematical truths could then be recaptured on purely syntactic grounds, and mathematics itself freed of all referents. Stated otherwise, Formalism asserts that formalization is always possible, and that it loses no truth. Thus the central question is: Is the Formalist program correct? The answer, as has been known for a long time, is "no". This is the import of the famous theorem due to Gödel (1931), who showed that Number Theory could not be syntacticized in the Formalist sense. No matter how we may program a mathematical machine, there are always propositions in Number Theory which are true, but cannot be entailed by (i.e., "caused" by) the inferential structure of the machine; such propositions can never be effects.

Such syntactically unentailed but true propositions are hence of a fundamentally semantic character, in the sense that they depend essentially on what Number Theory is about.

What Gödel showed, then, is that already in the realm of purely formal systems, there are inferential structures which cannot be simulated by purely syntactic means. We shall now argue that this situation has precise parallels in the world of natural systems too. The trick to exploring these parallels, as we indicated earlier, is to look at the models of a natural system, and to explore whether these models are subject to simulation or not. But before turning to this, we need to say a few further words about the entailment structure of natural systems, and their models.

IV. ENTAILMENT IN NATURAL SYSTEMS: CAUSALITY

The idea of the ancient Greek atomists, such as Democritus, that all of material reality could be analyzed down to structureless particles pushed around by impressed forces, is obviously a very syntactic idea. There is an obvious parallel between a structureless particle and a meaningless alphabet symbol; between a configured set of particles and a word; and the rearrangement of configurations by forces and the processing of words. Correspondingly, there is an equally obvious parallel between formalization and reductionistic analysis.

Atomistic ideas were resuscitated, and indeed made all-encompassing, by Newton in his Principia. So successful was this endeavor that, despite all the upheavals which have occurred in Physics in the past century, the basic epistemological framework enunciated by Newton has not changed. Far from it; the current preoccupation with elementary particles, and unified fields, are simply the most modern manifestation of this framework.

As has often been noted, the success of Newtonian mechanism in physics, and hence of the epistemology which underlies it, thenceforth completely changed the very character of the scientific enterprise itself. Prior to Newton, science had been what Aristotle said it was almost two millenia earlier. To Aristotle, science was precisely the study of causation, the study of causal entailment. Causation, in turn, was the inquiry into "the why of things"; the Aristotelian

Categories of Causation were in turn developed to express all the different ways of answering the question "why?". These Categories (material, formal, efficient, and final causation) were regarded as distinct, independent, inequivalent, each one necessary, and all together sufficient, for the understanding of the phenomena of material nature.

The Aristotelian approach to science thus begins with a partition between effects (i.e., between those things in the world for which we seek explanation) and their causes (the things in terms of which explanations are expressed). Newtonian science, on the other hand, starts with a partition between system and environment, which is a very different matter. The Newtonian partition is reflected exactly in the mathematical structure which Newton himself invented to reflect his view of material reality. Specifically, what is system is represented in terms of states (phases) and trajectories of state transitions; what is environment, on the other hand, is the seat of external forces, and also (equally fundamental but less clearly recognized) the seat of initial conditions; initial configurations, and initial velocities. There remains, of course, causality in the Newtonian scheme, but it is now a poor thing when compared to its Aristotelian predecessor. Specifically, causality is now restricted entirely to the state-transition sequence in the system. Stated otherwise: The only thing in the Newtonian scheme which can be an effect, the only thing which can be entailed, is the state of a system at an instant. In turn, this state is itself entailed by "causes" themselves unentailed; by initial conditions, and external forces, arising from an environment in which nothing can be an effect. From the very outset, then, the Newtonian scheme exempts the environment from the province of causality.

The partition of a natural system from its environment is roughly analogous to the distinction we drew earlier between hardware (system) and software (environment). In each case, all entailment is embodied in the hardware, and none is left for software.

Since these matters are crucial, we will discuss them in more detail. In the Newtonian epistemology, which as we have seen is of an ultimately syntactic character, every attribute of a system, every observable, can be dissected down to the configurations and phases of constituent structureless particles. Thus, every system attribute is already entailed if this state, or phase, is known. State, on the other hand, is entailed from previous state, according to the rule

$$(x_o, v_o) \rightarrow (x_o + v_o dt, v_o + F_o dt) \quad (1)$$

or, in another form,

$$x(t) = \int dt \int F(\tau)d\tau. \quad (2)$$

Here, F is the force, which mathematically depends on, or is a function of, (a) initial conditions, and (b) a number of parameters, which must also be independently specified, and which represent the couplings between system phases and environmental forces. This force F represents a description of the environment, just as phase represents a description of system.

Thus if, as we have indicated, we think of configuration x(t) of system as effect, we can elucidate its causes (i.e., the only possible answers to the question: "Why is the system in configuration x(t) at instant t?" as follows:

1. Because the system was initially in phase (x_o, v_o) (material causation).
2. Because of the specific parameter values which couple system to environment (formal causation).
3. Because of the forces impressed on the system, mathematically expressed by the operator $\int dt \int F(\ldots)\, d\tau$ (efficient causation).

We may note here that there is simply no room left in this picture for the category of Final Causation. Speaking roughly, this is because final causation must run in the "opposite direction" from the other causal categories. That is, in material, formal, and efficient causation, the effect is entailed by these causes. In final causation, on the other hand, must be something which is itself entailed by the effect. For this reason, final causation is often described in terms of the future acting on the present, and this is simply not allowed in the Newtonian picture. As we have seen, the Newtonian epistemology allows no entailment beyond the state-transition sequence, and these are fully accounted for by the three "forward" causal categories alone.

We can in any case notice that the bulk of the Newtonian epistemology consists of hypotheses about the non-causal environment, rather than about the system itself. We can, of course, always try to reach into this acausal environment, and pull further systems out of it. In this way, we can put some causality into this environment, in the form of states and state transitions for these other systems. Moreover, we can put these systems into interaction, by making the forces experienced by one of them depend, at least in part, upon the states of the others. In effect, we create thereby a new, bigger system, of which the original system is in some sense a subsystem. But, of course, this does not change the underlying epistemology at all; it merely shifts the focus of attention from the original system to the bigger one which contains it.

V. SIMPLE SYSTEMS

Pragmatically, the most important classes of systems are defined by negation. Thus, a nonlinear system is one which is not linear; an open system is one which is not closed. In each case, it has proved simpler to characterize a restricted class of system (e.g. linear, closed, etc.) and define the class of real interest as the complement of that restrictive class. Likewise, it is easier to define and circumscribe a class of systems which are simple; a complex system will then be any system which is not simple.

Heuristically, we shall consider a natural system to be simple if the Newtonian epistemology, which we have outlined in the preceding section, applies to it. We have already seen how little entailment there actually is in such a system. Furthermore, we have seen that what entailment remains in such a system is of a very special form. This, together with the syntactic character of atomistic analysis of Newtonian systems, actually serves to determine almost completely the form of a mathematical system which can be a model of a simple natural system.

This form is, of course, a very familiar one and it can be briefly described. For instance, since all of the entailment is to be embodied in trajectories of state transitions, there must above all be a state set, in which these trajectories lie. The trajectories themselves are generated by infinitesimal differential operations like (1), or by equivalent integral operators like (2). These operators embody the forces which generate the state transitions, and they contain the specific parameters which serve to couple the system to the forces which move it.

Thus, each of the three categories of causation which are allowed in the Newtonian epistemology are embodied by, or encoded into, a specific mathematical image which is fixed once and for all. The state set embodies material causation; the parameter set embodies formal causation; the integral or differential operation which encodes into equations of motion embodies efficient causation. There is no other form which a model of such a system can take. Our only real free choice in this situation is the choice of state set, and it is only in this single degree of freedom that such a system allows apparently distinct models to be generated.

However, the syntactic character of atomistic analysis takes away most of even this freedom. For it is implicit in the Newtonian mode of analysis that for every natural system there exists a "biggest", universal model, from which all others can be generated by purely formal means. This universal model must therefore contain every shred of material reality embodied in the system. In mathematical terms, it is like a free structure in the set of all models of the system; every other model is essentially a quotient of this free structure. Above all, it is a model <u>of the same type</u> as all the others, and above all, it is purely syntactic in nature. It is the ultimate goal of Newtonian science to construct this free universal model, which makes every other model superfluous and redundant.

All of this can be summed up, and systematized, in the following way: <u>A natural system is simple if every model of it can be simulated on a mathematical machine</u>. And the upshot of the entire Newtonian epistemology can then be distilled into a very simple assertion about the material world: <u>Every natural system is simple</u>.

These hypotheses about the external world are of course very attractive in several ways. Not the least of their charms is that they serve in a real sense to democratize science; they render every problem in natural science a purely technical one; to be solved by mere labor, governed by explicit and well-defined rules. Nowhere has this hypothesis been more attractive than in biology, where the "machine metaphor" originally propounded by Descartes has never been more powerful than it is today. It was Descartes who took the breathtaking step of identifying life with mechanism (= simple system), several generations before Newton wrote his <u>Principia</u>, without any clear notion of either life or of mechanism, largely on the basis of superficial similarities between organisms and various kinds of clockwork automata. And, as has been discussed elsewhere (cf. Rosen, 1985b) it has been this "machine metaphor", together with the apparent rarity of organisms in the class of all material systems (i.e., the postulated non-genericity of organisms) which has above all estranged biology from the world of theoretical physics and its quest for "universal laws". We shall have more to say about this in a moment.

VI. COMPLEX NATURAL SYSTEMS

We have already remarked that the Formalist program, which is the formal analog of the Newtonian hypothesis that all natural systems are simple, fails. The question is whether it likewise fails in the world of natural systems. As we have indicated, one strategy for investigating this problem is to translate it back into the formal world through the concept of a model; all we have to do is produce a natural system which has a non-simulable model.

Motivated initially by quite different considerations, I had earlier introduced (cf. Rosen, 1979) a class of system models which turned out to be non-simulable on a variety of grounds. We shall briefly describe them here, as an illustration of what a complex system might look like. The point of departure for the earlier discussion was an attempt to establish a relation between the technical vocabularies of physics (comprising terms like energy, force, work and the like) and biology (comprising "informational" terms like coding, translation, recognition, specificity and the like), using the Rosetta stone of <u>stability</u>. In turn, that discussion was based on an observation of Higgins (1967), that the stability of a system of first-order rate equations

$$dx_i/dt = f_i(x_1, \ldots, x_n), \tag{3}$$

which as we have seen is already a very Newtonian kind of thing, can be discussed entirely in terms of the Jacobian coefficients

$$\begin{aligned} u_{ij}(x_1, \ldots, x_n) &= \partial/\partial x_j(dx_i/dt) \\ &= \partial f_i/\partial x_j. \end{aligned}$$

In fact, the important property of these coefficients u_{ij} is only its <u>sign</u>; whether they are positive or negative at a state.

As Higgins pointed out, each of these n^2 quantities u_{ij} could be given an <u>informational</u> interpretation. Namely, if u_{ij} is positive in a state, it means that an increase in x_j also increases the rate at which x_i grows (or, alternatively, that a decrease in x_j also decreases that rate). Thus we can rigorously speak of x_j as an <u>activator</u> of x_i in that state. Conversely, if u_{ij} is negative, then an increase in x_j decreases the rate at which x_i grows, and hence we can speak of x_j as an <u>inhibitor</u> of x_i in that state. Thus, specification of the n^2 quantities u_{ij} gives us an alternate description of the system (3), which I called an activation-inhibition pattern.

The above ideas can clearly be iterated further. For instance, we can form the n^3 quantities

$$u_{ijk}(x_1, \ldots, x_n) = \partial/\partial x_k(u_{ij}) = \partial/\partial x_k(\partial/\partial x_j(dx_i/dt)).$$

These, too, have an informational interpretation. Namely, if u_{ijk} is positive in a state, it means that x_k potentiates the effect of x_j on x_i; we can thus call x_k an agonist of x_j with respect to x_i. Conversely, if u_{ijk} is negative, then x_k attentuates the effect of x_j on x_i; thus x_k is an antagonist of x_j with respect to x_i. The n^3 functions u_{ijk} provide another description of the original system, which we may call an agonist-antagonist pattern.

We can obviously keep iterating this procedure, obtaining at each stage a new description of the original system with an obvious informational interpretation.

It will be noted that we began our discussion with a given set of rate equations (3), from which we produced the subsequent layers of informational descriptions. The question then immediately arises: If we were to start with, say, a given activation-inhibition pattern (i.e., a set of functions u_{ij}), could we recover an overarching set of rate equations, of the form (3), which generates precisely the given pattern?

The answer, of course, is "not in general". In order to pass backward from a given activation-inhibition pattern to a set of rate equations, the functions u_{ij} must satisfy an extremely strong property. Namely, we must require that each of the differential forms

$$\omega_i = \sum_{j=1}^{n} u_{ij} dx_j$$

be exact. For then we can write

$$\omega_i = df_i$$

for some function $f_i = f_i(x_1, \ldots, x_n)$, and then put

$$dx_i/dt = f_i.$$

If any of the forms ω_i are not exact, then there is no set of rate equations (3) which generates the given activation-inhibition pattern u_{ij}.

The same considerations hold at each iteration. At the level of agonist-antagonist patterns, for example, the condition of exactness translates into

$$u_{ijk} = u_{ikj}$$

for all indices i, j, k. In other words, the condition requires complete symmetry between agonism/antagonism and activation/inhibition. This is clearly no way to "process information".

Clearly, if we start from a set of rate equations like (3), the exactness conditions we have indicated must hold at every step of the iteration. All the descriptions collapse back to the original rate equations, from which they are generated by purely formal means (differentiation and integration allows us to pass from any one of them to any other). On the other hand, if exactness does not hold anywhere, then (a) all these different descriptions become independent of each other, and (b) there is no set of rate equations governing the system at all.

It is clear that a system of the kind we have described possesses a potentially infinite number of independent descriptions, each of which violates the Newtonian epistemology at every step. It is likewise clear that such systems cannot be simulated. Thus, such systems are complex in our sense.

Biology is replete with information-handling systems for which the u_{ij}, the u_{ijk}, ... are the basic things to observe. The central nervous system, the immune system, and the endocrine systems are obvious examples, but so too are cellular control mechanisms and ecosystems. What am I bet that, among these systems, we cannot find many which will violate the canons of Newtonian epistemology?

VII. SOME IMPLICATIONS

If there are truly complex systems in nature (and I believe that biology provides an abundance of examples) then the entire epistemology of the past three centuries of natural science is fundamentally inadequate. Adherence to it immediately puts biology completely out of reach.

Such considerations make it immediately clear why theoretical physics, the most advanced science of material nature, has had so inordinately little to say about organisms. It is not, as has often been suggested, that we have been insufficiently clever in bringing physical universals to bear, or that we lack data, or that living systems are as yet incompletely characterized. It is that contemporary physics is, and remains, a science of simple systems, and thus, our inherited conceptual tools are fundamentally inadequate.

The situation is analogous to the one in which physics found itself at the end of the last century; phenomena of atomic spectra, radioactivity, chemical bonding and the like simply could not be dealt with on the basis of the existing conceptual structure, and required a massive transmutation of physics itself. In my view, biology, and complexity in general, pose exactly such questions, except that the required transmutations will be even more radical.

The considerations developed above suggest that the natural world cannot be completely syntacticized, any more than the formal world can. Complexity in our sense means precisely that there is an essential semantic component to material reality which cannot be ignored. Stated another way: Purely syntactic models of material nature are too empoverished in entailment to mirror the causal sequences which can occur in nature, just as they are too empoverished to encompass all of natural language, or even of Number Theory.

The question can then be raised: If contemporary physics is indeed the science of simple systems or mechanisms, why has it been so successful in dealing with the material world? One answer, I believe, will shed some light on the relation between complexity and simplicity, may metaphorically be put as follows. In the early days of cartography, mapmakers were very successful in making planar maps of local areas. Indeed, it was found that a good map could be made of any local area. However, when one attempted to extend such maps to larger areas, it was discovered that (a) planar maps became increasingly inaccurate as the region mapped grew in extent, and (b) planar maps of remote local areas could not be combined into a single larger such map. Ultimately, of course, it was recognized that these difficulties arose because the surface to be mapped was not itself a plane; it was an object of a new topological type; it was a sphere. One could then regard the discrepancy between a planar map and the surface it imaged as measuring a new or emergent property, namely, curvature. The new spherical surface could be regarded in a sense as a limit of all its local planar maps, but there could be no single universal or overarching planar map to which this limit could converge.

This is rather the situation we face in dealing with complexity. If we think of Newtonian models, in the broad sense in which we have been using this term, as the analogs of planar maps, and complexity as the analog of the surface of a sphere, we have a metaphor for the structure of the set of models of a complex system. Accordingly, a complex system can be modelled locally, and for a circumscribed time, by an approximating simple system. Sooner or later, however, the discrepancy between the behavior of the complex system and its local simple approximation will become too great; the model must then be abandoned and replaced by another. The growing discrepancy between the two is the essential root of emergent novelty in biology, and of error in other areas (cf. Rosen, 1985a); this is the analog of the "emergence" of curvature in our cartographic analog.

As with cartography, physics is successful to the extent that it restricts itself to situations in which discrepancies are small, i.e., to analogs of the local regions which can successfully be imaged by planar maps. But biology does not allow this luxury.

Complexity has many other ramifications, which we cannot enter into in this short space. For one thing, the absence of a single overarching ("free") model means that we cannot fix a "set of states" which can be fixed once and for all. There is thus no fixed system-environment boundary which can be established a priori. More generally, the causal sequences which can occur in complex systems are very different from those to which we have become accustomed; in complexity, the categories of causation become themselves mutable, and can interchange in novel ways. Above all, we can introduce ideas of final causation in a perfectly rigorous and non-mystical way, and open up a true theory of anticipation (cf. Rosen, 1985a).

We appear to lose a lot in giving up the accepted epistemology of three centuries, but I believe we obtain ample compensation for doing so. Above all, in dealing with questions of natural science, we must let nature tell us what to do. In human law, it may be true that precedent is everything. But in Natural Law, this had better not be the case.

REFERENCES

1. Gödel K. 1931. "Über formal unentscheidbare Sätze der Principia Mathematica und verwandte Systeme". Monatshefte der Mathematik und Physik 38, 173-198.

2. Higgins J. 1967. "Oscillating Reactions". J Ind & Eng Chem 5, 18-62.

3. von Neumann J. 1951. "The General and Logical Theory of Automata". In: Cerebral Mechanisms in Behavior (Jeffress LA, ed.), 1-49. John Wiley, New York.

4. Rosen R. 1962. "Church's Thesis and its Relation to the Concept of Realizability in Biology and Physics". Bull Math Biophys 24, 375-393.

5. Rosen R. 1969. "Hierarchical Systems in Biology". In: Hierarchical Structures (Wilson D, Whyte LL, eds.). Elsevier, New York.

6. Rosen R. 1977. "Complexity as a System Property". Int J Gen Systems 3, 227-232.

7. Rosen R. 1979. "Some Comments on Activation and Inhibition". Bull Math Biophys 41, 427-445.

8. Rosen R. 1985a. Anticipatory Systems. Pergamon Press, London.

9. Rosen R. 1985b. "Information and Cause". In: Information Processing in Biological Systems. (Mintz SL, Perlmutter A, eds.). Plenum Press, New York.

Part I

Pattern Formation and Pattern Recognition in Multistable Systems

PATTERN RECOGNITION AND PATTERN FORMATION AS DUAL PROCESSES

by H. Haken and A. Fuchs
Institut für Theoretische Physik und Synergetik
Universität Stuttgart
Pfaffenwaldring 57
D-7000 Stuttgart 80
West Germany

For quite a while pattern recognition and associative memory realized by machines were problems treated by the computer sciences. In digital as well as in syntactic pattern recognition the main task was to develop algorithms for pattern recognition and associative memory to be implemented on computers [1]. Some time ago in the frame of the synergetics enterprise one of us suggested to consider pattern recognition as a process dual to pattern formation, which can be performed by dynamic systems [2]. A typical example for pattern formation by a dynamic system is the formation of rolls or hexagons in fluids - the well-known Bénard problem [3], schematically shown in fig. 1. It is easy to show that pattern formation by a fluid can be interpreted as the result of a parallel computation where the individual volume elements of the fluid act as individual computers. The simplest dynamical system is one in which the equations of motion can be derived from a potential [4]. Indeed, the formation of rolls in a fluid obeys such potential conditions. Each minimum of the potential corresponds to a specific pattern.

Fig. 1: If a fluid layer is heated from below coherent convective motion sets in above a certain threshold.

An important step towards the realization of pattern recognition by a dynamic system was done by Hopfield [5] who introduced the spin-glas model. A generalization of this concept was given more recently by Banzhaf [6].

In this paper we shall discuss a different kind of approach [7], namely one which exhibits the analogy between pattern recognition and pattern formation quite explicitly. To this end we decompose a pattern into specific features which are lumped together into a vector, $\mathbf{v}$. If there are several prototype patterns, we will distinguish them by an index k so that we introduce the vectors of the prototype patterns, $\mathbf{v}^{(k)}$. The individual features, which are the components of the vectors $\mathbf{v}^{(k)}$, can be chosen for instance as different degrees of grey in a picture, but there are many other possibilities also. For instance, the pattern can be evaluated in a system of orthonormal functions, where $\mathbf{v}^{(k)}$ would then be given by the expansion coefficients of the pattern. We further introduce the transposed vector $\bar{\mathbf{v}}^{(k)}$ and assume that $\mathbf{v}^{(k)}$ is normalized according to

$$\bar{\mathbf{v}}^{(k)} \mathbf{v}^{(k)} = 1 \quad . \tag{1}$$

The prototype vectors need not be orthogonal. We rather introduce the adjoint vectors $\mathbf{v}^{(k)+}$ with the property

$$\mathbf{v}^{(k)+} \mathbf{v}^{(k')} = \delta_{k,k'} \, . \tag{2}$$

For what follows we will construct the adjoint vectors as superpositions of the original vectors

$$\mathbf{v}^{(k)+} = \sum_{k'} a_{kk'} \mathbf{v}^{(k')} \, . \tag{3}$$

Note that the dimension of the space we are constructing is determined by the number of features N, whereas the number of prototype patterns may be $M \leq N$.

We assume that the vectors $\mathbf{v}^{(k)}$ are linearly independent. We then construct a potential function according to

$$V = -\frac{1}{2}\,\mathbf{q}^{+}\sum_{k=1}^{M}\lambda\,\mathbf{v}^{(k)}\cdot\mathbf{v}^{(k)+}\,\mathbf{q}$$

$$+\frac{1}{4}\,B\sum_{k\neq k'}^{M}(\mathbf{v}^{(k)+}\,\mathbf{q})^{2}\,(\mathbf{v}^{(k')+}\,\mathbf{q})^{2} \tag{4}$$

$$+\frac{1}{4}\,C\left\{\sum_{k=1}^{M}(\mathbf{v}^{(k)+}\,\mathbf{q})^{2}\right\}^{2}$$

were we define $\mathbf{q}^{+}$ so that

$$d_k = (\mathbf{v}^{(k)+}\,\mathbf{q}) = (\mathbf{q}^{+}\,\mathbf{v}^{(k)}) \tag{5}$$

holds. A typical behaviour of such a potential function is shown in fig. 2, where the minima correspond to the prototype patterns $\mathbf{v}^{(k)}$.

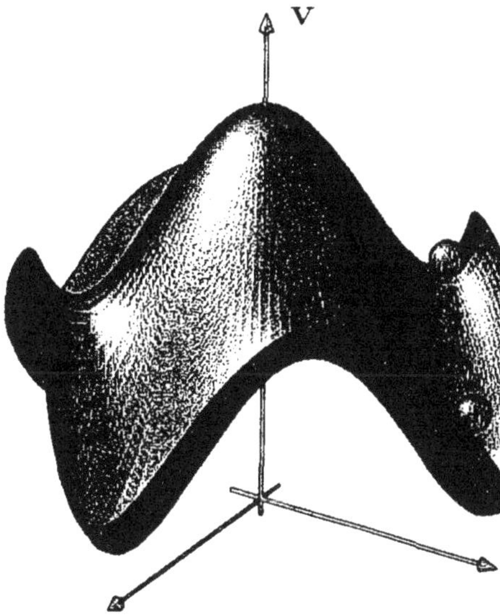

Fig. 2

When a pattern, $\mathbf{q}(0)$, will be offered to the system two tasks must be fulfilled:

1. If it is incomplete it must be completed so that it coincides with one of the prototype patterns, or that it is rejected altogether.

2. The machine must distinguish between different prototype patterns with which $\mathbf{q}$ is to be identified.

One may show by some little analysis that the minima of V are those in which $\mathbf{q}$ is parallel to one of the prototype patterns. Thus we have to seek for such a minimum of V whose $\mathbf{q}$ is closest to the initially given $\mathbf{q}(0)$. The search for minima of a potential function V is a widespread task which is tackled for instance by so-called evolution strategies [8], [9]. We decided to use the usual gradient strategy by analogy with classical mechanics (overdamped motion of a particle in the potential field V). Therefore, we find $\mathbf{q}$ as a time dependent function which obeys the equations

$$\dot{\mathbf{q}} = -\frac{\partial V}{\partial \mathbf{q}^+} + \mathbf{F} \tag{6a}$$

or

$$\dot{\mathbf{q}}^+ = -\frac{\partial V}{\partial \mathbf{q}} + \mathbf{F} \tag{6b}$$

respectively. Here we have added fluctuating forces $\mathbf{F}$ which may enhance the speed with which the minimum is achieved. For our practical calculations we chose

$$\mathbf{F} \equiv \mathbf{0} \tag{7}$$

however. One may show [10] that the only minima are those in which $\mathbf{q}$ is parallel to $\mathbf{v}^{(k)}$, i.e. there are no local minima which are irrelevant, provided $B > C$. The equations (6a) or (6b) describe a dynamic system in which the original state vector $\mathbf{q}(0)$ is pulled into a final state $\mathbf{v}^{(k)}$.

$$\mathbf{q}(0) \rightarrow \mathbf{q}(t) \rightarrow \mathbf{v}^{(k)} \tag{8}$$

Multiplying the equations (6a) by $\mathbf{v}^{(k)}$, we readily obtain the equations

$$\dot{d}_k = \lambda\, d_k - B \sum_{k \neq k'} d_{k'}^2\, d_k - C \sum_{k'} d_{k'}^2\, d_k \,. \tag{9}$$

The equations (9) are quite typical equations obtained in synergetics [4], [11] when we treat the formation of patterns by dynamic systems. In it d_k plays the role of an order parameter. Equations of type (9) appear for example in fluid dynamics or laser physics. Let us discuss the case of fluid dynamics more closely. To this end we interpret the vector $\mathbf{v}^{(k)}$ as a continuously varying quantity according to the relationship

$$\mathbf{v}^{(k)} \rightarrow N \exp\{ i(k_x\, x + k_y\, y)\} \tag{10}$$

These functions then describe standing plane waves which are just the projections of a role pattern of the fluid on to the horizontal plane. The scalar product must be replaced according to the rule

$$(\mathbf{v}^{(k)+}\, \mathbf{q}) \rightarrow \int dx\, dy\, \exp\{-i(k_x x + k_y\, y)\}\; q(x,y) \,. \tag{11}$$

As a detailed analysis shows, the equations (9) together with (10), (11) and (5) describe the formation of rolls in a fluid. All orientations of these rolls are possible, but in each specific experiment only one orientation is realized. In other words the fluid acquires a structure which is determined by initial conditions, for instance initially present fluctuations. But then it enters a stable state with a local minimum of the corresponding potential. Equation (9) can be interpreted in such a manner that competition among the various order parameters d_k takes place and only one survives.

Precisely the same mechanism holds in pattern recognition which is illustrated by means of an explicit example. Here 5 faces were photographed and given to the computer in a digitized form with 60 rows and 60 columns (fig.3). Then the vectors of the prototype patterns were constructed and the equations (6a) established.

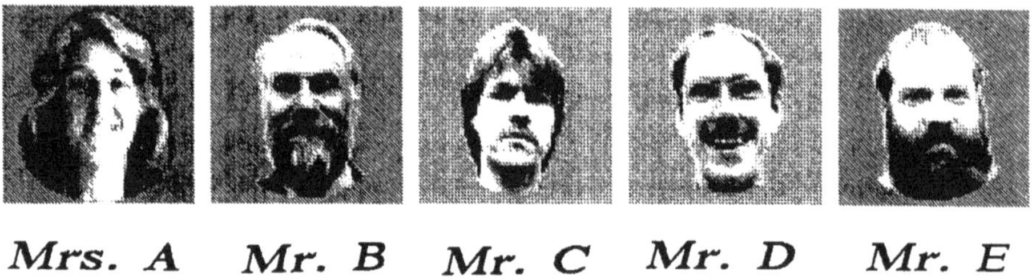

Fig. 3

They were solved on a digital computer and we immediately see the capability of the system to act as associative memory (fig.4).

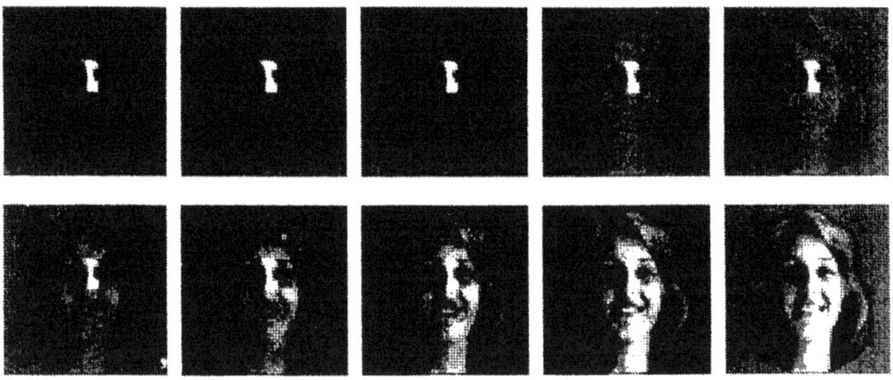

Fig. 4

In the same way it acts as a pattern recognizer when we supplement each pattern by an index and then eventually read off the index (fig. 5).

Fig. 5

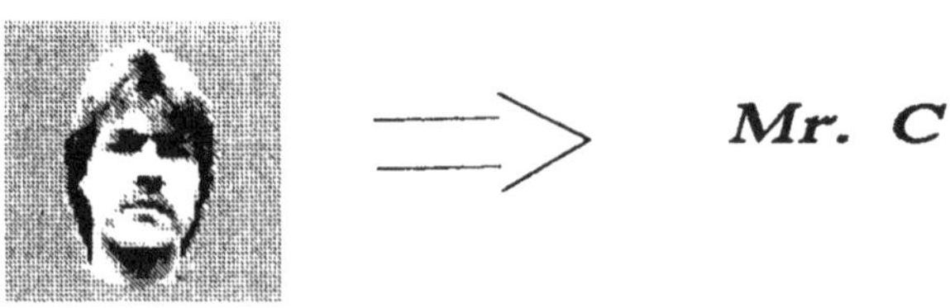

Our above results show quite clearly that there is a one-to-one correspondence between pattern recognition and pattern formation. This analogy opens new ways for the construction of computers for pattern recognition, namely computers which are dynamical systems from the outset. For instance more recently one of us [12] suggested a chemical parallel computer using these principles. The equations (6) and (9) can be interpreted from the point of view of the cognitive sciences. In (6) jointly with (4) the prototype patterns $\mathbf{v}^{(k)}$ are prescribed and the signal from the outer world given by $\mathbf{q}(0)$ is pulled into one of the prototype patterns. In this way the outer world is assimilated by the system. On the other hand the equations (9) can be interpreted as if the order parameters, d_k , are competing with each other, or in other words the internal state of the system is adapted according to the initially given value of the individual d_k . Though we solved the equation (6) on a conventional digital computer, the full power of our approach will become available on a parallel computer (i.e. neurocomputer). To this end we only need to write the equation (6) in the form

$$\dot{q}_j = \sum_l J_{jl}\, q_l - \sum_{l,m,n} T_{jlmn}\, q_l\, q_m q_n \tag{12}$$

where J_{jl} and T_{jlmn} describe the synaptic connections.

Our approach can be generalized to patterns which are shifted, rotated or deformed with respect to the prototype patterns $\mathbf{v}^{(k)}$. In this case we replace $\mathbf{v}$ by $v(\mathbf{x})$, $\mathbf{x} = (x,y)$ and $\mathbf{q}$ by $q(\mathbf{x})$. The scalar product $(\mathbf{v}^+\ \mathbf{q})$ must be replaced by the integral

$$\int v^+(\mathbf{x})\ q(\mathbf{x})\ d\mathbf{x} \tag{13}$$

We now replace

$$v^{(k)+}(\mathbf{x}) \quad \text{by} \quad w^{(k)+}(\mathbf{x}) = v^{(k)+}\ (\underline{\underline{\mathbf{S}}}\ \mathbf{x} - \xi) \tag{14}$$

where $\underline{\underline{\mathbf{S}}}$ denotes rotation and scaling, and ξ a shift.

The matrix $\underline{\mathbf{S}}$ which may be parametrized by the rotation angle φ and scaling factor s, and the displacement vector $\boldsymbol{\xi}$ will be formally denoted by a vector η. Thus V becomes a functional of $\mathbf{q}$ and of η. Pattern recognition amounts to a search for those $\mathbf{q}$ *and* η for which V becomes a minimum. First numerical tests as shown in fig. 6 for the case of scaling have substantiated the feasibility of this approach.

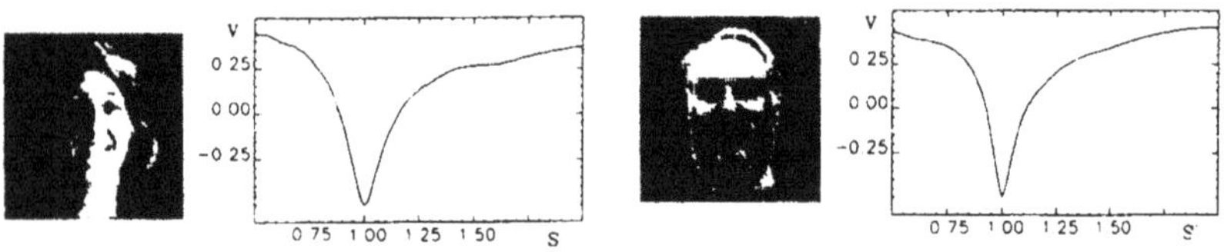

Fig. 6: Behaviour of the potential V due to a scaling of the pattern

Acknowledgement

We thank B. Hölle for valuable discussions, and K. Marx for the plot of the potential (fig. 2).

References:

[1] T. Kohonen, *Associative Memory*, Springer, Berlin (1977)
K. S. Fu, Ed., *Digital Pattern Recognition*, Springer, Berlin (1976)
K. S. Fu, Ed., *Syntactic Pattern Recognition, Applications*, Springer, Berlin (1977)

[2] H. Haken in *Pattern Formation by Dynamic Systems and Pattern Recognition*, H. Haken, Ed., Springer Series in Synergetics Vol. 5 (1979)

[3] S. Chadrasekhar, *Hydrodynamic and Hydromagnetic Stability*, Clarendon Press, Oxford (1961)

[4] H. Haken, *Synergetics, an Introduction*, 3rd. ed. (Springer 1983)

[5] J. J. Hopfield, *Proc. Natl. Acad. Sci. 79*, 2554 (1982)

[6] W. Banzhaf, to be published

[7] H. Haken in *Computational Systems, Natural and Artificial*, H. Haken, Ed., Springer, Berlin (1987)

[8] I. Rechenberg, Evolutionsstrategien, in *Med. Informatik und Statistik Vol. 8*, B. Schneider, U. Ranft, Ed., Springer, Berlin (1978)

[9] H. P. Schwefel, *Numerical Optimization of Computer Models*, Wiley, Chichester (1981)

[10] H. Haken, to be published

[11] H. Haken, *Advanced Synergetics*, 2nd. corr. printing, Springer (1987)

[12] H. Haken, *Journal de Physique Chimique*, in press

Holovision: a Semantic Information Processor for Visual Perception

H. SHIMIZU, Y. YAMAGUCHI AND K. SATOH

Faculty of Pharmaceutical Sciences, University of Tokyo
Hongo, Tokyo 113, Japan

ABSTRACT

The self-organization of semantic information, namely, how a figure emerges from pieces of elementary information with semantic separation, is studied in the case of visual perception. The figure is regarded as a kind of relational order self-organized by cooperative activities of holons, synergetic unit processors with many selectable internal modes. Holons are able to behave as local rule-generators by cooperative actions, a property that makes autonomic information processing with flexibility and creativity possible. Holovision, a semantic parallel processor for visual perception with three levels is composed of limit cycle oscillators. Holovision has memory that stores abstract information as pattern concepts. As a result, it has the ability to recognize pattern independent of size, position, and orientation and free from rigid similarity in shape. Some applications of holovision are shown to demonstrate the flexible and creative ability of the memory.

INTRODUCTION--VISUAL PERCEPTION

Information processing in the brain is characterized by parallel processing in a self-organized neural network. Two kinds of neuronal connections exist, static and dynamic. The static connections are concerned with learning processes while the dynamic connections are related to the perception of an instantaneous feature of the external world.

Connectionists [1] treat psychological processes using models with parallel distributed processors. This is a meaningful approach to understand neural networks--an area that has been treated traditionally from rather mathematical and technological standpoint as is also the case in classical network model. Connectionist models, usually employ static connections among the unit processors in quasi equilibrium. One important issues of their works is the back propagation method proposed with the Boltzmann machine [2] as a computational theory on a hierarchical network.

Malsburg [3] pointed out that cognitive processors are related by synaptic modulations acting on a short time scale due to synchronization of unstable neuronal activities with temporal disorders. In fact, modifications of a network due only to the classical Hebbian mechanism are insufficient to explain perception that occurs within about 0.1 second. The importance of network dynamics has been pointed out by many. For example, a limit cycle was noticed by Hopfield [4] though it occurs as an exceptional case in his model.

As studied by Cowan and Ermentrout [5] in their work on hallucination, a kind of dissipative structure--self-organized in a neural network of the visual field--might be perceived as a visual pattern. From a dynamical point of view, parallel processing by distributed synergetic elements is crucial. The idea of a synergetic computer, working in a state far from equilibrium was proposed by Haken [6]. In the same spirit, an autonomic processing system for visual perception called "holovision", was proposed by the authors [7,8] using oscillators as synergetic elements. Information is self-organized in holovision by entrainment among relevant oscillations. Based on the knowledge of neuronal networks in the brain, we [9] have shown that entrainment of periodic changes in neuronal activities explains important features of information processing at a psychological level.

In addition to information processing, holovision has an important property that is indispensable for perception and cognition. That is, self-organization of *semantic* information occurs when input signals are interpreted in terms of concepts stored in the long-term memory. As expressed in Kant's philosophy [10] it is impossible for man to know an object itself. It is only possible to perceive its perceptual image (expressed by neuronal activities in brain, as we would say). Although perception reflects detailed features of an input signal from the object, only a part of the signal is perceived as "figure" and the remaining part is compressed as "background". The "figure" part comprises semantic information expressed by the image. Semantic separation between figure and background may be regarded as "the breaking of semantic symmetry of information".

Normally, recognition of an object is linked to perception, and meaning is given to the image in the process of recognition. As pointed out by Wittgenstein [11], recognition may be viewed as a kind of interpretation of an object, (or more precisely, an image,) in terms of concepts stored in the memory as semantic items with articulated meanings. In other words, the meaning of the image is found when the image is related with one of the items. Thus, the self-organization of the image may be regarded as the self-organization of semantic information from elementary information, normally given as an assembly of meaningless signals picked up by a sensory organ. For the sake of simplicity, we call such meaningless signals, elementary signals.

Grossberg [12] suggested that the functional unit of cognitive coding is an adaptive resonance, or amplification of neural activities in a hierarchical neural network system. Perception is regarded as a matching process between the afferent data and the efferent expectancies based on a process he terms adaptive resonance. Hierarchical levels were needed in the system in order to encode these different kinds of information. There is a similarity between Grossberg's and our approach which will be explained below. Visual perception linked with pattern recognition will be presented in the case of holovision with examples illustrating its functional capabilities. Holovision has a three-level structure, a retinal plane or R-plane, an image synthesizer corresponding to the visual fields of the brain, and a long-term memory where various items are stored as concepts. The image synthesizer is a kind of synergetic computer whereby a spatio-temporal pattern is self-organized using parallel computation among synergetic unit processors. Semantic information is organized when the pieces of elementary information, so-called elementary signals, are connected with each other according to what we call relevant relations. The meaning of the organized semantic information not only depends on the elementary signals but also on their connections. Holovision has an autonomic mechanism for self-organizing a visual image, a sort of semantic information, from many pieces of elementary information forming a complex pattern.

BASIC IDEAS OF HOLOVISION

It is very hard to describe the mechanism of visual perception by taking a single neuron as a unit of information processing. As pointed out by Marr [13], it is rather important to find a level of study relevant to the problem. Heuristic approaches based on a model with suitable functional units, are a useful way to study the fundamental principles of information processing in visual perception. If such a model is capable of simulating characteristic features of visual perception, some of the principles may be contained in the model. For this purpose the model should be as simple as possible. The following facts seem to be important in finding a basis for such models.

(1) Input signals from the external world are received by retinal sensory cells, which means that the signals evoke a mosaic structure composed of various degrees of excitation of the sensory cells arranged in the retinal plane. Excitation in sensory cells is then transmitted to the visual field of the brain through the lateral geniculate nuclei. In this process the signals are properly assembled into a number of pieces of elementary information, i.e. elementary (visual) signals. (2) The visual image of the figure as an object is self-organized from a part of the neuronal group coding these elementary signals, while the other part is suppressed as the background for the figure. (3) The self-organization of the image is governed by the concept of an item aroused in memory.

From the above (1)-(3), a model can be proposed based on the self-organization of visual semantic information, in which an image of the object is formed from an input signal.

The input signal is decomposed into an assembly of elementary signals in the R-plane (receptive plane), maintaining topographical identity with the input signal. The elementary signals are coded one by one to synergetic unit processors (groups of neurons) in the image synthesizer. The mutual relations among the elementary signals are uncertain at this stage. In the image synthesizer, such relational uncertainty is removed by the unit processors according to a strategy given by one of the items stored in the memory. This process is associated with the autonomic formation of an image by making connections of some parts of the elementary signals (e.g. in specific directions) and compressing others as background. This semantic separation is connected with the self-organization of a spatio-temporal pattern (a dissipative structure), coding the image into the dynamics of the synergetic unit processors (holons) under dynamical constraints given by the referred concept.

CONDITIONS FOR AUTONOMIC CONNECTIONS OF ELEMENTARY SIGNALS

Let us illustrate the information processing needed for the self-organization of an image in the image synthesizer with the following example. When a photograph printed in a newspaper or a journal is enlarged many times with a photocopy machine, one can see a number of dots of various sizes that are arrayed on a square lattice as shown in Fig. 1. The dots may be tentatively regarded as elementary visual signals given in a "syntactic plane". (The syntactic plane is a part of an information space representing the apparent features of information.) If one wants to see anything in the assembly of the dots, they must be relevantly related. Qualitatively, this process is based on either connecting some of the dots in relevant directions or on compressing other dots. It should be noted that the connected dots become the elements of the lines composing a figure. The meaningless dots properly acquire their elementary meanings within the meaning of the whole figure; e.g. they are regarded as a line element the direction of which is determined by the direction of the line to which they belong.

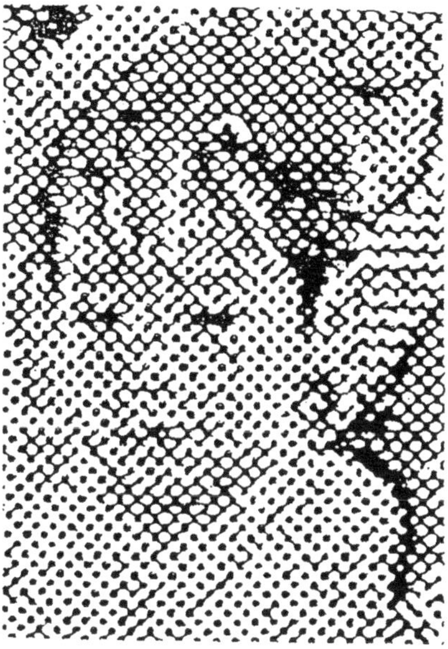

Figure 1 A photograph enlarged with a photocopy machine.

How can one obtain relevant connections among, and compressions of, the dots. The relevance depends on a figure to be self-organized from the dots, and, on the other hand, the figure depends on the connections among the dots themselves. Thus, we have a kind of logical circle which does not converge. For the self-organization of the figure in the syntactic plane, it is therefore crucial to have a method that makes the logical circle spontaneously convergent. A number of local rules would be present for connecting the dots, but a specific one will be selected out of the possibilities when the figure is determined. Moreover, different rules may be applied to different parts of a figure, since each part has its own specific features. Selection of the relational rules by means of a memorized item is crucial to determine the relevance of such features.

As long as we confine ourselves to information processing in the syntactic plane, a number of different rules may be given to the unit processors. However, for the spontaneous selection of *specific* rules without confusion, it is necessary to reduce the "entropy" of different rules: Any change on a syntactic plane must be related to only a single rule. This means that the original information space of the syntactic plane must be split to many subplanes each of which has only one specific rule. We shall call such subplanes hyperplanes. Note that the syntactic space is degenerate with respect to the many relevant connections a point can make. The self-organization of an image is impossible unless a new dimension is introduced to the space to resolve the degeneracy. The new dimension is used to express a specific rule which provides specific meanings to the elementary signals, in this case meaningless dots.

It is impossible for the rules themselves to select the rule that is most relevant. This is one of the reasons for the emergence of the logical circle. An additional law is necessary to compare the relative relevance of the rules. This law is of higher order than the relational rules for the connections among dots because it governs these rules. Pattern concepts may be regarded as such higher order laws.

SELF-ORGANIZATION OF DYNAMICAL RELATIONS AMONG HOLONS

We try to identify the minimal essentials of an autonomic mechanism that finds a figure in an assembly of dots given in a plane (the receptive plane) with a square lattice structure. For the sake of simplicity, we assume that the input signal is represented by a mosaic pattern composed of dots and blanks of the same size corresponding to unit cells of the planar lattice. An image is self-organized from the mosaic pattern in an image synthesizer (such as the visual field of the brain) composed of synergetic unit processors; the structure of the synthesizer is topographically identical to that of the receptive plane. Such unit processors are called ***holons*** because they are individual components whose behavior is determined by a holistic order. (The name holon was originally given by the English writer, Arthur Koestler [14], though we use the same word in a modified sense.) When a dot is present in a given unit cell the corresponding holon receives an elementary signal. In reverse the holon does not receive any signal when a dot is absent in the corresponding unit cell. We refer to this process as the coding of elementary information in the image synthesizer. Elementary signals are therefore coded in the dynamics of the holons.

It cannot be overemphasized that only those dots that constitute a figure should be perceived. We will therefore assume the following: Although the holons that receive an elementary signal are brought to an excited state, this is not a sufficient condition that their activities have further consequences. Initially, they are brought only to a weakly excited state with activity below a threshold. (As explained below, the excitation of a holon means the excitation of its internal state.) Then the image of a figure is self-organized cooperatively by all the holons with and without a coded dot. When dynamical coherence is self-organized in some of the holons, the dots coded by those holons are assumed to be connected. In this coherent state the activities of the holons become stronger than in the weakly excited state. However, for unconnecetd holons with activities below a threshold value, the coded dots are assumed to be in a compressed state.

For the self-organization of an image, relevant rules (or local interactions) must be chosen by the holons to connect some of the dots according to the meaning of the figure. To provide autonomic ability, we assume that a holon has an internal state with many distinguishable modes. Moreover, the holon obeys different rules depending on its internal state, that is, on which mode was selected. Thus, the holon is able to chose a rule by selecting its internal state. To determine the elementary meaning of a coded dot (e.g. as a line element in a specific direction) by connections, a specific meaning of the dot is encoded in each of the internal modes. Thereby, the selection of an internal mode is associated with the selection of a specific meaning. More concretely, the internal state of a holon specifies the direction of a line element that is associated with a coded dot.

The internal state of a holon depends on the interactions with its neighbors through a relational rule. Positive (or excitatory) interactions are assumed only in the direction specific to a hyperplane while interactions in the other directions are negative. (See Fig. 3.) For example, if a holon interacts with neighboring ones only in the specific direction, the internal mode of the holon encoding a line element of that direction is excited. To select only one internal mode, negative (or inhibitory) interactions are introduced among different internal modes of a holon. Thus, the excitation of the holon's internal states is based on competition.

When an elementary signal is coded into a holon, all the internal modes of the holon are weakly and equally excited. This "homogeneous excitation" of the internal state means that the holon initially has uncertainty in its connections with neighbors. The initial uncertainty is partly reduced due to competitive excitations among the internal modes of the holons in the self-organization of local order. Finally, a single or a few modes are selected in those holons constitute a self-organized figure, which means that the initial uncertainty is compressed. In general, a figure is composed of various lines the orientations of which are not generally the same. This means that *different* relational rules are applied to various parts of the same figure. In this respect, the spatio-temporal pattern--dynamic order--self-organized by holons, that code the image of a figure, differs from the dissipative structures appear in uniform systems [15]: in contrast to the former, the latter is governed by a single relational rule [16].

Competitive selection of internal modes cannot completely compress the initial uncertainty. In other words, the slaving principle [16] is only partly applicable; partial frustration might appear among local orders because they are generally governed by different relational rules. This indicates that the local patterns cannot be organized into an image because of remaining uncertainty; a suitable mechanism must be provided to the image synthesizer to suppress such frustrations and to further remove uncertainty. Moreover, there is no specific mechanism yet in the image synthesizer to separate figure and background. What is self-organized in the synthesizer might be a mixture of some parts of many different figures and background. Such problems are related to the logical circle between semantic information and its components.

For semantic separation according to a concept, one needs to have constraints to limit self-organization to be consistent with a one of the memorized concepts. To escape from the logical circle, the following processes will be sufficient: 1) self-organized local connections evoke the carriers of concepts (constraints) that are compatible with at least one of the local orders; 2) each of the concepts must excite only holons that code local orders that are compatible, while the activities of holons coding incompatible local orders are suppressed. Thus, competition starts in the evocation of concepts with positive and negative feedback loops; 3) eventually, the most relevant concept is left, which leads to semantic separation in the local patterns.

THE STRUCTURE OF A SEMANTIC PARALLEL PROCESSOR FOR VISUAL PERCEPTION

Based on the above discussion, a three-level model of visual perception is proposed that is schematically illustrated in Fig. 2. The lowest level is a plane composed of a square lattice of receptive

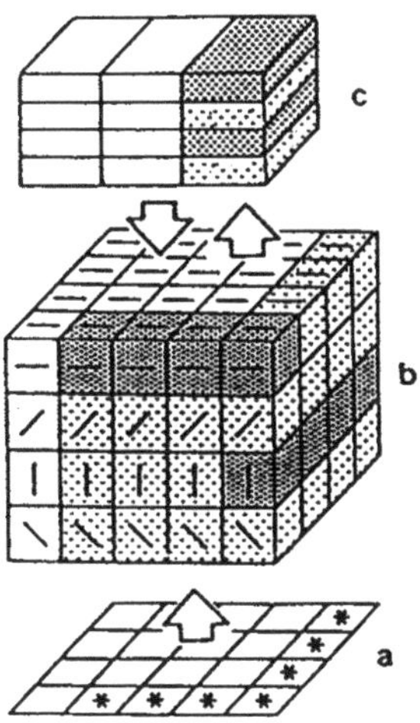

Figure 2 Schematic representation of the original three-level model of holovision: the original Holovision model. (a) R-plane, (b) the image synthesizer, and (c) the memory processor for concepts of figures.

cells. The middle level consists of the image synthesizer--an assembly of holons arranged in a syntactic space. The holons correspond here to the "hypercolumns" proposed by Hubel and Wiesel [17]. Each holon has three distinguishable states: a resting state, a weakly excited state with a uniformly excited internal state leading to uncertain relations with its neighbors, and a strongly excited state with a nonuniformly and strongly excited internal state giving rise to a specific relation with some of its neighbors. The latter two states have various degrees of excitation. Concepts of various figures are stored in the highest level encoded into memory holons.

In the image synthesizer shown in Fig. 3, each simple cell in a hypercolumn (i.e. in a holon) corresponds to an internal mode of the holon. We will refer to these as *line processors*. The orientation of a line element, which is the elementary semantic information to be given to a dot, is explicitly shown in each line processor of a holon. To separate each of the relational rules, we introduce hyperplanes, horizontal intersections on which the line processors for

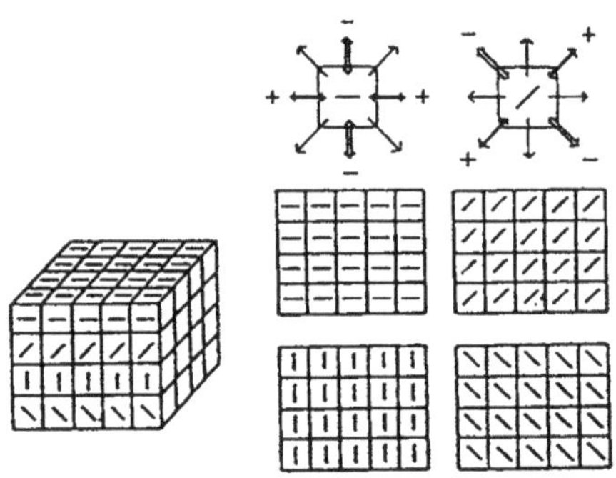

Figure 3 The image synthesizer composed of line processors is illustrated on the left. A unit cube denotes a line processor. The set of cubes in the vertical direction is called hypercolumn (holon) and the set in the horizontal plane is called a hyperplane. The excitatory and inhibitory connections between line processors in each hyperplane are shown on the right. The orientation of encoded elementary information activated by the local interactions is indicated by a small line in each cube.

the same orientation are arranged. The relational rule differs from hyperplane to hyperplane as indicated in Fig. 3. The structure of the image synthesizer differs from that of the R-plane on the lowest level in that an additional axis, the semantic axis, is introduced to resolve the degeneracy of the relational rules. The presence of many hyperplanes in the image synthesizer may be visualized as the coexistence of many sets of cellular automata obeying different rules.

Note that the new three-dimensional information space, (a syntactic-semantic space) is composed of two types of coordinate axes. One provides a syntactic space, i.e. here a plane that is topographically identical with the receptive plane. The other type, the semantic axis is perpendicular to the two axes spanning the syntactic space. A spatio-temporal pattern is self-organized in the three-dimensional information space of the image synthesizer, simultaneously determining syntactic and semantic relations among the elements of the figure.

NONLINEAR OSCILLATORS AS RELATIONAL RULE-GENERATORS

The image synthesizer is composed of holons with a limited number of line processors (which may be a realistic assumption for the brain). In this sense there are static connections between the processors in the synthesizer. Here one problem arises. Namely, why the brain is able to find lines in directions which are not encoded in the holons. This problem becomes important, for instance, when *curved* lines are found in the input signal. This problem is solved, however, holons are rule-generators in the sense that they are able to create new semantic information by themselves by properly modifying semantic information that is encoded statically. Such flexibility is crucial for creative information processing. We will discuss this property, which is also an important property of memory holons, in more detail later. The presence of unit processors in the synthesizer with rule generating properties is crucial for visual perception of varying and moving objects, as is memory itself.

As we shall explain, rule-generators in the above sense can be realized by using nonlinear oscillators as line processors. The crucial point for making a rule-generator is to construct a holon from a set of elements with excitability and entrainment capability: in holovision rule-generators are composed of limit cycle oscillators whose phases are locked by phase-shift interactions with a reference oscillator. The elementary semantic information (such as a line element with a specific orientation) is encoded in the phase of each oscillator. The relative phases of the oscillators are attracted towards each other due to entrainment and are locked. But the relative phases may also be modified from the locked values due to interactions between holons. Such deviations lead to modifications of the elementary meaning encoded in the holons. In any case, modifications of the encoded information is required for semantic information processing with flexibility. (The presence of the reference oscillator is not crucial for holovision. It was introduced to make the interactions among various line processors simple.)

SEMANTIC MEMORY

To self-organize an image with semantic separation, the most relevant concept in memory is needed to evoke and to guide the self-organization according to its *meaning*. For such control, information must be transmitted from the synthesizer to the memory both to evoke memory holons and to select the most relevant one. The reverse transmission from memory to the synthesizer is also needed. Thus, one has a positive feedback loop exists for the information flow between the synthesizer and memory. The guidance by the memory must be flexible and independent of the detailed features of objects. This means that the stored information influences the self-organization of the image in its semantic part. Since the semantic part of self-organized information is coded in the phases of the oscillators, semantic memory must control the phases of the oscillators in the synthesizer. On the other hand, the information transmission from the synthesizer to the memory carries also syntactic information. The connection between semantic and syntactical information is determined through the positive feedback loop in order to induce maximal values in the excitation of holons both in the memory and the synthesizer, i.e. the maximal feedback strength. When many semantic items are evoked, the one with maximal feedback strength will be finally selected. Thus, the logical circle will converge.

ORIGINAL HOLOVISION MODEL

Original holovision is based on the above three-level model. The holons are composed of a set of van der Pol oscillators, present in the memory as well as in the image synthesizer as line processors. Information is encoded in the phases (and amplitudes) of the oscillators of the same frequency. Originally, holovision had a structure as shown in Fig. 2. In addition to these three components, a reference oscillator locks the relative phases of the line processors by means of phase-shift interactions.

An image--self-organized in the image synthesizer as a spatio-temporal pattern of collective oscillations of line processors--is transmitted to the memory in the form of a set of rhythmic pulses. The oscillations of the line processors are transformed to pulses to eliminate contributions from line processors whose amplitudes are below a threshold value. The orientations of lines are coded to the positions (phases) of the pulses relative to the phase of the reference oscillator. The output of the reference oscillator works as a "carrier wave" for the pulses. A semantic concept is encoded in a specific memory holon as a set of lines constituting the figure of the concept. The unit oscillators code a line one by one in their phases which are locked by phase-shift interactions with the reference oscillator relative to the reference oscillation. The carrier wave excites all the line processors in the memory holons. The output from the memory holons is sent back to the synthesizer as thythmic pulses. The pulses activate only those oscillators that are in the same phase. Thus, we have a positive feedback loop, a *semantic loop,* between holons in the synthesizer with memory holons encoding various concept. The memory holons that encode various kinds of concepts compete with one another. This competition is governed by the amount of coincidence between the pulses from the memory holons and those from the synthesizer. According to that amount of coincidence a "gate variable" of a memory item, that governs its access to the carrier wave is controlled.

NEW HOLOVISION MODEL WITH FOUR LAYERS

In this original type of holovision, information on lines self-organized in the synthesizer was expressed by means of the phases of the pulses, and no information on the positional relations of the lines was sent to the memory, where lines were encoded also without information on their mutual relations. The absence of information on the positional relations of lines in the communications between the synthesizer and memory, as well as in the information stored in the memory restricts the applications of the original holovision as pointed out by Haken [18]. In the present paper, a new type of holovision is proposed in which relational information of lines is included both in the communications and in the memory. However, because of its simplicity, the original type of holovision may be used for the study of the fundamental features and principles of semantic information

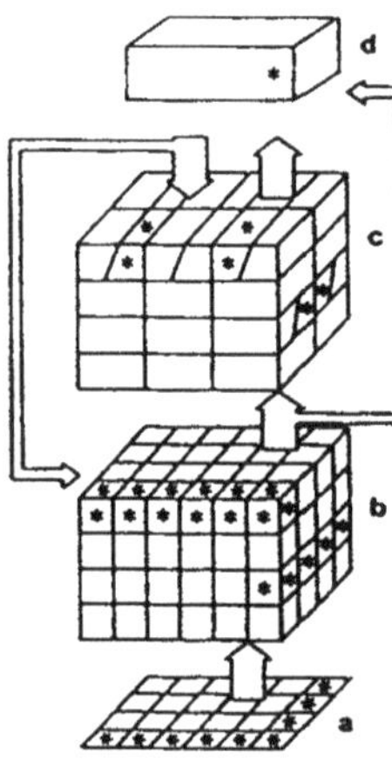

Figure 4A A schematic representation of the new Holovision with four layers. (a) R-plane, (b) layer A as the first part of the image synthesizer, (c) layer B as the second part of the image synthesizer, (d) memory.

processing. No detailed explanation of the original holovision will be given here because it has published elsewhere. [7,8]

The new holovision [19] has a structure that is schematically illustrated in Fig. 4.

a. R-plane

The input information is introduced a binary arrays (1,-1) to the unit cells of a square lattice, the R-plane.

b. Image Synthesizer

The image synthesizer has two layers. The first layer--called layer A is a line synthesizer as explained above. Another synthesizer, an angular relation or angle synthesizer, is placed in layer B. The spatial relations among lines, self-organized in the line synthesizer, are input to the angle synthesizer and described in terms of the directions of a set of vectors that point from a line end to any other line end. The functional units of the angle synthesizer are also holons which are different from those of the line synthesizer. These two kinds of holons will be called angle and line holons, respectively.

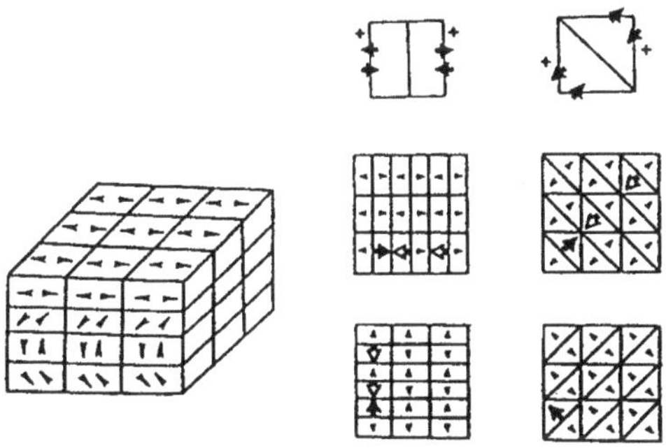

Figure 4B An illustration of layer B in the new version of Holovision. On the left a unit cube denotes a pair of angle processors. On the right the connections and the resulting angle selectivity for each angle processor are illustrated for every hyperplane, where the processor denoted by a large closed symbol has mutually excitatory connections with another one denoted by a open symbol.

The angle holon consists of unit processors, so-called angle processors, corresponding to the line processors of line holons. A pointing vector indicating a direction from the position of the angle holon is encoded in an angle processor. The direction of a pointing vector encoded in each angle processor is illustrated in the angle synthesizer in Fig. 4B together with a vector pointing in the opposite direction that corresponds to another angle processor of the same angle holon. What is important is that these paired processors are placed in the same hyperplane of the angle synthesizer. The syntactic plane of the angle synthesizer where angle holons are arrayed has a lattice structure that is topographically identical with that of the R-plane.

The input information sent from layer A reports the positions of characteristic points of lines (self-organized in the line synthesizer) as pieces of elementary information. These, then, are elementary signals for layer B: Characteristic points are defined as the positions of the line ends or of the points at which curved lines have a curvature much stronger than those at other positions. A line end is detected by comparing the activities of two line processors at neighboring sites in the direction of a coded line. Information on the position of a line end is transmitted from the line holon situated at that site, to the angle holon at the corresponding site in the angle synthesizer.

Analogous with our previous discussion, the angle holon is weakly excited with initial uncertainty when it receives an elementary signal from the line holon. If it can detect a partner, (belonging to another weakly excited angle holon in the same hyperplane), it is brought to a strongly excited state. The partner is an angle processor with a vector pointing in the opposite direction. Such vector pairs of pointing in opposite directions in syntactic space indicate the relations of line ends by means of the orientation of a line connecting two arbitrary ends. The detection of the pair, i.e. the self-organization of (semantic) information on positional relations of characteristic points, results from cooperative actions of angle holons coding the elementary signals. Such cooperativity is a typical

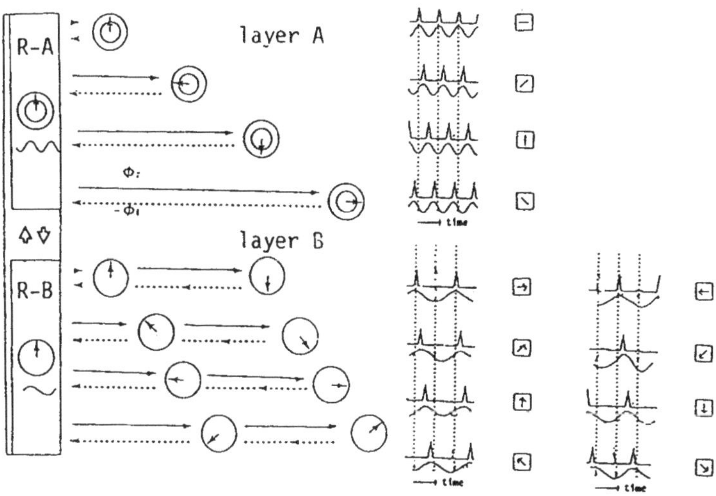

Figure 5 Phase coding of geometrical information in the image synthesizer. An oscillator in the reference processor shown on the left hand side interacts with every unit processor in layer A. The amount of phase-shift in input/output reflects the line orientation coded by the line processors. The phase-shift interaction in layer B is equivalent to that in layer A.

characterictic of holons in creating new (semantic) information.

c. Rhythmic Activities of Unit Processors in the Image Synthesizer

The unit processors in either layer A or B are assumed to possess intrinsic rhythmic activity. The utilization of the phase of the unit oscillators for coding information might be an interesting method also for more general applications. In the present case, quantities such as the direction of a vector or the orientation of a line are coded to the phases because there is a one to one correspondence between the spatial and temporal periodicities. As schematically shown in Fig. 5, the unit oscillators in the line synthesizer have the same frequency which is twice the frequency of the unit oscillators belonging to the angle synthesizer. Correspondingly the frequency of the reference oscillator for the line synthesizer is twice the frequency of the corresponding reference oscillator for the angle synthesizer. Thus the two kinds of reference oscillators are entrained with 2:1 phase relation. The orientations of line elements and the directions of vectors pointing to each other are coded as illustrated in Fig. 5.

Encoded information can only be kept stable when the relative phases of the unit oscillators are stable. Entrainment of the oscillators with a common reference oscillator (with phase-shift interactions) are utilized for this purpose. To describe the rhythmic activity of the unit processors a van der Pol equation is used. Entrainment of a unit oscillator with a specific phase-shift is given by means of an interaction term with a rotational transformation with a given shift value as shown in the Appendix (where the fundamental equations used in the new holovision are given.) In addition to the phase-shift interactions, the unit oscillators in layers A and B oscillating with the same phase in the same hyperplane have positive interactions only in a specific direction and negative interactions in the other directions, leading to cooperative activations and inhibitions. Interactions are also introduced between unit processors belonging to different hyperplanes. As mentioned earlier, such interactions are used 1) to select a single mode from many internal modes after initial uncertainty; and 2) for the self-modification of elementary meanings encoded to the locked phases of unit processors with that of the reference oscillation.

The particular connections between various oscillators serve to fix pathways of interaction between various processors in such a way that the introduction of phase locking provides a "frame" for the self-organization of information in the image synthesizer. Such connections will be called "static connections" of the processors. On the other hand, the encoded information is modifiable within the frame of the static connections by cooperative activation and inhibition of unit processors with dynamic coherence. Coordination among the processors with dynamic coherence are called "dynamic connections". Dynamic connections depend on the detailed features of the input information to the synthesizer. The ability of the image synthesizer to self-organize dynamic connections gives flexibility to the information processing.

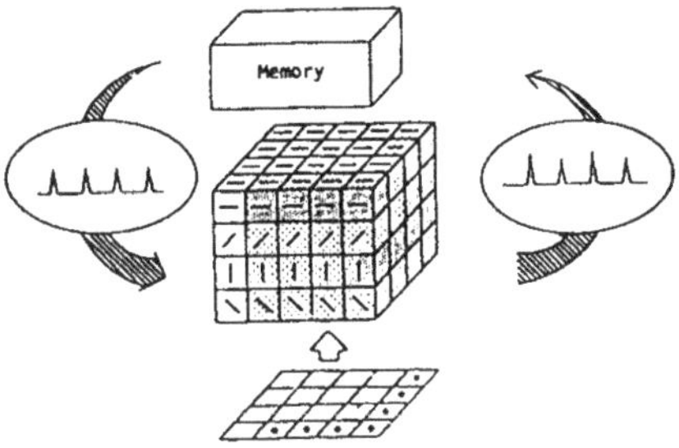

Figure 6 A schematic representation of the semantic loop for communications between the image synthesizer and the memory through one channel.

All the communications between the image synthesizer and the memory is sent by pulses transmitted from holons without use of channels specific to spatial positions of holons in the syntactic space. This is essentially because only semantic information encoded to the phases of holons is communicated. As a result pattern recognition is independent of size, position and orientation. A saturation level is introduced to the pulses output from the unit processors, and all the output pulses are superposed producing trains of rhythmic pulses as shown in Fig. 6.

d. Dynamical Memory

In the new holovision, the long-term memory is encoded in the phases of the unit (memory) processors of memory holons. Each memory holon encodes an item of semantic information. One of the important features of such memory holons is their ability to encode abstract and general information as a concept. The autonomic ability of the memory holons to form dynamic connections within a frame of static connections (that correspond to the conceptual frame of the item) makes the interpretation of a variety of input signals possible. Within the frame of the concept the encoded information is modified flexibly in the memory holon according to pulses transmitted from the synthesizer. Thus the new holovision has the ability of flexible recognition based on pattern concepts in addition to its independence in pattern recognition of size, position and orientation. For instance, triangles and tetragons are respectively recognized as triangle and tetragon, independent of shape (see below).

The extent of flexibility in the dynamics of holons is related to the degree of abstraction of encoded information. If the memory holon is unable to modify encoded information, it can only respond rigidly to input information.

In contrast to the line and angle holons in the image synthesizer, each memory holon (encoding a semantic item) has its own characteristics because the semantic item has its particular meaning. The oscillators, unit memory processors are not phase-locked to the reference oscillators and constitute a memory holon. They have, however, preferred phase relationships among each other due to static connections particular to the holon. A semantic concept is defined by pointing vectors and geometric relations among feature points. It is stored by means of pairs of unit oscillators with reversed phases (coding the pointing vectors) and static connections among the corresponding oscillators (that code the geometric relations).

e. The Semantic Loop; Dynamic Feedback between Image Synthesizer and Dynamic Memory

A damped unit oscillator in the resting state is excited by a rhythmic pulse and is entrained to the pulse. Hence, when a set of rhythmic pulses is transmitted to the memory in the resting (damped) state, some of the memory holons will be evoked due to entrainment with the pulses. The phases of the unit memory oscillators in the memory holon may be changed due to entrainment to the transmitted pulses. A match or fitness may be defined that measures the amount of coincidence of the output pulses of a memory holon with the pulses transmitted from the image synthesizer. The extent of the evocation or activation of memory holons is an increasing function of this fitness. The maximum fitness will be obtained when the set of output pulses from the unit oscillators of a memory holon completely coincides with the pulses transmitted from the image synthesizer. To assure the selection of only a single memory holon some form of competition among memory holons must be introduced. One simple method to do this is the introduction of a dynamic gating variable for each memory holon. The gate variable is an increasing function of the fitness and controls the amount of output from the memory holon. This is somewhat similar to Eigen and Schuster's approach [20] in their logistic equation for the competition among various species in biological evolution.

The output pulses from the memory holons induce changes in the activation of unit oscillators in the line and angle synthesizers according to the degree of fitness or match with the memorized information. These changes lead to semantic separation on the basis of the most relevant memory item. Thus, through dynamic feedback of semantic information between image synthesizer and memory, the most relevant memory item with the best fitness is selectively evoked in the memory, while self-organization of an image occurs in the synthesizer with semantic separation according to that memory item.

APPLICATIONS AND DISCUSSIONS

We exhibit a number of examples to demonstrate the properties of holovision. Specific attention is given to show how holons as autonomic rule-generators provide flexible and creative memory dynamics.

a. Effects of Semantic Feedback in the Original Holovision

Fig. 7A demonstrates the perception of fuzzy information by the original holovision. The left figure of Fig. 7A(a) indicates the elementary signals given in the R-plane. The middle figure shows an image self-organized in the image synthesizer, i.e. the line synthesizer, in the absence of any information in the memory, (no semantic loop). The pattern is decoded from the phases and amplitudes of the unit oscillators of the synthesizer. The amplitude encodes elementary information on the length of a line element. Many dots due to the excited holons which received no signal are also seen. No semantic separation occurs in the image because of the absence of a semantic loop. The right figure illustrates the image that is perceived by the memory holons. This is the part of the self-organized image that is compatible with information encoded in the evoked memory holon. In this case no pattern appears because no memory holons are present.

Fig. 7A(b) shows similar patterns in the presence of three kinds of semantic items in the memory encoded by memory holons. The time courses of various outputs are also shown in Fig. 7B(a) from top to bottom: output from the reference oscillator utilized as a carrier wave of output from the image synthesizer, outputs from the image synthesizer and from memory holons for triangle, tetragon and octagonal. Curved lines on each of the output signals indicate the opening size of the dynamic gate. The dynamic gate assures that

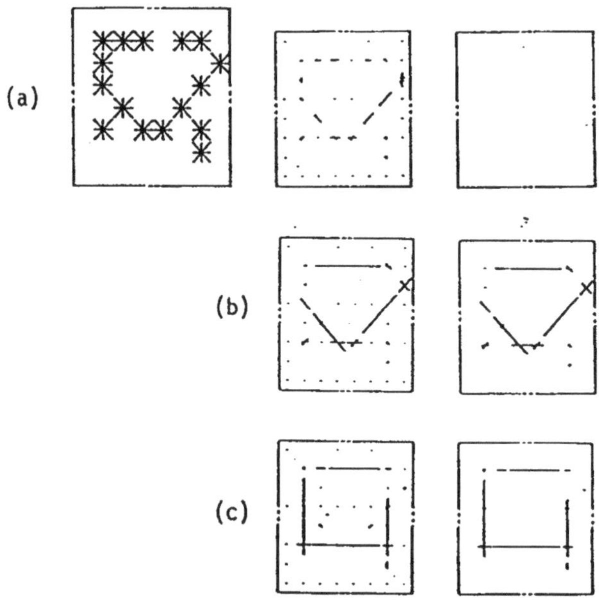

Figure 7A Perception of a fuzzy input pattern by the original holovision. The input pattern is shown on the left hand side in (a). The result of perception is represented by two kinds of images in three cases: in the absence of memory (a), after spontaneous selection of a triangle among three items (b) after forced evocation of a tetragon (c). In each case the left image denotes the organized image in the image synthesizer and the right one the perceived image. In the latter the line segment is represented only when the feedback gain from the processor is not zero.

the synthesizer is not saturated by the feedback signals from the memory. As an image is self-organized in the synthesizer three kinds of items are evoked simultaneously. Then, due to the competition of memory holons only the memory holon encoding the triangle is recognized. The self-organization of the image is influenced by the evoked memory. For instance, the image has a continuous line at the top side of the triangle of the image even though an elementary signal is absent there in the input information. Regardless, the input signal is recognized as a triangle.

Fig. 7A(c) demonstrates the perception of holovision when it expects to see a tetragon in the same input information. The selective value for the gate of the tetragon is increased in the initial stage. The self-organization of the image is thus influenced to be consistent with a tetragon. The time courses corresponding to Fig. 7A(c) are shown in Fig. 7B(b).

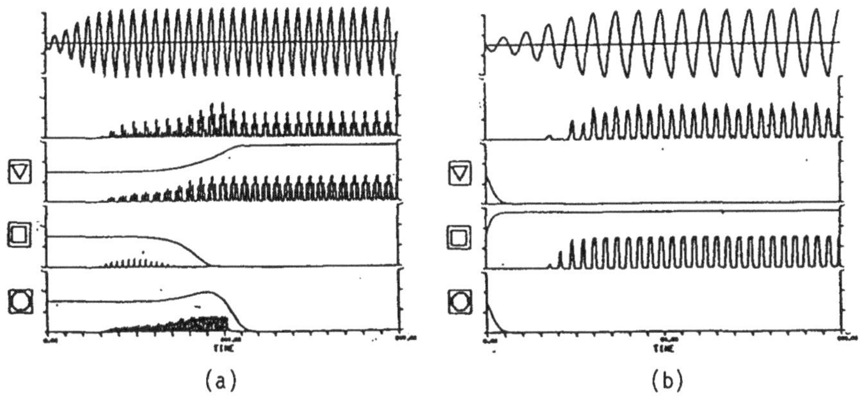

Figure 7B Time evolution of various processors in pattern recognition processes by holovison. The conditions used in (a) and (b) are the same as those in Fig. 7(b) and Fig. 7(c), respectively. Top-to-bottom: The reference oscillator, the image synthesizer and three kinds of items in memory (shown by small figures on the left).

b. Creative Dynamics of Memory Holons

Man is often able to discover a new meaning when new input signals are given. For this ability the creation of a new concept is important, because preunderstanding or acquiring a concept is necessary before cognition. The discovery of new concepts differs from a simple learning process governed by the Hebbian law. It would be related to a discontinuous change in recognition (visualizable as an inspiration to find something in a signal). At least partly, it would be related to the creation of a new concept based on associations with existing concepts.

The ability of the holon to generate rules makes such a creative process possible in the simplest case. We store two kinds of semantic items, triangle and square, to the original holovision as long-term memory and give a fuzzy input signal as shown in Fig. 8a (an image self-organized in the image synthesizer in the absence of any semantic item is also shown). Now, in the presence of the memory item, holovision has self-organized two kinds of images as

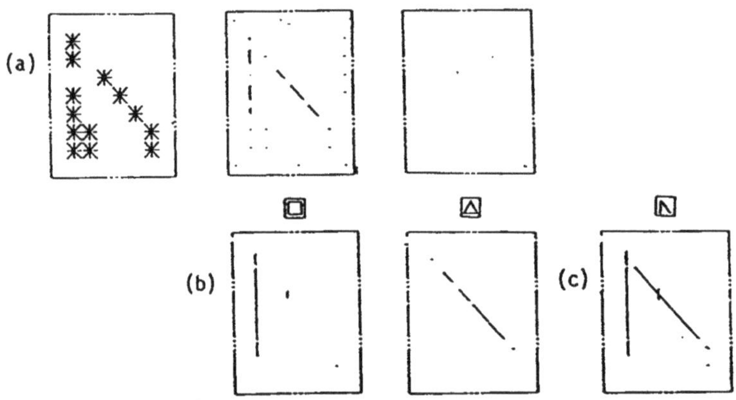

Figure 8 Perception of fuzzy input signals under various conditions of the memory. (a) Left-to-right. The input signal in the R-plane, the self-organized image and the perceived image in the absence of any memory item, respectively. (b) The perceived image under forced evocations of an memory item. The evoked item is shown in a small square on top of each image. (c) The perceived image organized after creation of the new item as indicated in the little square above the image.

demonstrated in Fig. 8b. each of which is self-organized under the influence of a single item, corresponding to a square and triangle, respectively. However, neither of these concepts is capable of extracting a large amount of information from the signal. In other words these two concepts are only partially relevant, and we cannot say which is the more relevant one of these two.

For such cases of only partially relevant memory items we introduce a mechanism to holovision that allows for the self-organization of new semantic items. As illustrated in Fig. 9. a memory holon without encoded information participates in the semantic

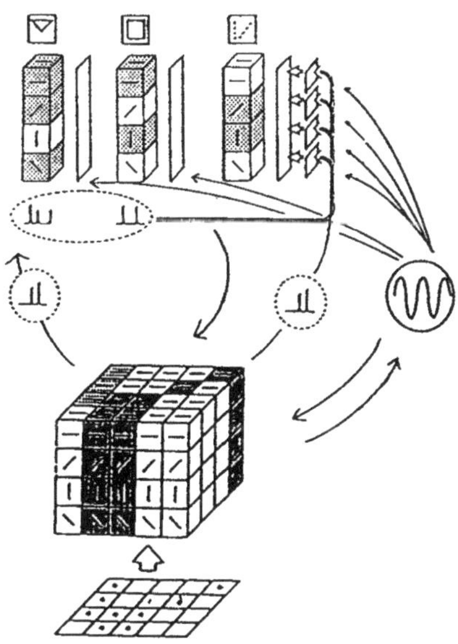

Figure 9 A schematic representation of the creation process of a new item in the memory in original holovision. An "empty" holon additionally prepared is shown in the upper right. When no item can be uniquely selected, two kinds of pulse trains are sent to the empty holon. One is from the image synthesizer and the other one from all old items. In the empty holon the timing coincidence between the two kinds of pulse trains opens the gate at the column whose pulse also has timing coincidence. The creation of the new item is completed by opening its gates. Pulses emitted from the new item are sent to the image synthesizer and also enter the competition among items.

loop with the other two memory holons and spontaneously acquires relevant information from the semantic loop.

Fig. 8c illustrates a new item and the image self-organized in the synthesizer under the influence of the new item. Fig. 8d demonstrates the time course of the changes in the synthesizer and the memory. It clearly shows that two kinds of only partially relevant memory holons, corresponding to triangle and square, are evoked in the initial stage. Neither is able to win the mutual competition. Then, after some period of time, a new item is self-organized, that wins against the old ones. The amplitudes of the rhythmic pulses from the synthesizer becomes larger after that period because a relevant item was created in the memory.

c. Memory of Pattern Concepts in the New Holovision

In Fig. 10 a memory holon for a straight line is schematically indicated. It is composed of two oscillators damped with inhibitory interaction. The activation of the two oscillators by a rhythmic pulse leads to two oscillations with reversed phases. These oscillations describe a line by two vectors at the line ends pointing toward each other.

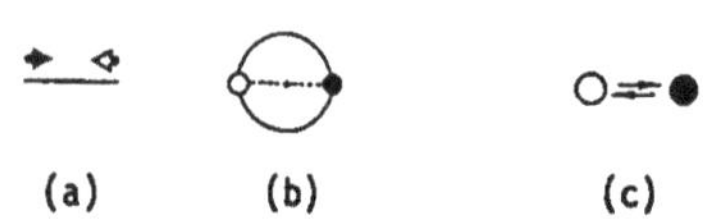

Figure 10 The memory holon of a straight line. (a)The geometry of the information, (b)The pattern of phase distribution for (a) where black and white circles represent unit oscillators. (c)Coupling of oscillators. The two oscillators denoted by black and white circles are negatively coupled.

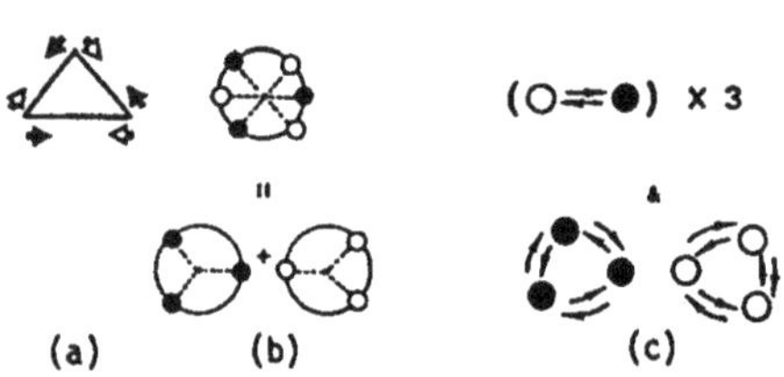

Figure 11 The memory holon for the triangle is composed of three pairs or six unit oscillators. The three black and the three white oscillators have negative interaction leading to phase dispersion among the set of either black or white oscillators as shown in the figure.

Fig. 11 illustrates the oscillators of a holon encoding the memory of a triangle. As shown in this example, the memory holon for a n-polygon has n pairs of oscillators whose phases are indicated by the position of small circles on a larger phase circle. If we number these circles from 1 to 2n and introduce inhibitory interaction among all the odd numbered oscillators as well as among the even numbered oscillators, all the oscillators in each set are activated stably distributed in phase on the phase circle such that they do not overlap. The phase relations of the oscillators of the two sets are mirror images of each other with respect to the center of the phase circle.

Fig. 12 illustrates how triangles of various shapes are related with a memory holon encoding a triangle. The input signal in the R-plane, an image in the line synthesizer and the spatial relations of the characteristic points of the self-organized image with dynamic connections among oscillators of the memory holon, are shown by means of the phase positions of the oscillators. One of the interesting aspects of our method of encoding a pattern concept to the memory holon is that a figure like Fig. 12 is consistent with theorems of Euclidean geometry for a triangles i.e. the phases of the circle and the angles of the triangle always sum to 2π. The time

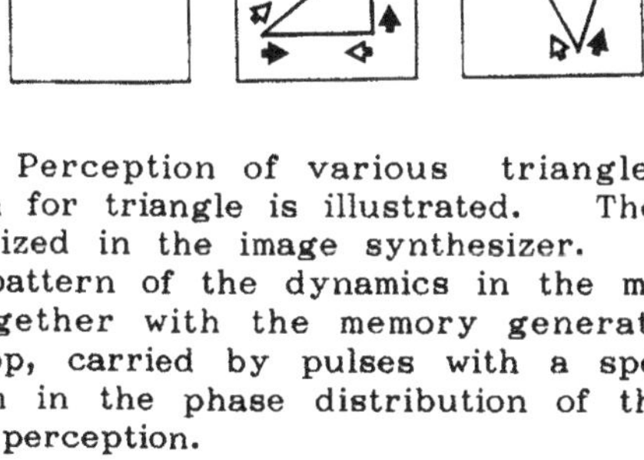

Figure 12 Perception of various triangles in the presence of a memory item for triangle is illustrated. The lower part denotes the image organized in the image synthesizer. The upper part denotes the phase pattern of the dynamics in the memory. The image synthesizer together with the memory generate a feedback loop, the semantic loop, carried by pulses with a specific phase distribution. The freedom in the phase distribution of the unit oscillators allows for flexible perception.

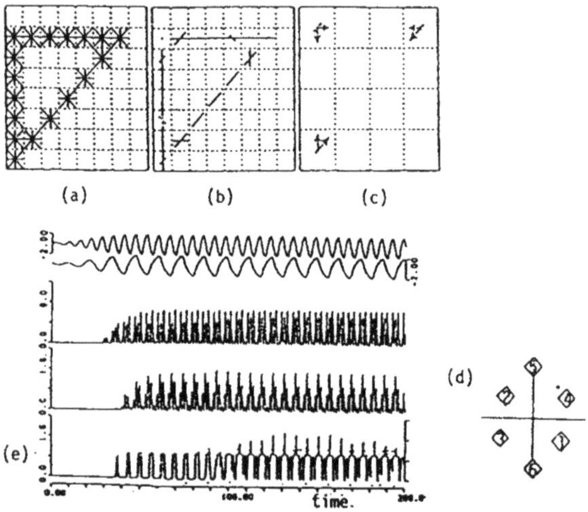

Figure 13 Computer experiments by the new holovision in the presence of a memory item for triangle. (a) - (c) show in this order the input signal, the organized line image in layer A, and the organized angle image in layer B at the stationary stage. These images are decoded from the activities of unit processors. In (d) the oscillations of the memory unit processors are represented by numbers 1 to 6 in their phase plane. The reference oscillation is also indicated by a star. The phase relation among oscillators shows the pattern of phase-locking caused by mutual interactions and by signals from the image synthesizer. Pairs of anti-phase oscillators are observed there. The time evolution of processors is shown in (e). From top, the reference oscillations for layers A and B,a pulse train transmitted from the line synthesizer in layer A, from the angle synthesizer in layer B, and from the memory, respectively, are plotted.

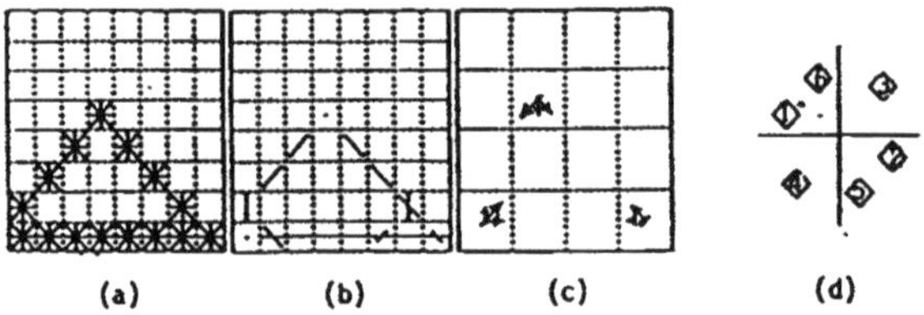

Figure 14 The self-organized image for another input signal. All the conditions except the input signal are the same as those in Fig.13.

course of rhythmic pulses carrying information are also shown. Fig. 12 also illustrates the flexibility of the recognition by memory. Thus, the phase distribution pattern of the coupled oscillators that represents an item varies according to the entrainment by the signals from the image synthesizer. It is obviously independent of the absolute size or position of the input information in the R-plane.

Fig. 13 shows an example of computer experiments by the new Holovision. For the purpose of demonstrating the dynamics of memory, only the concept of triangle is stored. The self-organized information in the image synthesizer, the orientations of the contour lines, and the angles denoting positional relations among the line ends are shown. An instantaneous distribution of the phases of the memory oscillators in the stationary state is depicted on the right. Note that three pairs of oscillators have an anti-phase relation. The behaviors of the holons in the memory and the synthesizer become temporally coherent through a semantic loop, the feedback loop carried by rhythmic pulses as shown in Fig. 12(e).

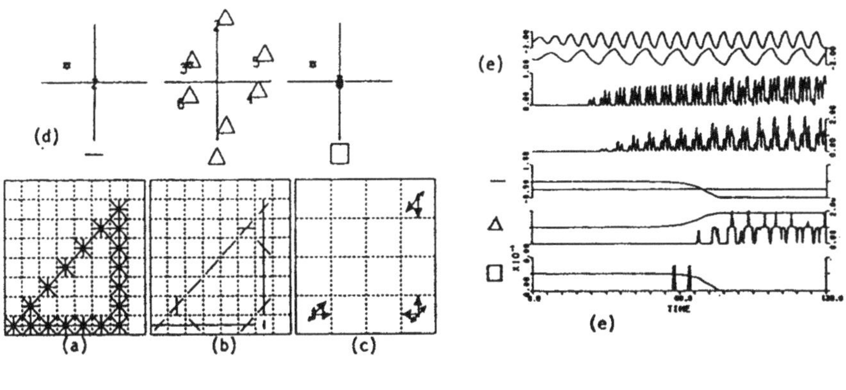

Figure 15 Perception in the presence of three kinds of memory items. The input pattern, the self-organized images in layer A and B at the stationary state are shown in (a)-(c). The dynamics of the memory holons coding each of the three items are shown in (d) in the phase plane. The size of symbol at the each number denotes the pulse amplitude of the numbered oscillator. The time evolution of each holon is represented in (e) in the same way as Fig. 13. The magnitude of the dynamic gate for each holon is also shown as a measure of competition among memory items.

Fig. 14 shows the flexibility of the memory holon. An input pattern which significantly differs from that in Fig. 13 is given for the sake of comparison. The two different types of input patterns are both recognized as triangles, showing that holovision has the capability to recognize patterns independent of size, position, orientation and detailed difference in shape.

In the next computer experiments three kinds of semantic item are prepared in the memory. The spontaneous connection of an input signal to one of the items is examined in Fig. 15. The items used here are the linear line, the triangle and the tetragon. The self-organized image is obviously a triangle. The time evolution of the memory processors shows that due to the mutual competitions among these three concepts only the triangle is evoked at the final stage.

CONCLUSION

Holovision, a dynamical model of visual perception, has three-levels. The basic elements of holovision are the so-called holons, a kind of synergetic processor with many internal modes in which various kinds of elementary semantic information are encoded. Holons can be conceived of as rule-generators; they generate local rules to modify the encoded information according to the "syntactic text" of the input information. Because of the special features of holons, holovision is able to self-organize an instantaneous image of the external world. This image, the "figure", that is endowed with semantic information, is self-organized from a relevant part of the elementary information, concomitantly with information compression of the irrelevant part as background. The self-organized figure is also influenced by an evoked memory holon encoding a relevant concept through the feedback of semantic information. This semantic separation of figure and background is necessary for perception associated with recognition, and is a scientific and technological realization of the "interpretation" process of visual information. This is one of the important and necessary steps to treat the dynamics of semantic information in science and technology.

Furthermore, holons, synergetic elements working in states far from equilibrium, endow holovision with unique properties: (1) Dynamical memory is flexible within the frame of a concept defined by static connections. (2) An ability to recognize patterns is not only independent of size, position and orientation of an object but also free to recognize the concept in a variety of shapes. (3) Creative memory: In order to interpret a new kind of input signal with sufficient relevance, holovision is capable of creating a new concept by extracting the essential features of the input signal based on stored items that are only *partially* relevant to the input signal. Thus, we conclude that holovision is capable of simulating fundamental features of visual perception along with pattern recognition.

The authors wish to acknowdedge Professor J. A. S. Kelso and Dr. G. Schoener for their kind suggestions for the improvement of the manuscipt.

REFERENCES

[1] Waltz, D. L., (executive ed.), Cognitive Science **9**,(1) Special Issue: Connectionist models and their applications. (1985).

[2] Rumelhart, D. E., Hinton, G. E. and Williams R. J.,in Parallel Distributed Processing, (eds.), Rumelhart, D. E., McClelland, J. L. and The PDP research group, vol.1, 318, MIT Press (1986).

[3] Malsburg, Ch. von der and Schneider, W., Biol. Cybern. **54**, 29-40 (1986).

[4] Hopfield J. J., Proc. Natl. Acad. Science USA **79**, 2554 (1982).

[5] Cowan, J. D. and Ermentrout, G. B., in Pattern Formation by Dynamic Systems and Pattern Recognition. Springer Verlag, **122** (1979).

[6] Haken, H., in Pattern Formation by Dynamic Systems and Pattern Recognition. Springer Verlag, 2,(1983).

[7] Shimizu, H., Yamaguchi, Y., Tsuda, I. and Yano, M., in Complex Systems - Operational Approaches, Haken, H., (ed.),Springer Verlag, 225 (1985).

[8] Shimizu, H. and Yamaguchi, Y., Physica Scripta, **36**, 970 (1987).

[9] Koerner, E., Tsuda, I. and Shimizu, H., preprint.

[10] Kant, E., Kritik der reinen Vernunft. (1781).

[11] Wittgenstein, L., Philosophical Investigation. Anscobe, G. E. M. and Rhees, R., ed., Basil Blackwell (1953,1958,1967).

[12] Grossberg, S .(ed.), The Adaptive Brain I-II, Advances in Psychology **42, 43** North-Holland (1987).

[13] Marr, D., Vision. W.H.Freeman and Company (1982).

[14] Koestler, A., The Ghost in the Machine. Huchinson & CO LTD (1967).

[15] Nicolis, G. and Prigogine, I., Self-Organization in Nonequilibrium Systems. John Wiley & Sons. Inc. (1977).

[16] Haken, H., Synergetics -- an Introduction. Springer, 3-rd ed. (1983).

[17] Hubel, D. H. and Wiesel, T. N., Proc.R.Soc.Lond. **B 198,** 1 (1977)

[18] Haken, H., in Complex Systems - Operational Approaches. Haken, H., (ed.), Springer, 1 (1985).

[19] Yamaguchi, Y and Shimizu, H., Technical Report of IEICE, in Japanese, MBE **86-136,** 309 (1987).

[20] Eigen, M. and Schuster, P., Hypercycle--A principle of Natural Self-Organization, Springer (1977).

APPENDIX

The mathematical formulation used in holovision is briefly explained in this section. In the absence of interactions or input signals the unit processors are commonly assumed to show damped oscillation. The evolution equation is given by the following differential equations.

$$(d/dt)\, X = F(X) \tag{1}$$

where

$$X = \begin{pmatrix} x \\ y \end{pmatrix} \tag{2}$$

and

$$F(X) = \begin{pmatrix} y \\ -\omega^2 x + \varepsilon(d - x^2)y \end{pmatrix} - g\begin{pmatrix} x \\ y \end{pmatrix} \tag{3}$$

with ω and ε respectively denote the native angular frequency and the magnitude of the nonlinearity. The value of ω is 1.0 in layer A and 0.5 in layer B and $\varepsilon \ll 1$ is always assumed. The value of d is defined by the input data as a positive or negative value in the image synthesizer. The last term in the right hand side of eq.(3) leads to damping of the oscillations with a damping coefficient g. Eq. (1) coincides with van der Pol equation when $g = 0$.

Interaction terms in eq.(1) of the unit processors are introduced as follows.

$$(d/dt)X = F(X) + G_r + G_n + G_m, \tag{4}$$

where G_r denotes the input from the reference processor, G_n the neighboring interaction among unit processors and the G_m input from the memory processors.

The reference processor entrains the unit processors with a fixed phase difference ϕ. The explicit formulation is:

$$G_r = b\begin{pmatrix} \cos\phi & -\sin\phi \\ \sin\phi & \cos\phi \end{pmatrix} \begin{pmatrix} x_r \\ y_r \end{pmatrix} \tag{5}$$

where x_r and y_r denote the reference oscillation and the coefficient matrix is an usual rotation matrix with a rotation angle ϕ. The value ϕ is determined as the desired value of phase shift for each unit processor so that a line orientation or angle is coded in the phase. The mutual interaction with reciprocal phase-shift stabilizes the entrainment with a phase difference ϕ. The interactions in the same layer are restricted only to nearest neighbors for the sake of simplicity. A threshold function of y_i produces pulses which serve in the interaction of image synthesizer and the memory. They endow the system with phase-specific amplification in a semantic loop. The connections between the image synthesizer and the memory are controlled by a dynamic gate defined for each memory item. The dynamic gate evolves obeying a kind of logistic equation depending on a selective value. One with the largest "selective value" overwhelms the others. The evolution of the selective value also depends on the temporal correlation between the two pulse sequences from the image synthesizer and from the relevant memory item.

The unit oscillators in the memory processor are also described by equations of the type of eq. (4). In the original holovision the specific pattern of the phase distribution of the memory is generated by entrainment of individual memory oscillators with the reference oscillator. In the new holovision the phase relation is generated through the interconnection of memory oscillators that are not coupled with the reference oscillator. In the memory for a straight line, for example, two coupled oscillators are defined as follows.

$$(d/dt)X_i = F(X_i) - cX_j + G_i \tag{6}$$

where i,j = 1,2 and the second term on the right hand side denotes the negative interaction with the other oscillator. The third term is the superposition of pulses generated in the image synthesizer. It is also multiplied by the threshold function depending on the activity of the memory oscillator itself. The magnitude of the dynamic gate in the new holovision is used as the value of the damping coefficient g. For the memory of a n-polygon, the second term is a superpositon of the connected oscillators.

Dynamic patterns of biological coordination: Theoretical strategy and new results

G. Schöner and J.A.S. Kelso

Center for Complex Systems
Florida Atlantic University
Boca Raton, Florida 33431 U.S.A.

Abstract

It is well-established that spatial, temporal and functional patterns can arise spontaneously in nonequilibrium physical, chemical and biochemical systems. According to Haken's slaving principle [1] when environmental conditions, as expressed by control parameters, reach a critical point, the behavior of such systems is governed by only a small number of degrees of freedom. It is not clear whether the functionally coordinated behavior of complex, biological systems can be understood in a similar fashion. In biology, the definition of an appropriate microscopic level is often vague and the collective, macroscopic state unknown. Using a strictly operational approach in which theory and experiment go hand in hand, we show that temporally coordinated movement patterns (their stability and change) may arise in a purely self-organized fashion. Many of the features of observed behavioral patterns, including synchronization, phase-and frequency-locking, multistability, loss of stability etc., are common also to neuronal patterns generated by many species. By linking functionally specific dynamics at different scales of observation, our approach offers a way to relate macroscopic behavioral levels to more microscopic physiological levels (e.g., neural networks). We elaborate the theoretical strategy to include the influence of perception, memory and intention on behavioral patterns. In each case information is viewed as *meaningful* and *specific* to the extent that it contributes to the collective dynamics attracting the system toward a (environmentally-specified, memorized, intended) behavioral pattern. The theory accounts for recent observations on environmentally-specified and learned patterns of coordination, as well as new switching time experiments by Kelso and Scholz showing that *intentional switching* is governed by the relative stability of behavioral patterns.

1. Introduction

Pattern formation has been extensively studied in physical, chemical, and to a certain extent, biological systems. A central theoretical theme of this work is the enormous reduction of number of degrees of freedom in such systems that occurs at or close to the critical points at which structure is formed spontaneously. This means that the large number of microscopic degrees of freedom is governed by only few collective variables or order parameters and their dynamics. That insight was mathematically formulated for a large class of systems [1,3] and is variously referred to as the *slaving principle*, the center manifold theorem or adiabatic elimination.

In a few biological systems structure formation has been treated in a related fashion. Well-known examples are the pre-pattern formation models of morphogenesis [4,5] and the aggregation patterns of slime mold cells [6,7]. However, one may raise the question of whether it is not merely the more *physical* aspects of biological pattern formation that have been addressed with the tools of nonequilibrium theory rather than *essentially* biological ones. Without attempting completeness let us clarify those *essentially biological* features of pattern formation that need to be addressed.

1. Functional coordination: Many components relate in an ordered fashion to perform a biological function. (In the study of behavior, physiology or neurobiology recognizable patterns are, in fact, often named by their function.)
2. Stability of function: Biological function is sustained under changeable environmental conditions and a variety of perturbations.
3. Flexibility of patterns: The same biological function may be realized by many different, flexibly assembled patterns.
4. Informationally open character: One form of flexibility is the ability of biological systems to adjust their functional patterns to environmental conditions. We may term this aspect the informational openness of biological systems.
5. Goal-directedness: Functional patterns may be actively sought. In humans, behavioral patterns are often said to be *intended*. The underlying concept is that of purpose. It is important to

realize however, that the purpose of a behavioral pattern is not absolute, but relative to its function and subject to change.

6. Learning and adaptation: Such processes occur on a time scale larger than the adjustment to environmental conditions mentioned in (4) above. Thus, functional patterns may change according to environmental demands in such a way that the change persists.
7. Development: On an even slower time scale, marked changes of functional patterns occur during development.
8. Evolution: More globally yet, ecological systems and evolutionary processes involve even larger time scales and larger systems.

Solutions to the problems outlined above involve a step beyond the mere application of the theory of pattern formation in nonequilibrium systems. First, it is often not clear how to characterize the functionally significant patterns. No unique and biologically relevant description of the patterns and their dynamics exists. Relatedly, the path from microscopic to macroscopic is not easily accessible in functioning biological systems. Often we do not have a theoretical description on any functionally relevant scale.

In this paper we report on our attempts to address some of these essentially biological features of pattern formation. This work developed from the study of movement coordination, in which a close relation of theory and experiment was crucial in arriving at the conceptual framework. We would like to stress the *operational character* of the theory, which requires that all theoretical constructs must bear an explicit relation to measurement. The paper is structured as follows: First we review the theoretical strategy that addresses points 1 to 3 of the *essentially biological* aspects of pattern formation listed above (Sect. 2). We propose a number of general theoretical propositions for behavioral patterns and illustrate them by referring to our standard model system that involves rhythmic movement coordination (rhythm as a general case). In Sect. 3 we lay out the generalization of this theory to incorporate environmental information (cf. point 4 above). Again both general theory and a model system are discussed. Next we show that memorized information can be treated in a similar fashion (Sect. 4). Finally we report some new results that show how intentional behavior may be understood in terms of behavioral dynamics (Sect. 5).

2. Dynamic theory of behavioral patterns

The main idea is to view behavioral patterns in terms of their dynamics. The key step is to characterize the behavioral patterns by collective variables. Thus the dynamics of the patterns can be determined as dynamics of these collective variables.

The following set of theoretical propositions [8,9] summarizes the dynamic theory of behavioral patterns:

1. Behavioral patterns on any level of observation (e.g. kinematic, muscular, neuronal,...) are characterized by low-dimensional collective variables (or order parameters).
2. Observable (i.e., reproducible, stationary over a certain time scale) behavioral patterns are mapped onto attractors of the order parameter dynamics.
3. Biological boundary conditions (e.g., environmental context, task variables) act as parameters on the collective dynamics. These parameters may be quite unspecific to the resulting stable behavioral patterns, in which case we might call the order parameter dynamics *intrinsic dynamics*.
4. Fluctuations determine various time scales (local and global relaxation times). The relations among these account for the observed metastability and switching among multiple behavioral patterns. Loss of stability leads to switching of behavioral pattern and the switching process is governed by stochastic order parameter dynamics.
5. If behavioral patterns are thus characterized at different levels of description, these levels may be related mathematically without introducing additional concepts.

Propositions (1) to (3) may be used as a strategy to obtain a dynamic pattern description of a behavior. Note that identifying collective variables cannot take place without addressing their dynamics as well. When a consistent description in the sense of propositions (1) to (3) has been reached, one has also chosen a level of observation. It is important to realize that the linkage of levels of observation in the sense of proposition (5) is possible only if a dynamic pattern description is available on both levels in question. Proposition (4) affords crucial tests of the dynamic pattern theory because it allows for predictions, as we show below.

For later reference we give a general mathematical formulation for the intrinsic dynamics. For a given level of observation, an order parameter $\mathbf{x}_t$ will be subject to

$$\dot{\mathbf{x}}_t = \mathbf{f}_{intr}(\mathbf{x}_t, \ \text{noise}) \tag{1}$$

where $\mathbf{f}_{intr}$ is in general nonlinear and is defined by its attractor structure.

Before we illustrate dynamic pattern theory by reviewing our "standard" model system of coordinated, rhythmic movement in humans, we would like to point out its relationship to other approaches. For example, Shimizu and Haken [10] have developed a similar view on a more microscopic level. They treat dynamic cooperativity of many identical components (e.g., actin-myosin sites) that interact weakly, and find that a functionally relevant macroscopic behavior emerges (e.g., muscle contraction). A new approach to physiological systems that stresses dynamics and their change (see e.g., the respective work of Goldberger & Rigney in cardiology [11] or Rapp in neurophysiology [12]) is related to our approach. From the dynamic pattern view the goal is to find well-defined collective variables, obtain measurements of pattern stability, time scales and noise, etc. Finally, it seems feasible to understand pattern formation in neurobiological systems in dynamic terms. The ubiquity of patterns of temporal order (e.g., synchronization, phase-locking), relaxation phenomena, collective effects, pattern switching and other features of neural networks, suggest that they too may best be understood by a dynamic pattern theory (see [13] for a discussion; cf. also [14] and [15]).

In order to illustrate the main points of the dynamic pattern approach we now turn to our model system on human bimanual movement coordination. The basic phenomenon is the following [16]: subjects are asked to move two limbs rhythmically at a common frequency. Two patterns of coordination (in-phase and anti-phase) are found to be stably performed at various frequencies. However, when frequency of movement is increased, a spontaneous switch from the anti-phase to the in-phase pattern of coordination is observed, while the in-phase pattern remains stable at all reachable frequencies.

To obtain a theoretical description, we identify relative phase, ϕ as a collective variable characterizing the behavioral pattern (this choice also determines the level of description). The observed stable behavioral patterns are mapped onto point attractors of relative phase at $\phi = 0$ and $\phi = \pm\pi$. The simplest dynamics of ϕ that accomodates these (and also fulfills basic symmetry requirements) are [17]:

$$\dot{\phi} = -\frac{dV}{d\phi} \tag{2}$$

where $V(\phi) = -a\cos(\phi) - b\cos(2\phi)$. This model captures the observed phase diagram: for $a < 4b$ two stable states $\phi = 0$ and $\phi = \pm\pi$ exist, while for $a > 4b$ only $\phi = 0$ remains stable.

In such a theory the spontaneous switch from anti-phase to in-phase coordination is due to a loss of stability (namely of the anti-phase pattern). One may argue that we have simply found a compact way to describe patterns. However, because the central concept of stability is measurable, the theory provides novel predictions that can be checked experimentally. In particular, when fluctuations are taken into account [18]:

$$\dot{\phi} = -\frac{dV}{d\phi} + \sqrt{Q}\xi_t \tag{3}$$

(where ξ_t is gaussian white noise) predictions about the behavior near the transition are possible. For example, theory predicts that the variability of the anti-phase coordination pattern (but not the in-phase pattern) should increase as the transition is approached. Such *enhanced flucuations* were observed experimentally in several studies ([19, 20]). Another measure of stability, *relaxation time* (the time it takes to return to the coordination pattern after a small perturbation) was shown to increase upon approaching the transition. In experiments in which perturbations were mechanically induced during the coordinated movement, relaxation time was measured and such enhancement was found [21]. Another aspect of the theory [18] is that local and global relaxation times have to fulfill certain relationships for multiple patterns to coexist. In experiments [21] it was indeed found that the switching occurred as these relationships were violated. Finally, characteristic features of the actual transient switching process (like the mean duration of this process, the so-called mean switching time, or the distribution of the switching times) were predicted [18] and found experimentally to be in striking agreement with theory [21].

The above discussion demonstrates–in a concrete experimental system– that propositions (1) to (4) of the dynamic pattern theory are valid. To illustrate proposition (5), we briefly discuss the component level in the same system. To characterize the individual rhythmically moving limb as a dynamic pattern we may choose end effector position x and velocity $\dot{x}$ as collective variables–collective now with respect to underlying neuromuscular activities. The observed pattern can then be mapped onto a limit cycle attractor of the $(x, \dot{x})$ dynamics. The simplest model that captures experimentally observed amplitude-frequency-velocity relations is [17,22]:

$$\ddot{x} + f(x, \dot{x}) = 0 \tag{4}$$

with $f(x, \dot{x}) = \alpha\dot{x} + \omega^2 x + \beta\dot{x}^3 + \gamma\dot{x}x^2$ where $\alpha < 0, \beta > 0, \gamma > 0, \omega > 0$ are constants. Experimental consequences of this description (stability, autonomy, stable frequency-amplitude relation, dimensionality) were tested [22,23] with positive results. Thus we have a dynamic pattern description also on the component level. Haken, Kelso & Bunz [17] showed that a nonlinear coupling between two such component systems leads to the observed relative phase dynamics. For example, from the coupling structure

$$\begin{aligned} \ddot{x}_1 + f(x_1, \dot{x}_1) &= g_{intr}(x_1, \dot{x}_1; x_2, \dot{x}_2) \\ \ddot{x}_2 + f(x_2, \dot{x}_2) &= g_{intr}(x_2, \dot{x}_2; x_1, \dot{x}_1) \end{aligned} \tag{5}$$

with $g(x_1, \dot{x}_1, x_2, \dot{x}_2) = k(\dot{x}_1 - \dot{x}_2) + \ell(\dot{x}_1 - \dot{x}_2)(x_1 - x_2)^2$ the dynamics of relative phase (Eq. (2)) were derived in the limit $|\alpha| \ll \omega$.

A significant feature of this formulation is that the experimentally observed phase transition arises – because of the nonlinear coupling structure – through a simple change of the oscillation frequency, ω.

3. Dynamic theory of environmental information

Many behavioral patterns adjust continuously to the environment. As examples we may think of prey seeking or pursuit behaviors, catching, interception and tracking. A simple example of temporal behavior is the synchronization of a rhythmic movement with a metronome, i.e., a periodically structured environment. How does dynamic pattern theory accommodate this ability to adjust behavior continuously to the environment ? In the following we address this question [25].

In a first attempt we treat the environment in a parametric fashion, i.e., as that aspect of the physical environment that influences the behavioral patterns without itself being influenced by the system. In order for the physical environment to become meaningful to the biological system it is necessary that a dynamic relation exists between the system and its environment. One may conceive this process as the formation of a perception-action pattern (cf. also [26] for a dynamic model that addresses how a part of the physical environment becomes meaningful to a system). More specifically we conceptualize environmental information as a *required* behavioral pattern. It is thus *specific* information rather than the unspecific environmental conditions included in the parameters of the intrinsic dynamics.

The following assumptions define a dynamic theory of environmental information:

1. The order parameter dynamics in the absence of environmental information (so-called intrinsic dynamics) persist in the presence of such information;
2. Environmental information perturbs the vector-field of the order parameter dynamics (1), that is, it forms a part of the environmentally-modified behavioral dynamics. This perturbation is defined such that in the absence of intrinsic dynamics,

the order parameter dynamics have attractors corresponding to the required behavioral patterns.

A general mathematical formulation in the notation of Equ. (1) reads:

$$\dot{\mathbf{x}}_t = \mathbf{f}_{intr} + c_{env}\mathbf{f}_{env}(\mathbf{x}_t, t) \tag{6}$$

where c_{env} is the strength of the perturbation $\mathbf{f}_{env}$. The latter function is defined such that in (6) it attracts the collective variable $\mathbf{x}_t$ toward the required behavior. For illustration, consider the case where the *required* behavioral pattern is a fixed point $\mathbf{x}_{req}$ of the collective variable dynamics. Locally around $\mathbf{x}_{req}$ we can then generally linearize $\mathbf{f}_{env}$ as ($c_{env} > 0$):

$$c_{env}\mathbf{f}_{env}(\mathbf{x}_t, t) = -c_{env}(\mathbf{x}_t - \mathbf{x}_{req}) \tag{7}$$

Our definition of environmental information purely in terms of behavioral pattern dynamics may help to clarify the term intrinsic dynamics as used above. Intrinsic dynamics refers here to the dynamics of the behavioral patterns in the absence of a specific behavioral requirement. This does not imply that the intrinsic dynamics are necessarily more "hard-wired" than, for example, the environmentally specified dynamics. Rather, the division into intrinsic and environmentally-specified dynamics depends on which environmental aspect requires an adjustment of the behavioral pattern.

Although formulated quite generally, already a few predictions emerge from such a dynamic theory of environmental information. First, the presence of intrinsic dynamics in (6) will be felt as a systematic deviation of the *actual* behavioral pattern from the *required* behavioral pattern. This deviation will be in the direction (in terms of the collective variable space) of the stable states of the intrinsic dynamics. Second, qualitative effects of environmental information can be seen when the corresponding perturbation in (6) breaks a symmetry of the intrinsic dynamics. Then asymmetric solutions may be stable, or different solutions that are related by the symmetry operation may differ in their stability. Finally, another important qualitative effect of environmental information is the possibility of phase transitions in the order parameter dynamics as the environmental information changes. Such phase transitions exist if the intrinsic dynamics (i.e. the limit $c_{env} \to 0$ of (6)) are qualitatively different from the purely environmentally-specified dynamics (i.e. the limit $c_{env} \gg 1$ of (6)). For example, multiple patterns may be intrinsically stable, while only one–the required pattern–may be stable for large c_{env}. We will clarify this case in the following example.

The experimental model system that guided the formulation of the present theory again involves bimanual, rhythmic movement coordination, but now in the presence of two metronomes (Tuller & Kelso [27, 28]) The two metronomes have the same frequency, well below the typical transition frequencies in Kelso's experiments [16] and their relative phase, ψ (the required behavioral pattern) is varied experimentally. The basic experimental results can be summarized as follows [28]: First, the performed relative phase deviates systematically from the required relative phase. The performance is attracted either to the in-phase or to the anti-phase pattern. Second, the variability of the performed relative phase is markedly lower in the in-phase and anti- phase pattern than at the intermediate required relative phase conditions. A typical subject's data are shown in the left half of Fig. 1.

To theoretically understand this experimental system we proceed as follows [24]: On one level of analysis we choose relative phase, ϕ as the collective variable. The dynamics (3) established for the case of rhythmic movement coordination without a metronome (or with a single metronome) are now viewed as intrinsic dynamics. The environment is characterized only by the parameter corresponding to the required relative phase, ψ. It is through the formation of perception-action patterns (of each finger with a metronome) that this parameter becomes available and meaningful to the movement system. The environmental information is now a perturbation of the intrinsic dynamics (3) that attracts relative phase toward the required relative phase. The simplest functional form fulfilling certain symmetry requirements [24] is

$$-c_{env} \sin\left(\frac{\phi - \psi)}{2}\right) \tag{8}$$

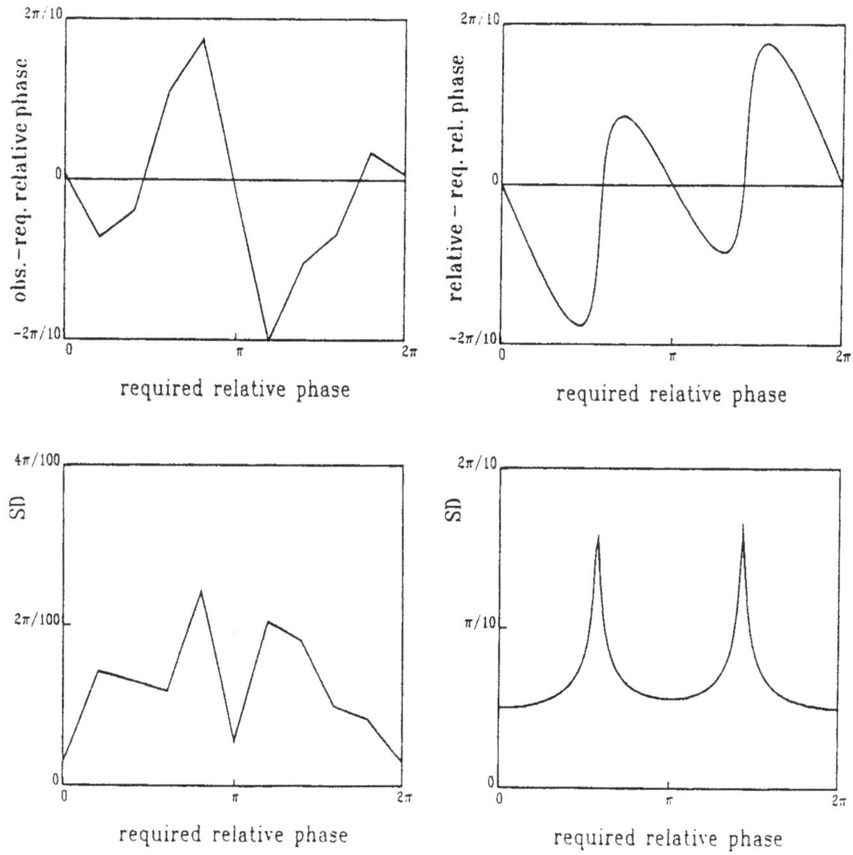

Fig. 1: Left: Deviation of performed from required relative phase (top) and variability of performed relative phase (bottom) as a function of required relative phase for a typical subject from Tuller & Kelso's experiment [28]. Right: The same measures as calculated from the model (9) (see text for details).

where $c_{env} > 0$ is a constant. Thus the complete model is:

$$\dot{\phi} = -a\sin(\phi) - 2b\sin(2\phi) - c_{env}\sin\left(\frac{\phi - \psi}{2}\right) + \sqrt{Q}\xi_t \qquad (9)$$

Note that the environmental term (8) breaks the $\phi \rightarrow \phi$ symmetry, so that left and right fingers are no longer exchangable, thus comforming to the experimental system.

From this simple model we can already obtain a number of interesting results. First, we can capture the experimental findings of Tuller and Kelso [28]. Using parameter values for the intrinsic dynamics as before [18] and adjusting the strength c_{env} of the perturbation we can determine the stationary states of (9) and their deviation from the required relative phase (see Fig. 1, right, top). Locally around that state we can determine the variability of relative phase, shown in the lower right of Fig. 1 (for details see [24]. (In fact positive statements can be made about the order-of-magnitude quantitative agreement with experiment [24]). Second, a testable prediction results if we examine the system above the phase transition of the intrinsic dynamics. There the attraction of performance to the anti-phase pattern should disappear and the differential variability be less pronounced at $\phi = \pm\pi$. Third, as discussed above, phase transitions are predicted as a function of environmental information. In the limit $c_{env} \rightarrow 0$ the dynamics have two stable states at $\phi = 0$ and $\phi = \pm\pi$. For large c_{env} only one state, slightly detuned from $\phi = \psi$ is stable. As c_{env} increases (which may be experimentally realizable through psychophysical techniques) one of the intrinsically stable patterns looses stability. This is shown in Fig. 2 for typical parameter values of the intrinsic dynamics. Another way of studying this phase transition is to vary the required relative phase at intermediate values of c_{env}. In such cases, as shown in Fig. 3, hysteresis effects in ϕ are predicted. In Tuller & Kelso's experiments an aspect of such hysteresis was, in fact, observed, namely a bimodal distribution of actually produced relative phase at intermediate required relative phases [28].

In the foregoing model system, the influence of environmental information may also be understood on the level of the component oscillators (cf. Eqs. (4) and (5)) [25]. The perturbation of the vectorfield in (5) takes the form of nonautonomous driving forces that attract the oscillators to the phase of their corresponding metronomes. Such a theory allows us to address the question of the nature of the perception-action patterns in which the physical environment is meaningful to the coordination system. Again the dynamics of relative phase, here Eq. (9), can be derived from the underlying

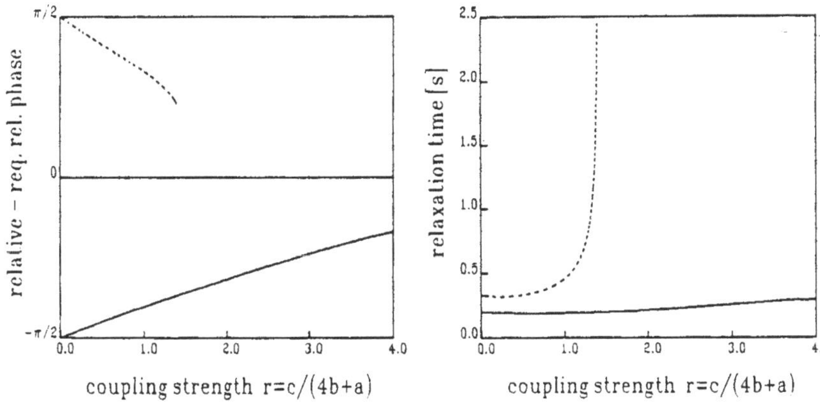

Fig. 2: The deviation of the stationary states of relative phase from the required relative phase (left) and the relaxation time of that stationary state (right) as a function of the strength r of the environmentally-specified perturbation (8) relative to the intrinsic dynamics (3). One of the two states existing at low r (dashed) looses its stability as r increases.

oscillatory dynamics. For more information we refer the reader to [25].

4. Dynamic Theory of memorized information

Our definition, in purely dynamic terms, of specific information that influences behavioral patterns is open to considerable generalization. In this section we show that the influence of memory can be viewed in this way. In Section 5 we will discuss how intentional information may also be captured dynamically.

As a first step, we study the limiting case in which a specific behavioral pattern has been practiced until learned and has been elicited (say, by an environmental event) or activated (directly from memory). In this case we can treat memory as a set of parameters that influences the behavioral dynamics, but is not influenced in return back by these dynamics. Nor are the memory parameters dynamic themselves. Thus, in this limiting case, we can proceed in close analogy to our dynamic theory of environmental information. We conceptualize the contents of memory as a *required, memorized*

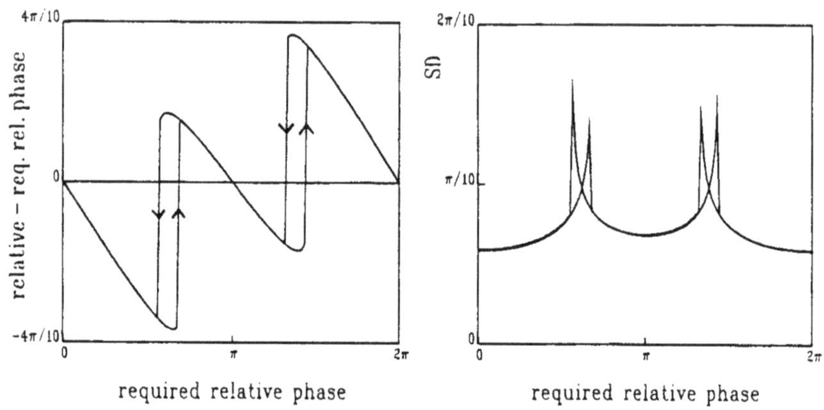

Fig. 3: The deviation of the stationary states of relative phase from the required relative phase (left) and the variability of those states (right) as a function of required relative phase at intermediate values of the strength r of the environmentally-specified perturbation (8) relative to the intrinsic dynamics (3). Close to $\pi/2$ and to $3\pi/2$ a bistable regime exists leading to hysteresis effects if the parameter (required relative phase) is scaled monotonically in time. The arrows indicate how the direction of such scaling differentiates the two states.

behavioral pattern. The following assumptions then define a dynamic theory of memorized information: (1) The intrinsic and, if present, the environmentally-specified dynamics persist. (2) Memorized information is a perturbation of the vector-field of the order parameter dynamics. The perturbation attracts the collective variables to the memorized required behavioral pattern. The strength of the perturbation is a measure of the relative strength of the intrinsic and the memorized dynamics. A general mathematical formulation (in the absence of environmental information) in the notation of Eq. (1) is:

$$\dot{\mathbf{x}}_t = \mathbf{f}_{intr} + c_{mem}\mathbf{f}_{mem}(\mathbf{x}_t) \tag{10}$$

where c_{mem} is the strength of the perturbation and $\mathbf{f}_{mem}$ is defined by its attractor structure. Again we may illustrate this in the case in which the memorized pattern is a fixed point of the collective variable dynamics. Then a linear approximation of $\mathbf{f}_{mem}$ that is local to the memorized pattern $\mathbf{x}_{mem}$ is:

$$c_{mem}\mathbf{f}_{mem}(\mathbf{x}_t) = -c_{mem}(\mathbf{x}_t - \mathbf{x}_{mem}) \tag{11}$$

with $c_{mem} > 0$.

In analogy to the dynamic theory of environmental information, three general predictions follow: (1) The presence of intrinsic dynamics leads to systematic deviation from the memorized pattern toward the intrinsically stable patterns. (2) The symmetry of the intrinsic dynamics can be broken by the memorized information. (There is an interesting analogy here with the breaking of the symmetry between figure and background by memory in Shimizu, Yamaguchi and Satoh's dynamic model of perception and recognition [26], a phenomenon they call *semantic separation*.) (3) The dynamics of the behavioral patterns may change qualitatively as a function of memorized information. Thus, for example, a phase transition may occur as the strength of the memorized information increases from zero (no influence of memory) to large values (memory-dominated behavior). Such a prediction is of particular interest here, because a possible view of learning and recall within our theory is that such processes lead to increasing strength of the perturbation represented by memorized information. Thus, phase transitions during learning and recall are predicted [29].

We conceived of the parallel between environmental and memorized information when trying to understand the experimental model system discussed in Sect. 3. The striking fact was that similar experimental results to Tuller and Kelso [28] had been obtained earlier by Yamanishi, Kamato & Suzuki [30] under conditions in which the different patterns were produced from memory. (We note that Yamanishi and colleagues had a different theoretical approach and motivation, based on phase resetting analysis.) Briefly, Yamanishi and colleagues trained subjects in the dual metronome task described previously in the context of Tuller and Kelso's experiment [28] by providing feedback about deviations from the to-be-learned pattern. Then one of ten trained relative timing patterns was *elicited* by providing visual metronome information for ten cycles, following which the metronomes were turned off. The experimental results, obtained without the metronomes present, have the same basic features as Tuller and Kelso's, namely (1) Systematic deviation from the trained pattern in the direction of in-phase or anti-phase pattern; (2) Differential stability of different conditions: low in in-phase and anti-phase conditions, high at intermediate conditions.

The implementation of our theory is straightforward: On one level of observation we take relative phase again as a relevant collective variable and keep Eq. (3) as the intrinsic dynamics. The memorized pattern is described by a parameter, ψ_{mem}, corresponding to memorized relative phase. When a particular relative phase condition ψ_{mem} has been elicited, the memorized information is a perturbation of (3) that attracts relative phase toward ψ_{mem}. The simplest functional form fulfilling symmetry requirements is [24]:

$$-c_{mem}\sin(\phi - \psi_{mem}) \tag{12}$$

with $c_{mem} > 0$. Thus we have:

$$\dot{\phi} = -a\sin(\phi) - 2b\sin(2\phi) - c_{mem}\sin(\phi - \psi_{mem}) + \sqrt{Q}\xi_t \tag{13}$$

The results are obviously closely analogous to those obtained in Sect. 3. In particular we can capture Yamanishi et al.'s [30] results, and predict the behavior at higher frequencies. Also we predict a phase transition, now as a function of memory strength. Such phase transitions offer a very interesting window into studying learning and are a novel aspect of dynamic pattern theory [29]. In this context, it is worth mentioning that Yamanishi et al. [30] reported bimodal distributions of relative phase at intermediate relative phase conditions. Bimodality is here a signature of the hysteresis phenomenon predicted from (13) (cf. Fig. 3) and represents one aspect of the new phase transition that arises due to memorized information.

Finally, it is possible to achieve a complete theoretical understanding also on the level of the component oscillators [25]. In this case additional coupling functions act as perturbations onto the oscillators Eq. (5), attracting their phases to memorized relative phase.

The relative phase dynamics (13) can again be derived from the coupled oscillators. It is remarkable that functionally equivalent dynamics on the level of relative phase (Equ. (9) and (13)) result from very different underlying oscillatory dynamics (driven versus elastically coupled).

5. Intentional change of behavioral pattern

How can a biological system actively seek a behavioral pattern? Can intention be part of a dynamic pattern theory ? We conceptualize intentional information as an intended behavioral pattern, i.e., the *goal* of the intentional act. The dynamic theory is defined by the following assumptions: (1) The intrinsic and, if present, environmentally specified or memorized dynamics persist. (2) Intentional information is a perturbation of the order parameter dynamics that attracts the behavioral pattern toward the intended behavioral pattern. Mathematically, in the notation of Eq. (1) we may write:

$$\dot{\mathbf{x}}_t = \mathbf{f}_{intr}(\mathbf{x}_t) + c_{int}\mathbf{f}_{int}(\mathbf{x}_t) \tag{14}$$

where the constant c_{int} measures the strength of the influence of intention and $\mathbf{f}_{int}$ is defined by its attractor structure. So far we have, however, swept an important question under the rug: Which patterns can be intended? This is clearly a difficult question. At present we see three ways of giving it meaning. First, it seems reasonable to assume that the intrinsically stable patterns can be intended because these patterns are already attractors of the intrinsic dynamics. The perturbation by intention in (14) amounts to modifying the strength of this attractor. Thus, *switching among intrinsic patterns*, as we shall discuss below, may be addressed in this manner. Second, environmentally specified patterns may also be viewed as intended patterns. This amounts to a possible modification of the strength of a perturbation in (6) and is related to the interesting problem, briefly mentioned in Section 3, of how the biological system renders environmental information meaningful. Third, and relatedly, memorized behavioral patterns may be intended. This relates to the problem of activation of memorized information and can be addressed within the current theory [29].

It is clear from the foregoing discussion that, in a general sense, we have to introduce new degrees of freedom or non-autonomy to express the onset of intention. Such a step is, however, beyond the scope of the present contribution. Some general results follow, however, from the present formulation, which are: (1) That the intrinsic dynamics will influence the intentional switching among available behavioral patterns; (2) That intention, defined here as part of the dynamics, can change dynamic properties such as stability even without changing the behavior itself.

Kelso and Scholz have recently completed an experiment that demonstrates the influence of the intrinsic dynamics on intentional switching [31]. Again, the task was rhythmic bimanual movement starting either in an in-phase or an anti-phase coordination pattern. For the first ten cycles the movement was paced by a metronome, which was then turned off. The subject's instruction was to continue the oscillation at the initial frequency. After a further ten cycles a signal indicated to the subject that she or he should switch as quickly as possible to the other mode of coordination. The paradigm is illustrated in Fig. 4.

Using interactive computer displays the length of the actual transient between the two modes of coordination was determined (as also illustrated in Fig. 4). The mean of these durations, called mean switching time, is shown in Fig. 5 as a function of the various oscillation frequencies used. Clearly, switching from the in-phase to the anti-phase mode is slower by approximately a factor of 2 than switching in the opposite direction. Fig. 6 shows the experimental distributions of the switching times in the two directions. Note the longer tail and larger mean of the switching time in the anti- to in-phase condition.

A theoretical model is straightforward [32]. Using relative phase as the order parameter the intrinsic dynamics are Eq. (3). The intended pattern, ψ is either 0 (for anti- to in-phase switching) or $\pm\pi$ (for in- to anti-phase switching). The onset of the intention, t_0 may be viewed as the time when the transient starts. In a simplest approximation we assume a perturbation of the form:

$$-c_{int}\theta(t-t_0)\sin(\phi-\psi) \tag{15}$$

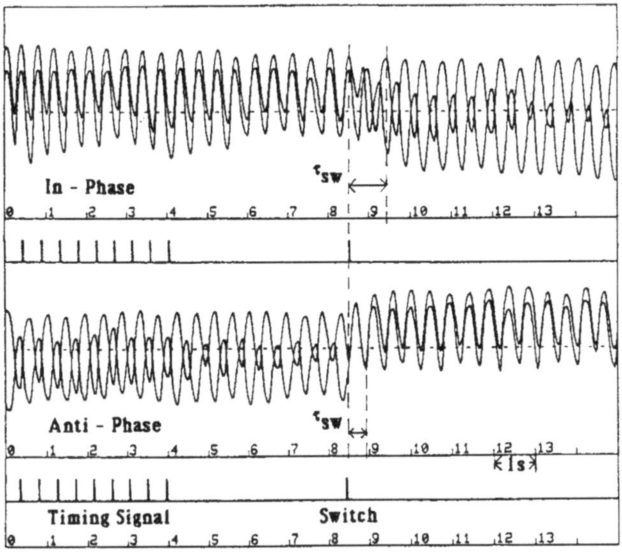

Fig. 4: Two typical runs of Kelso and Scholz's experiment. In each case the trajectories of left and right index finger are shown as well as the 10 metronome pulses that initially establish the oscillation frequency and the single pulse that cues the intentional switching process. Top: switch from in-phase to anti-phase; Bottom: switch from anti-phase to in-phase.

that attracts relative phase, ϕ to intended relative phase, ψ with strength $c_{int} > 0$. Here we assumed an abrupt onset of the intention

$$\theta(t) = \begin{cases} 0 & \text{for } t < 0; \\ 1 & \text{for } t > 0. \end{cases}$$

The complete model is thus:

$$\begin{aligned} \dot{\phi} &= -a\sin(\phi) - 2b\sin(2\phi) - c_{int}\theta(t - t_0)\sin(\phi - \psi) + \sqrt{Q}\xi_t \\ &= -\frac{dV}{d\phi} \end{aligned} \tag{16}$$

There are several methods for characterizing the transient switching process [18]. We may calculate the probability density

$$v(t) = \int_{\psi-\delta}^{\psi+\delta} \dot{P}(\phi, t)d\phi \tag{17}$$

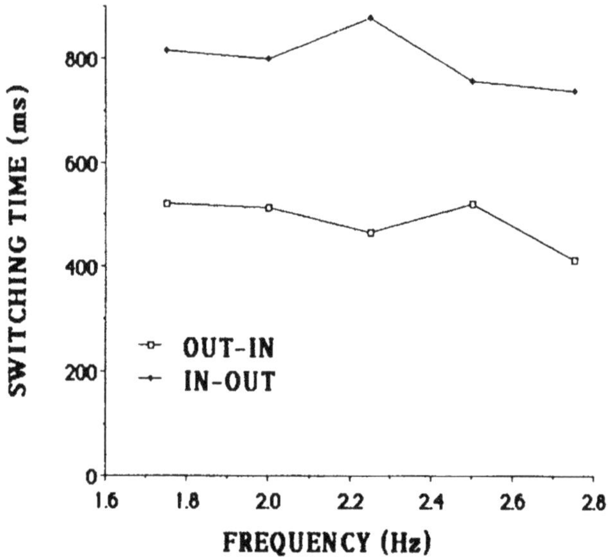

Fig. 5: The mean switching time (mean length of transients from one mode of coordination to other, mean over subjects and runs) as a function of oscillation frequency for the two directions of switching in Kelso and Scholz's experiment.

that switching is completed at time t. Here $P(\phi, t)$ is the time-dependent probability density of relative phase that may be determined by solving the Fokker-Planck equation associated with (16). The mean switching time is then

$$\tau_{switch} = \int_0^\infty t \quad v(t) \quad dt \tag{18}$$

We determined these measures numerically (see [32] for details) along a path in the parameter plane (a,b) that was established from earlier work, to correspond roughly to the frequency increase in the experiment ([18], Fig. 4 there). The results for τ_{switch} are shown in Fig. 7. On the left the attractors at $\phi = 0$ and $\phi = \pm\pi$ are equivalent and the mean switching times are very similar (different only due to the common initial condition used). As the anti-phase state becomes

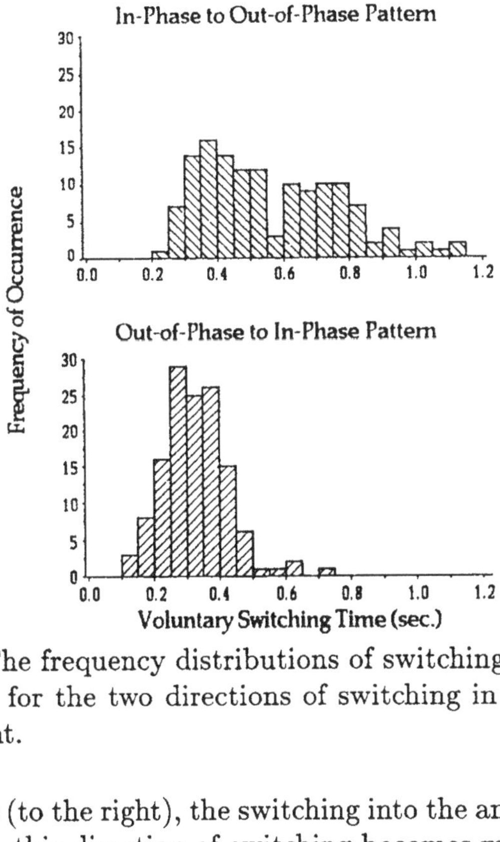

Fig. 6: The frequency distributions of switching times (all subjects and runs) for the two directions of switching in Kelso and Scholz's experiment.

less stable (to the right), the switching into the anti-phase state slows down, i.e., this direction of switching becomes more "difficult" than the opposite one. The distributions of switching time obtained for intermediate parameter values through Eq. (17) are shown in Fig. 8. The agreement with experiment is quite astounding if one considers that only one parameter c_{int} was adjusted to obtain the correct order of magnitude of only one of the times ! Another measure of switching time, the mean first passage time between the two states, yields highly consistent results [32].

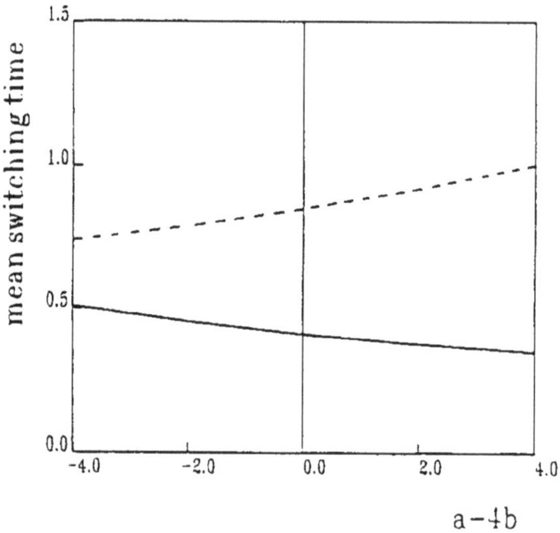

Fig. 7: The mean switchings times as calculated from the theoretical model Eq. (16) along a particular path in the parameter plane (a, b). Note that the mean switching time for switching from in-phase to anti-phase (dashed) is much larger than that for switching from anti-phase to in-phase (solid). This difference is due to a difference in stability: For $a < 4b$ both $\phi = 0$ and $\phi = \pm\pi$ are stable, but $\phi = \pm\pi$ less so. At $a = 4b$ the anti-phase state looses stability. For $a > 4b$ only $\phi = 0$ is stable.

6. Summary

We reviewed the dynamic pattern theory of biological coordination [8,9] which stresses collective variables and their dynamics, stability, fluctuations, time scales and linkages of levels of analysis. An experimental model system from human rhythmic movement coordination illustrated the close relation of theory and experiment that is necessary and possible in this approach. The main thrust of the present paper, however, was to elaborate the theory to show how information with meaning to the biological system may be understood from a dynamic pattern perspective. The key idea was to view meaningful information as part of the behavioral pattern dynamics that attracts the system toward an environmentally-specified, memorized

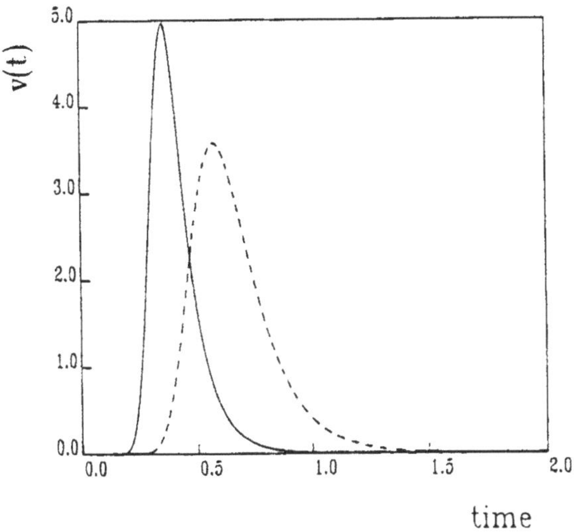

Fig. 8: The theoretical distributions of switching time (Eq. (17)) for switching from anti-phase to in phase (solid) and from in-phase to anti-phase (dashed). Note that for both figs. 7 and 8 only one parameter (c_{int}) was adjusted.

or intended behavioral pattern. As general results of such a theory, we found that the intrinsic dynamics (in the absence of meaningful information) influence the processes of perception, memory and intention. Also, qualitative change of the behavioral pattern dynamics is predicted due to these processes. In addition to making new predictions, now open to empirical test, we were able to show for each of the three types of information–environmental, memorized, and intentional–how the present theory accounts for observed behavior.

Acknowledgements

This work was supported by a NIMH (Fundamental Neurosciences Branch) Grant MH42900-01 and a joint contract N00014-87-G-0156 from the U.S. Office of Naval Research (Physics Division and Integrated Biology Program) and the Air Force Office of Scientific Research (Artificial Neural Networks Program). We express our warm thanks to Dr. J.P. Scholz for collaborating and sharing experimental details with us. We would also like to thank Prof. H. Shimizu and Dr. Y. Yamaguchi for stimulating discussions.

References

[1] Haken, H.: *Synergetics-An introduction*, 3rd ed., Springer Verlag, Berlin, Heidelberg, New York, 1983.

[2] Kelso, J.A.S., Schöner, G., Scholz, J.P., & Haken, H.: Phase-locked modes, phase transitions and component oscillators in biological motion, *Physica Scripta* **35**, 79-87 (1987)

[3] Haken, H.: *Advanced synergetics*, Springer Verlag, Berlin, Heidelberg, New York, 1983

[4] Gierer, A. & Meinhardt, H.: Biological pattern formation involving lateral inhibition. *Lectures on Mathematics in the Life Sciences* **7**, 163 (1974)

[5] Haken, H. & Olbrich, H.: Analytical treatment of pattern formation in the Gierer-Meinhardt model of morphogenesis, *J. Math. Biol.* **6**, 317 (1978)

[6] Gerish, G.: Chemotaxis in Dictyostelium, *Ann. Rev. Physiol.* **44**, 535 (1982)

[7] Segel, L.A.: *Modeling dynamic phenomena in molecular and cellular biology*, Cambridge Univsity Press (1984)

[8] Kelso, J.A.S. & Schöner, G.: Toward a physical (synergetic) theory of biological coordination, *Springer Proc. Phys.* **19**, 224 (1987)

[9] Kelso, J.A.S. & Schöner, G.: Self-organization of coordinative movement patterns, *Human Movement Science* (in press)

[10] Shimizu, H. & Haken, H.: Cooperative dynamics in organelles, *J. Theoret. Biology* **104**, 261, (1983)

[11] Goldberger, A.L. & Rigney D.R.: Sudden death is not chaos, this volume

[12] Rapp, P.E., Albano, A.M., & Mees, A.I.: Calculation of correlation dimensions from experimental data: Progress and porblems, this volume

[13] Schöner, G. & Kelso, J.A.S.: Dynamic pattern generation in behavioral and neural systems, *Science* (in press)

[14] Mpitsos G.J., Creech H.C., Cohan, C.S. & Mendelson, M.: Variability and chaos: neurointegrative principles in the generation of rhythmic motor patterns, this volume

[15] Selverston, A.I.: Invertebrate central pattern generators as computational data bases, this volume

[16] Kelso, J.A.S.: Phase transitions and critical behavior in human bimanual coordination, *Am. J. Physiol.: Reg. Integ. Comp.* **15**, R1000 (1984)

[17] Haken, H., Kelso, J.A.S. & Bunz, H.: A theoretical model of phase transitions in human bimanual coordination *Biol. Cybern.* **51**, 347-356 (1985)

[18] Schöner, G., Haken, H. & Kelso, J.A.S.: A stochastic theory of phase transitions in human hand movement, *Biol. Cybern.* **53**, 247 (1986)

[19] Kelso, J.A.S. & Scholz, J.P.: Cooperative phenomena in biological motion, in: Haken, H. (ed.): *Complex systems–operational approaches in neurobiology, physical systems and computers*, Springer, Berlin, Heideberg, New York, 1985)

[20] Kelso, J.A.S., Scholz, J.P. & Schöner, G.: Nonequilibrium phase transitions in coordinated biological motion: critical fluctuations, *Phys. Letts.* **A118**, 279 (1986)

[21] Scholz, J.P, Kelso, J.A.S. & Schöner, G.: Nonequilibrium phase transitions in coordinated biological motion: critical slowing down and switching time, *Phys. Letts.* **A123**, 390 (1987)

[22] Kay, B.A., Kelso, J.A.S., Saltzman, E.L. & Schöner, G.: Space-

time behavior of single and bimanual rhythmic movements: data and limit cycle model, *J. exp. Psych.: Hum. Perc. Perf.* **13**, 178 (1987)

[23] Kay, B.A.: Dynamic modeling of rhythmic limb movements: Converging on a description of the component oscillators, Ph.D. Thesis, Univ. of Connecticut, Storrs (1986)

[24] Schöner, G. & Kelso, J.A.S.: A synergetic theory of environmentally-specified and learned patterns of movement coordination. I. Relative phase dynamics, *Biol. Cybern.* (in press)

[25] Schöner, G. & Kelso, J.A.S.: A synergetic theory of environmentally-specified and learned patterns of movement coordination. II. Component oscillator dynamics, *Biol. Cybern.* (in press)

[26] Shimizu, H., Yamaguchi, Y., & Satoh, K.: Holovision: A semantic information processor for visual perception, this volume

[27] Tuller, B. & Kelso, J.A.S.: Coordination in normal and split-brain patients, paper presented at Psychonomic Society, Boston MA (1985)

[28] Tuller, B. & Kelso, J.A.S.: Environmentally-elicited patterns of movement coordination in normal and split-brain subjects, *Exp. Brain Research* (submitted)

[29] Schöner, G. & Kelso, J.A.S.: A theory of learning and recall in biological coordination, paper presented at conference on *Dynamic patterns in complex systems*, Fort Lauderdale, October 1987.

[30] Yamanishi. J., Kawato. M. & Suzuki. R.: Two coupled oscillators as a model for the coordinated finger tapping by both hands, *Biol. Cybern.* **37**, 219 (1980)

[31] Kelso, J.A.S. & Scholz, J.P. (to be published)

[32] Schöner, G. & Kelso, J.A.S. (to be published)

Editors' Note on R. Landauer's Paper
"Fluctuations in Bistable Tunnel Diode Circuit"
(J. Appl. Phys., **33**, *2209, 1962)*

Multistable systems, with more than one state of local stability, are widespread and appear in a number of papers in this volume. Moreover, all systems are subject to random influences that can affect the system's long-term behavior. Thus, as Landauer's paper points out, questions about relative stability, identifying the most likely state, in active dissipative systems far from equilibrium cannot, in general, be answered by appeal to deterministic (i.e., noiseless) equations of motion.

Noise in the tunnel diode circuit, discussed in this reprint of Landauer's pioneering work (done when such circuits were serious technological candidates), depends on the state of the system. The role of such nonuniform noise has become widely appreciated in recent years, and is now variously discussed in terms such as ***multiplicative noise, external noise, noise induced transitions*** (see e.g., W. Horsthemke and R. Lefever, *Noise Induced Transitions* (Springer Series in Synergetics, Vol. **15**), 1984). How to treat multistable, driven systems far from equilibrium, was not obvious a quarter century ago when the following paper first appeared. For these reasons, we reprint it here with the kind permission of Dr. Landauer and the American Institute of Physics. The interested reader can consult a broader, more conceptual discussion by Landauer (*Physics Today,* **31**, 23, 1978) that is closer to his actual talk at the conference.

Reprinted from JOURNAL OF APPLIED PHYSICS, Vol. 33, No. 7, 2209–2216, July, 1962

Fluctuations in Bistable Tunnel Diode Circuits

R. LANDAUER
Thomas J. Watson Research Center, International Business Machines Corporation, Yorktown Heights, New York
(Received January 24, 1962)

In a series of preceding papers connection has been made between the physics of the activated jump over a static potential barrier and related computing devices. This paper treats a more complicated system, namely, the tunnel diode, which stores information in one of two possible dissipative states. The activated jump between dissipative states is analyzed. An idealized physical model is used to find which of the two states is the really preferred one, and to also evaluate the rate at which fluctuations bring about an approach to this preferred distribution. To simplify the analysis, the case where the tunnel diode is in series with a vacuum diode, rather than a resistor, is emphasized. For typical germanium Esaki diodes, activated jumps are improbable if the junction cross section exceeds 10^{-11} cm^2.

1. INTRODUCTION

IN recent years it has become clear that there is a continuing search for smaller and less dissipative computer elements. This naturally poses the question: What are the ultimate physical limitations which determine the available progress in this direction? Three previous publications[1–3] have examined the effect of interactions of a bistable system with its thermal environment, emphasizing the type of system illustrated in Fig. 1, which holds information in one of two symmetrical and dissipationless states. The first of these discussions[1] pointed out that the thermal fluctuations determined an optimum size storage element; one which was large enough to be reasonably stable against loss of information due to the thermal fluctuations, but large enough for this purpose and no larger. This initial discussion evaluated the optimum-sized element through the use of information theory, thereby assuming that coding methods were available, permitting the use of redundancy in the memory to compensate for the existence of somewhat unreliable elements. A simpler approach would require the barrier height U of each element to be large enough so that the probability of thermally induced transitions from one well to the other during the information storage period, is extremely small. The simpler approach yields optimum volumes which are only slightly larger than those yielded by the use of information theory. In the second publication[2] the frequency factor ν, in the expression $\nu e^{-U/kT}$ for the probability, per unit time, of a thermally activated passage over a barrier, is analyzed, extending Kramers'[4] original one-dimensional treatment to many dimensions. The same interactions of a system with its thermal environments which cause fluctuation phenomena also account for dissipation, as is emphasized by the existence of dissipation-fluctuation theorems.[5] The effects of dissipation are discussed in the third paper[3] which shows that logical systems require dissipative effects; they are not just an inevitable nuisance.

These preceding discussions leave two obvious gaps. First of all, explicit recognition of quantum-mechanical processes in these discussions are somewhat limited. The most important quantum-mechanical effects can be expected to come from the quantization of signals exchanged by logical elements. A start in this direction has been made by the work of Lasher[6] who has extended the classical channel capacity theorem to the quantized case. This is, of course, limited to linear circuits. It is hoped that some of the concepts involved, however, are applicable to logical devices.

The other conspicuous gap in our earlier treatment arises from the limitation to devices which store information in a nondissipative state. Many of the simplest devices which hold information in a time invariant state are dissipative. Bistable vacuum tube and transistor circuits are in this category. Other examples involve relays and bistable circuits made up from combinations of electrically powered light sources and photoconductors.[7] This note is a first attempt to deal with thermal fluctuations in such devices. Our specific treatment is limited to the simplest example we could find, a somewhat idealized model of a bistable circuit using a tunnel diode. One circuit is shown in Fig. 2. The tunnel diode characteristic and the associated load line are shown in Fig. 3. Points A and C in Fig. 3 are stable points. Point B is unstable. These points are analogous to points A, B, and C in Fig. 1.

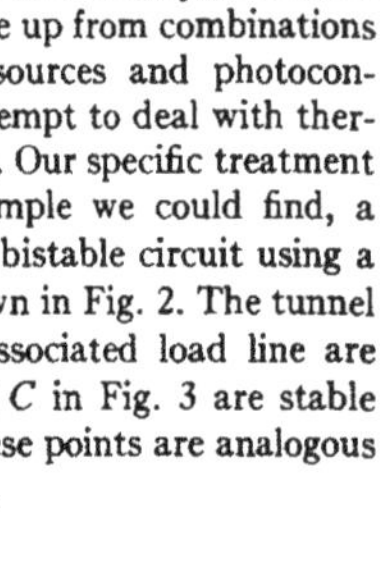

FIG. 1. Bistable potential well. x is a generalized coordinate representing quantity which is switched.

[1] J. A. Swanson, IBM J. Research Develop. 4, 305 (1960).
[2] R. Landauer and J. A. Swanson, Phys. Rev. 121, 1668 (1961).
[3] R. Landauer, IBM J. Research Develop. 5, 183 (1961).
[4] H. A. Kramers, Physica 7, 284 (1940).
[5] W. Bernard and H. B. Callen, Revs. Modern Phys. 31, 1017 (1959).
[6] G. Lasher, paper presented at Quantum Electronics Conference, Berkeley, California, *A Quantum Statistical Treatment of the Channel Capacity Problem of Information Theory*, to appear in the Conference Proceedings.
[7] G. Diemer and J. G. Van Santen, Philips Research Repts. 15, 368 (1960).

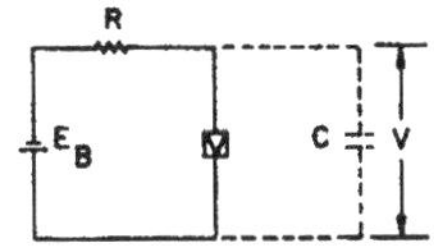

FIG. 2. Tunnel diode fed through series resistance R is diode capacitance C. V is the potential across the diode.

Despite the idealizations which are to be made it is nevertheless hoped that the physical concepts involved have a degree of generality, even if the detailed equations do not. Fluctuations in nonlinear circuits have been discussed many times before and we will only refer the reader to some of the latest discussions.[8–10] Lax,[11] in particular, emphasizes fluctuations about a state which is not necessarily one of thermal equilibrium and much of our content is implicitly treated in his general formulation. Most of the earlier treatments, however, emphasize the role of the frequent small fluctuations, and many are limited to expansions in the parameter characterizing the degree of nonlinearity. While our treatment refers to an idealized situation, and is at best of somewhat playful conceptual interest, the work suffers no more from these defects than the earlier literature on fluctuations in nonlinear circuits. Furthermore, we discuss a situation in which miniaturization is of real technological importance, though practical size reduction is not likely to be pushed to an extent that will make thermal fluctuations of real practical significance.

2. TUNNEL DIODES

Fluctuations in tunnel diodes have been treated in the literature.[12,13] Tiemann[13] points out that the current through a tunnel diode can be broken up into two terms

$$i=i_E+i_Z, \qquad (2.1)$$

where i_E is due to tunneling from the conduction band into the valence band, and i_Z is the "Zener" current due to valence electrons tunneling into the conduction band. In the forward direction, i_E is dominant and causes the negative resistance behavior; i_Z is of opposite sign to i_E. These two current streams contribute independent shot effect fluctuations, giving an effective noise current

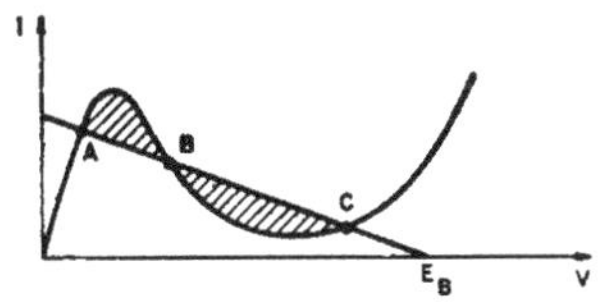

FIG. 3. Forward diode characteristic and load line.

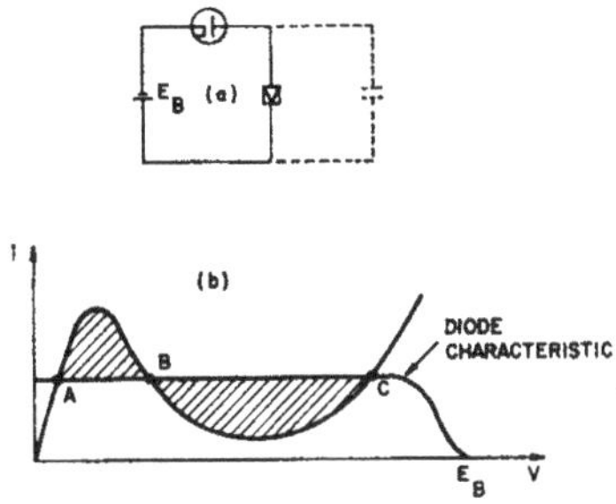

FIG. 4. Tunnel diode fed through diode drawing its full emission current. Circuit at top, characteristics at bottom.

generator in the bandwidth Δf,

$$\langle i^2\rangle=2e|i_E|\Delta f+2e|i_Z|\Delta f. \qquad (2.2)$$

In the subsequent discussions we will have three separate circuits in mind. One is that already illustrated in Fig. 2, and involves both shot noise sources as well as Nyquist noise generated in the resistor. It will turn out, however, that circuits which contain only shot noise circuits can be treated accurately in a much simpler fashion. Such circuits are illustrated in Figs. 4(a) and 5(a), with their respective characteristics shown in Figs. 4(b) and 5(b). In Fig. 4 we have the resistor replaced by a vacuum diode, which is drawing its full thermionic emission current, and contributing uncompensated shot noise. Figure 5 illustrates the "Goto twin" circuit.[14] For both Figs. 4 and 5 the additional capacitance, shunting the element which has replaced the resistor, can be lumped in parallel with the original tunnel diode capacitance, as far as the ac circuit behavior is concerned. The circuit of Fig. 4 will be explicitly discussed, but our remarks will apply with trivial modifications to the twin circuit of Fig. 5. The resistive load line case of Fig. 3 is the subject of an Appendix.

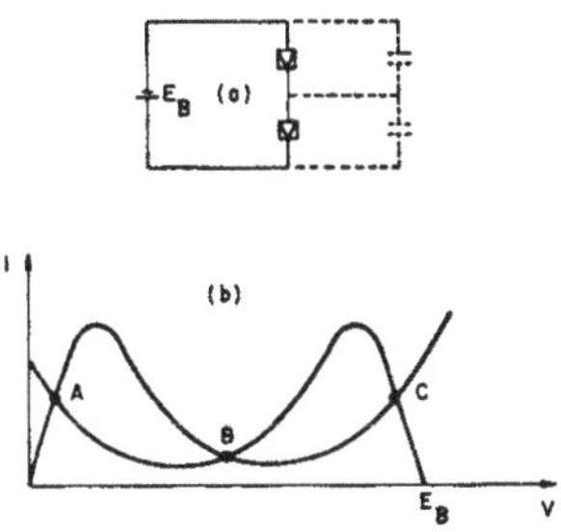

FIG. 5. Goto twin circuit and characteristic. Only one of the two diodes can be in a high voltage state, the other is in a low voltage state.

[8] W. Bernard and H. B. Callen, Phys. Rev. **118**, 1466 (1960).
[9] C. T. J. Alkemade and D. K. C. MacDonald, Conference on Statistical Mechanics and Irreversibility, held at Queen Mary College, London, December 19, 1960.
[10] N. G. van Kampen, J. Math. Phys. **2**, 542 (1961).
[11] M. Lax, Revs. Modern Phys. **32**, 25 (1960).
[12] R. A. Pucel, Proc. IRE **49**, 1080 (1961).
[13] J. J. Tiemann, Proc. IRE **48**, 1418 (1960).

[14] E. Goto, K. Murata, K. Nakazawa, K. Nakagawa, T. Motsoka, Y. Motsuoka, Y. Ishibashi, H. Ishida, T. Soma, and E. Wade, IRE Trans. on Electronic Computers **EC-9**, 25 (1960).

3. THE MASTER EQUATION AND THE FOKKER-PLANCK EQUATION

The stochastic variable in our considerations of the circuit of Fig. 4 is the charge q on the tunnel diode capacitance. Consider an ensemble of such tunnel diode circuits. Let $n(N)$ be the number of ensemble members for which the capacitive charge consists of N electrons. In that case $n(N)$ has a time dependence given by

$$\partial n(N)/\partial t=\Gamma(N+1,N)n(N+1)+\Gamma(N-1,N)n(N-1) -\Gamma(N,N+1)n(N)-\Gamma(N,N-1)n(N). \quad (3.1)$$

Here $\Gamma(i,j)$ represents the probability per unit time that a system with i electrons on the capacitance makes a transition to a state with j electrons. Since the charges are uncorrelated in their motion we can regard the probability of two or more charges arriving at the capacitor (or leaving it) in exact time coincidence, as negligible, and hence we take $\Gamma(i,j)=0$ unless i and j differ by unity.

The supply current of magnitude i_S coming through the diode tends to charge the capacitance, so does the Zener current i_Z; whereas i_E discharges the capacitance. Hence

$$\Gamma(N+1,N)=i_E(N+1)/e, \quad (3.2)$$

since only the Esaki current can reduce the value of N. The notation $i_E(N+1)$ is used to denote the Esaki current at the voltage associated with a capacitive charge of $N+1$ electrons. For simplicity we will take the diode capacitance as constant and independent of voltage. Similarly,

$$e\Gamma(N,N-1)=i_E(N),$$
$$e\Gamma(N,N+1)=i_S+|i_Z(N)|,$$
$$e\Gamma(N-1,N)=i_S+|i_Z(N-1)|. \quad (3.3)$$

We can now rewrite the master equation (3.1) using Eqs. (3.2) and (3.3). We shall also introduce a new notation $i_D=i_E$ and $i_C=i_S+|i_Z|$. Thus i_D (D for discharge) represents the current tending to reduce the capacitive charge and i_C (for charge) tends to increase the capacitive charge. Both quantities are taken as positive in the range of interest, with the actual capacitive charging determined by $\dot{q}=i_C-i_D$. This yields

$$e\partial n(N)/\partial t=n(N+1)i_D(N+1)+n(N-1)i_C(N-1) -n(N)\{i_D(N)+i_C(N)\}. \quad (3.4)$$

To obtain some physical appreciation of this equation as well as a form needed for later use, consider the special case where the quantities involved are slowly varying functions of N. Consider a set of terms in (3.4), such as

$$n(N+1)i_D(N+1)-n(N)i_D(N), \quad (3.5)$$

which is of the form $f(N+1)-f(N)$ and can be expanded.

$$f(N+1)-f(N)=f'(N)+\tfrac{1}{2}f''(N)+\frac{1}{3!}f'''(N)+\cdots. \quad (3.6)$$

If f is a sufficiently slowly varying function of N, then this is a rapidly converging sequence. If we expand all the finite differences in Eq. (3.4) in this fashion, and retain the first two terms in each expansion, we find

$$e\frac{\partial n(N)}{\partial t}=\frac{\partial}{\partial N}(ni_D)+\frac{1}{2}\frac{\partial^2}{\partial N^2}(ni_D) -\frac{\partial}{\partial N}(ni_C)+\frac{1}{2}\frac{\partial^2}{\partial N^2}(ni_C). \quad (3.7)$$

In terms of the charge $q=eN$, and the density distribution $\rho(q)\,dq=n(N)dN$, we find

$$\frac{\partial\rho}{\partial t}=\frac{\partial}{\partial q}[\rho(i_D-i_C)]+\frac{e}{2}\frac{\partial^2}{\partial q^2}\rho(i_C+i_D). \quad (3.8)$$

Now i_D-i_C, which occurs in the first right-hand term of Eq. (3.8), is the negative of the rate at which the capacitive charge q increases, if the system obeys macroscopic equations. In actual fact, of course, i_C and i_D are only average values determined by the charge q. Thus, Eq. (3.8) becomes

$$\frac{\partial\rho}{\partial t}=-\frac{\partial}{\partial q}(\rho\langle\dot{q}\rangle_{\rm av})+\frac{e}{2}\frac{\partial^2}{\partial q^2}\rho(i_C+i_D). \quad (3.9)$$

If all the diodes in the ensemble described by $\rho(q)$ were constrained to have $\dot{q}=\langle\dot{q}\rangle_{\rm av}$ then the equation

$$\frac{\partial\rho}{\partial t}=-\frac{\partial}{\partial q}(\rho\langle\dot{q}\rangle_{\rm av}) \quad (3.10)$$

would be the equation of continuity. Hence, the final term in Eq. (3.9) must represent the diffusive effects due to the fluctuations or departures from the macroscopic behavior. Equation (3.9) can be integrated to give a flow of ensemble density in q space:

$$j_q=\rho\langle\dot{q}\rangle_{\rm av}-\frac{e}{2}\frac{\partial}{\partial q}\rho(i_C+i_D). \quad (3.11)$$

Equation (3.8) and its equivalent, Eqs. (3.9) and (3.11) are usually called the Fokker–Planck equations, and we have rederived these equations for our particular case. We could instead, have referred to one of the standard treatments.[16] These consider the ensemble members given by a density $n(q)$ in q space, moving past a point q, in the time interval Δt. Let the mean displacement of these ensemble members be $\langle\Delta q\rangle_{\rm av}$ and

[16] S. Chandrasekhar, Revs. Modern Phys. 15, 1 (1943); and reference 11.

let $\langle \Delta q^2 \rangle_{av}$ be the mean squared displacement occurring in the same time interval Δt. Then the usual treatments lead to an equation

$$j_q = n(q) \lim_{\Delta t \to 0} \frac{\langle \Delta q \rangle_{av}}{\Delta t} - \frac{1}{2} \frac{\partial}{\partial q} \left\{ n(q) \lim_{\Delta t \to 0} \frac{\langle \Delta q^2 \rangle_{av}}{\Delta t} \right\}, \quad (3.12)$$

if the higher moments $\langle \Delta q^n \rangle_{av}$, for $n \geq 3$, go to zero more rapidly than Δt. Again it is clear that the first right-hand term represents the behavior which would be found if all ensemble members at a given point in q space behaved alike. Equation (3.12) is a generalization of an equation used in treating motion in an overdamped potential well, of the form shown in Fig. 1. For the latter case we find a form familiar from electronic transport problems:

$$j = -n(x)\mu \nabla V - D\nabla n, \quad (3.13)$$

where μ is the mobility, D the diffusion coefficient, and the Einstein relationship $D = \mu kT$ obtains. The equivalence between Eq. (3.12) and Eq. (3.13) is made readily apparent in the case where D and μ are independent of the position x. In that case $-\mu \nabla V$ represents a drift velocity, and is therefore equivalent to $\langle \Delta q \rangle_{av}/\Delta t$. Furthermore, in the presence of a constant diffusion coefficient D the mean square diffusive displacement of a particle, in time t, is given by $2Dt$. Hence $\langle \Delta q^2 \rangle_{av}/2\Delta t$ in Eq. (3.12) is equivalent to the diffusion coefficient in Eq. (3.13). In the case where D and μ are functions of x, the correspondence between Eq. (3.12) and (3.13) becomes less direct, and does not remain a term by term equivalence.

4A. STEADY-STATE SOLUTION

The steady-state ensemble distribution is primarily of interest because it indicates whether A or C in Fig. 4 is the preferred state. We shall first give a treatment based on Eq. (3.11), which has the form

$$j_q = \rho \langle \dot{q} \rangle_{av} - \frac{\partial}{\partial q}(D_q \rho), \quad (4.1)$$

with an effective diffusion coefficient

$$D_q = e(i_C + i_D)/2. \quad (4.2)$$

This diffusion coefficient represents the combined effect of all the shot noise sources. In the steady (thermally relaxed) state, $j_q = 0$. Hence, Eq. (4.1) integrates to

$$\rho = \frac{C}{D_q} \exp\left(\int \frac{\dot{q}}{D_q} dq \right), \quad (4.3)$$

with C a constant of integration which is adjusted to give the correct total ensemble population.

Let us continue to follow the analogy with the case of the particle in an overdamped potential, as shown in Fig. 1. In connection with Fig. 1, take $D = \mu kT$ as constant and $\dot{x} = -\mu \nabla V$. In this overdamped case of Fig. 1 we can write the thermal equilibrium distribution in a form

$$\begin{aligned} \rho(x) &= \exp(-U/kT) = \exp\left(-\int \frac{\nabla V}{kT} dx \right) \\ &= \exp\left(-\int \frac{\mu \nabla V}{\mu kT} dx \right) = \exp\left(\int \frac{\dot{x} dx}{D} \right), \end{aligned} \quad (4.4)$$

where the equivalence to Eq. (4.3) becomes obvious.

It is clear that $\int_B{}^A \dot{q} dq/D_q$ and $\int_B{}^C \dot{q} dq/D_q$ take the place of the potential barrier heights $V(B) - V(A)$ and $V(B) - V(C)$. It is these exponential factors in Eq. (4.3) which probably determine the favored state, rather than the multiplying coefficient D_q^{-1}, which is likely to have a more limited variation. The integrals which have to be evaluated differ from the simpler forms $\int_B{}^A \dot{q}\, dq$, $\int_B{}^C \dot{q}\, dq$, which are just equal to the cross-hatched areas in Fig. 4(b), by the presence of the factor $1/D_q(q)$ in the integrand. Thus if the two cross-hatched areas in Fig. 4(b) are equal, then it is the high-voltage state which is favored, since both i_E and $|i_Z|$ are less in the high-voltage region, giving a smaller value of D_q between B and C than between A and B.

The steady-state solution we have just discussed stems from the Fokker–Planck equation (4.1), and we have done this in some detail, since an interesting correspondence with Fig. 1 is involved. Let us now examine the validity of the Fokker–Planck equation. This requires that $\rho(q)$ be a slowly varying function of N, the number of electronic charges on the capacitance. Equation (4.3) tells us that

$$\begin{aligned} \frac{\rho(N+1)}{\rho(N)} &\sim 1 + \frac{\dot{q} e}{D_q} = 1 + \frac{(i_C - i_D)e}{\frac{1}{2}(i_D + i_C)e} \\ &= 1 + 2\frac{i_S + i_Z - i_E}{i_S + i_E + |i_Z|}. \end{aligned} \quad (4.5)$$

The final ratio of currents in Eq. (4.5) has $\dot{q}$, the total capacitive charging current in the numerator, and the sum of all current magnitudes in the denominator. When this ratio is small, the Fokker–Planck equation applies. $\dot{q}$ vanishes at points A, B, and C, and it is clear, therefore, that Eq. (4.3) correctly describes the variation of ρ near these points. In between, however, the ratio is not likely to be small, unless we have a tunnel diode whose characteristic is unusually flat, and which stays close to the characteristic of the charging diode. To see this in more detail, consider large voltages for which $i_Z = 0$, and near a pronounced tunnel diode valley, giving $i_E \ll i_S$, Eq. (4.5) then yields $\rho(N+1)/\rho(N) \sim 3$, which is indeed very far from the Fokker–Planck condition.

The correct solution must be obtained from Eq. (3.4). In the steady state we can actually obtain the condition $j_q = 0$ by requiring the net transition rate from $q = Ne$ to $q = (N+1)e$ to vanish. This gives, instead of Eq. (3.1),

$$\Gamma(N+1, N) n(N+1) = \Gamma(N, N+1) n(N), \quad (4.6)$$

or

$$i_E(N+1)n(N+1)=[i_S(N)-i_Z(N)]n(N), \quad (4.7)$$

which can be written

$$i_D(N+1)n(N+1)=i_C(N)n(N). \quad (4.8)$$

Therefore, finally,

$$n(N+1)/n(N)=i_C(N)/i_D(N+1). \quad (4.9)$$

A formal solution for $n(N')/n(N)$, for arbitrary N' is obtained by multiplying the ratios given by Eq. (4.9). In the typical case we have in mind, N will change by a large number in going from A to B to C. Therefore i_C and i_D are smooth functions of N. The ratio $n(N+1)/n(N)$, while it need not be close to unity, is also a slowly varying function of N. Using this fact we shall be able to represent the following sum in terms of integrals:

$$\frac{n(N')}{n(N_0)}=\exp\sum_{N_0}^{N'-1}\log[n(N+1)/n(N)]. \quad (4.10)$$

Or, taking the logarithm of (4.10) yields

$$\log n(N')/n(N_0)=\sum_{N_0}^{N'-1}\log i_C(N) - \sum_{N_0}^{N'-1}\log i_D(N+1). \quad (4.11)$$

The sums in Eq. (4.11) can then be approximated by integrals:

$$\sum_{N_0}^{N'-1}\log i_C(N)\simeq e^{-1}\int_{(N_0-\frac{1}{2})e}^{(N'-\frac{1}{2})e}dq\,\log i_C(N), \quad (4.12)$$

and similarly

$$\sum_{N_0}^{N'-1}\log i_D(N+1)=\sum_{N_0+1}^{N'}\log i_D(N)\simeq e^{-1}\times\int_{e(N_0+\frac{1}{2})}^{e(N'+\frac{1}{2})}dq\,\log i_D(N). \quad (4.13)$$

The integrals involved in Eqs. (4.12) and (4.13) can be reduced, approximately, to the same interval of integration:

$$e^{-1}\int_{e(N_0-\frac{1}{2})}^{e(N'-\frac{1}{2})}dq\,\log i_C\simeq e^{-1}\int_{eN_0}^{eN'}dq\,\log i_C - \tfrac{1}{2}\log i_C(N')+\tfrac{1}{2}\log i_C(N_0), \quad (4.14)$$

$$e^{-1}\int_{e(N_0+\frac{1}{2})}^{e(N'+\frac{1}{2})}dq\,\log i_D\simeq e^{-1}\int_{eN_0}^{eN'}dq\,\log i_D + \tfrac{1}{2}\log i_D(N')-\tfrac{1}{2}\log i_D(N_0). \quad (4.15)$$

Equations (4.14) and (4.15) can now be used to replace the sums in Eq. (4.11), yielding,

$$\log n(N')/n(N_0)=\int_{eN_0}^{eN'}\frac{dq}{e}(\log i_C-\log i_D) + \log(i_C(N_0)i_D(N_0)/i_C(N')i_D(N')^{\frac{1}{2}}. \quad (4.16)$$

Equation (4.16) can be rewritten

$$n(N')/n(N_0)=(i_C(N_0)i_D(N_0)/i_C(N')i_D(N')^{\frac{1}{2}} \times\left[\exp\int_{eN_0}^{eN'}e^{-1}dq(\log i_C-\log i_D)\right]. \quad (4.17)$$

Consider the rapidly varying exponential factor in Eq. (4.17). This has the same general character as Eq. (4.3), exhibiting maxima at A and C, and a minimum at B. In the neighborhood of each of these points, Eqs. (4.17) and (4.3) give the same variation of $n(q)$. Between A and B, for example, $i_S-i_Z<i_E$, and the difference of logarithms in Eq. (4.17) is negative. Typically, however, in the situation shown in Fig. 4, $i_S-i_Z=i_C$ is not very large compared to i_D. Hence,

$$n(B)\simeq n(A)e^{-\alpha[q(B)-q(A)]}, \quad (4.18)$$

with α smaller than unity, but comparable to it. In going from B to C, $i_S-i_Z=i_C$ exceeds i_E, and near the minimum of i_E can be a large number. Thus,

$$n(B)\simeq n(C)e^{-\beta[q(C)-q(B)]}, \quad (4.19)$$

with β likely to be much larger than α. With the typical broad valley, which will make $q(C)-q(B)>q(B)-q(A)$ and the fact that $\beta>\alpha$, it follows that

$$n(C)/n(B)\gg n(A)/n(B). \quad (4.20)$$

Hence C is likely to be the favored state.

4B. STEADY-STATE BEHAVIOR NEAR A, B, C

At A, B, and C, the capacitive charging current $\dot{q}=i_C-i_D$ vanishes. Near A (similarly for C) $\dot{q}$ is proportional to $-(q-q_A)$, causing a circuit to return to A, if it is forced away from there. Near B, $\dot{q}$ is proportional to $+(q-q_B)$, causing this to be an unstable point. In each case the proportionality can be written in terms of a time constant

$$\dot{q}=-(q-q_A)/\tau_A \quad \text{near } A,$$
$$\dot{q}=(q-q_B)/\tau_B \quad \text{near } B,$$
$$\dot{q}=-(q-q_C)/\tau_C \quad \text{near } C. \quad (4.21)$$

These time constants will have an order of magnitude typical of tunnel diode switching behavior. In particular $\tau_B=R_nC$, where R_n is the magnitude of the tunnel diode negative resistance at the point B.

Near A, B, and C, Eq. (4.3) is valid. The integral in

Eq. (4.3) becomes, through the use of Eq. (4.21),

$$\int_A^q \frac{\dot{q}dq}{D} = -\frac{1}{2D_A\tau_A}(q-q_A)^2,$$
$$\int_B^q \frac{\dot{q}dq}{D} = +\frac{1}{2D_B\tau_B}(q-q_B)^2, \qquad (4.22)$$
$$\int_C^q \frac{\dot{q}dq}{D} = -\frac{1}{2D_C\tau_C}(q-q_C)^2,$$

where we have taken D as sensibly constant within each short interval of integration. This gives us an ensemble density near A, of the form

$$n(q) = n(A)\exp(-q^2/2D_A\tau_A), \qquad (4.23)$$

and a total population near A,

$$n(A) = \int_{\text{near } A} n(q)dq \sim \int_{-\infty}^{+\infty} n(A)\exp(-q^2/2D_A\tau_A)$$
$$= n(A)(2\pi D_A\tau_A)^{\frac{1}{2}}. \qquad (4.24)$$

Similarly for point C.

5. RELAXATION TO THE STEADY STATE

If we have an ensemble of systems whose distribution differs from Eq. (4.3) or (4.17) a relaxation toward the steady-state distribution will take place. In analyzing this we will be essentially adapting the arguments of reference 2, where the over-damped well of Fig. 1 was treated, and we shall not repeat the arguments here in full detail. We shall assume that even in the absence of the steady state most of the systems are near A or C, and that near A and C the steady-state distribution holds. It is only the ratio $n(A)/n(C)$ that is adjusted in the relaxation process.

Consider the number of systems which are changing from $q = Ne$ to $q = (N+1)e$. In the notation of Eq. (3.1) the net rate is

$$\Gamma(N,N+1)n(N) - \Gamma(N+1,N)n(N) = J_{N,N+1}, \qquad (5.1)$$

or

$$i_C(N)n(N) - i_D(N+1)n(N+1) = eJ_{N,N+1}. \qquad (5.2)$$

In our case $J_{N,N+1}$ is effectively zero near A and C, giving a multiple of the steady-state solution in these neighborhoods. Near B we have a current which is approximately constant, and we shall therefore write $J_{N,N+1} = J$.

As in reference 2 we shall solve Eq. (5.2) by assuming $n(N) = \beta n_0(N)$, where $n_0(N)$ satisfies the steady-state equation

$$i_C(N)n_0(N) - i_D(N+1)n_0(N+1) = 0. \qquad (5.3)$$

The multiplier β is then determined by

$$i_C(N)\beta(N)n_0(N) - i_D(N+1)\beta(N+1)n_0(N+1) = eJ. \qquad (5.4)$$

Multiplying Eq. (5.3) by $\beta(N+1)$ and subtracting it from Eq. (5.4) yields

$$i_C(N)n_0(N)[\beta(N) - \beta(N+1)] = eJ, \qquad (5.5)$$

or

$$\beta(N+1) - \beta(N) = -eJ/i_C(N)n_0(N). \qquad (5.6)$$

Now $n_0(N)$ in the denominator on the right-hand side of Eq. (5.6) is not necessarily a slowly varying function of N, as we have seen in Sec. 3. At the point B, however, n_0 reaches a minimum, and is therefore not rapidly varying. Furthermore, we shall show that as we move away from B, $n_0(N)$ becomes large compared to its value at C, before it loses its smooth variation with respect to N. Hence in the region where β has its variation, $n_0(N)$ is essentially continuous. This reduces Eq. (5.6) to a differential equation, as in reference 2.

We still have to dispose of the point connected with the smooth variation of $n_0(N)$. In the vicinity of B, $1/n_0(N)$ is given by an equation similar to Eq. (4.21) as

$$1/n_0(N) = \exp[-(q-q_B)^2/2D_B\tau_B], \qquad (5.7)$$

and we wish to show that when this becomes very small, compared to its maximum value at B, it is still a smooth function of $N - N_B$. Equation (5.7) can be rewritten

$$1/n_0(N) \sim \exp[-(N-N_B)^2e^2/2D_B\tau_B]. \qquad (5.8)$$

Now $D_B = \frac{1}{2}e(i_C+i_D)$, so that we have

$$1/n_0(N) \sim \exp[-(N-N_B)^2e/(i_C+i_D)\tau_B]. \qquad (5.9)$$

The expression i_C+i_D in the exponent of Eq. (5.9) is a macroscopic current, and is in fact large compared to the actual net charging current $i_C - i_D$, in the vicinity of B. τ_B is also a macroscopic time, in which typically a great many electrons flow in and out of the diode capacitance. Hence

$$e/\tau_B(i_C+i_D) \ll 1, \qquad (5.10)$$

and therefore $(N-N_B)$ must become very large, before the exponent in Eq. (5.9) becomes equal to about 5 or 10, permitting $1/n_0$ to drop appreciably below its value at B. The smoothness of $1/n_0(N)$ is indicated by a small logarithmetic derivative with respect to N. We have

$$n_0^{-1}(dn_0/dN) = 2(N-N_B)/e(i_C+i_D)\tau_B. \qquad (5.11)$$

This quantity is equal to the exponent, (5 or 10) divided by $(N-N_B)/2$, and hence small compared to 1.

We have thus reduced the problem to that of the continuous case, and Eq. (5.6) becomes

$$d\beta/dq = -J/i_C n_0(q). \qquad (5.12)$$

Integrating this through the point B gives

$$\beta(C)-\beta(A)=\int_A^C dq(d\beta/dq)\simeq\int_{\text{near }B} dq(d\beta/dq). \quad (5.13)$$

Taking $n_0(N)$ in this integration of the form

$$D_B^{-1}\exp[-(q-q_B)^2/2D_B\tau_B] \quad (5.14)$$

determines the normalization for n_0, and yields

$$\beta(C)-\beta(A)=-(i_C)^{-1}JD_B \times\int_{-\infty}^{+\infty}\exp[-(q-q_B)^2/2D_B\tau_B]d\tau q. \quad (5.15)$$

Using the fact that $D_B=\frac{1}{2}(i_C+i_D)$ and that $i_C=i_D$ at the point B, we find

$$\beta(C)-\beta(A)=-J(2\pi D_B\tau_B)^{\frac{1}{2}}. \quad (5.16)$$

To evaluate a relaxation rate we need to know the extent to which the ensemble population deviates from the steady-state one, as given by Eq. (4.17). The total ensemble population near A is given by $N(A)=n(A)(2\pi D_A\tau_A)^{\frac{1}{2}}$. The ensemble density at A, $n(A)$, in turn is given by $\beta(A)n_0(A)$. The normalization of n_0 is defined by Eq. (5.14) which gives $n_0(B)=1/D_B$. The ratio $n_0(A)/n_0(B)$ can be obtained from Eq. (4.17). Thus

$$N(A)=\beta(A)(2\pi D_A\tau_A)^{\frac{1}{2}}[n_0(A)/n_0(B)]D_B^{-1}. \quad (5.17)$$

If we define a fictitious population associated with the unstable point B,

$$N_0(B)=(2\pi D_B\tau_B)^{\frac{1}{2}}n_0(B), \quad (5.18)$$

then Eq. (5.17) can be written

$$N(A)=\beta(A)[N_0(A)/N_0(B)](2\pi\tau_B/D_B)^{\frac{1}{2}}. \quad (5.19)$$

Let the ratio $N_0(A)/N_0(B)$ be denoted by $e^{\Delta A}$, where ΔA is then somewhat equivalent to a free energy of activation. The main contribution to ΔA will be the argument of the exponential function in Eq. (4.17). The logarithm of the pre-exponential factors in Eq. (4.17) will vanish in the special case of Fig. 4, where i_C and i_D are the same for all three points A, B, and C. In the case of Fig. 5, however, this logarithm would contribute to ΔA. The remaining term in ΔA is the logarithm of the factor $(2\pi D_A\tau_A)^{\frac{1}{2}}/(2\pi D_B\tau_B)^{\frac{1}{2}}$ arising from the fact that we are dealing with integrated populations, not just population densities. In our new notation,

$$N(A)=\beta(A)(2\pi\tau_B/D_B)^{\frac{1}{2}}\exp\Delta A, \quad (5.20)$$

$$N(C)=\beta(C)(2\pi\tau_B/D_B)^{\frac{1}{2}}\exp\Delta C. \quad (5.21)$$

To find the deviation from the steady state we must subtract from the above populations, a steady-state distribution: $N_0(A)$, $N_0(C)$ with the same total population:

$$N_0(A)+N_0(C)=N(A)+N(C).$$

The relaxation rate $1/\beta$ will then be given by

$$1/\tau=J/[N(A)-N_0(A)].$$

Upon carrying out this procedure we find

$$1/\tau=(2\pi\tau_B)^{-1}[\exp(-\Delta A)+\exp(-\Delta C)].$$

Thus the factor τ_B^{-1} related to the speed of deviation from unstable equilibrium, takes the place of the frequency factor.

6. SUMMARY

We have followed, in some detail, the similarities between the thermally activated jump between dissipative states and the more commonly considered problem of a thermally activated jump in a potential field. The probability distribution for the dissipative system after it has had a chance for complete thermal relaxation was found by an inverse of the consideration usually used to derive the Einstein relations between induced and stimulated emission. In deriving the Einstein relations one starts with the known distribution $\exp(-U/kT)$ and invokes detailed balance to obtain information about the kinetics of the system. We have instead assumed a complete knowledge of the kinetics, and found the form of the time independent distribution. The barrier U/kT is replaced, in our problem, by complicated quantities of the form αN, where α is not too terribly different from unity and N is the change in the number of electronic charges on the tunnel diode capacitance in the jump. (Under typical conditions, a tunnel diode of 10^{-11} cm^2 cross-sectional area would have a charge increase of about 25 electrons in going from A to B.) The quantity equivalent to the usual frequency factor ν, in giving the jump probability $\nu\exp[-U/kT]$, turns out to be the reciprocal of a relaxation time τ, closely associated with the switching characteristics of a tunnel diode and often used as a measure of its quality.

ACKNOWLEDGMENTS

An earlier version of this manuscript concentrated on an awkward analysis of the circuit of Fig. 3, and the associated Fokker-Planck equation. The present form of the work is largely due to suggestions made by Gordon Lasher and Peter Price.

APPENDIX. THE RESISTIVE LOAD LINE, FIG. 3

We shall here indicate the modifications required to treat the case where not only shot noise but Nyquist noise is also present. For all of our cases we can write a generalization of Eq. (3.12) in the form

$$j_q=n\langle\dot{q}\rangle+\text{effects of fluctuations}. \quad (A1)$$

The sources of shot noise and of Nyquist noise contribute independently to the second right-hand term in Eq. (A1). In thermal agitation noise the independent event is the mean free flight of a carrier, the charge transport associated with this is of the order el/L, where

l is the mean free path and L the length of the resistive current path. Hence, we cannot limit our considerations to integral multiples of the electronic charge on the capacitor. The probability for a transition from q to $q\pm e$, due to the tunnel diode currents i_Z and i_E, will, however, be the same as before. The explicit analytical representation of this effect, however, in Eq. (A1) will be more complicated now. The ensemble members making a transition through q can come from anywhere within the range $q-e$ to $q+e$. If we permit ourselves to make the Fokker-Planck approximation, then this becomes of no importance, and the shot effect fluctuations contribute exactly as in Eq. (3.13). It turns out, however, that the criterion for the validity of the Fokker-Planck equation is the same restrictive one as found before in Sec. 3; namely, that the load line and the tunnel diode characteristic must lie very close to each other.

The remainder of this Appendix is concerned with the Nyquist contributions to Eq. (A1). We shall first assume that these give only terms of the Fokker-Planck type, as shown in Eq. (3.12), and subsequently justify this point of view. A short-circuited resistor has current fluctuations, in the bandwidth Δf, given by

$$\langle i^2\rangle_{\Delta f}=4kT\Delta f/R. \tag{A2}$$

Our resistor is not short circuited. The physical electron scattering mechanisms which produce current fluctuations in the resistor are, however, independent of the external circuit connections. If the resistor is not short circuited, as in Fig. 2, the spontaneous current fluctuations cause charges to be piled up on the diode capacitance, which in turn modify the subsequent resistor current and diode durrent. This latter effect is, however, not a deviation from the average behavior, and is already described by the first right-hand term of Eq. (A1). It is therefore adequate, when evaluating $\langle\Delta q^2\rangle_{av}$ to assume that the resistor generates current fluctuations as if it were short circuited. (We have, here, simply argued for the usual noise representation of a resistor as a perfect resistor in parallel with a constant current noise generator.)

The current, given Eq. (A2) integrated over a time Δt, gives

$$\langle\Delta q^2\rangle_{\Delta f}=(8kT\Delta f/R\omega^2)(1-\cos\omega\Delta t). \tag{A3}$$

In computing the result (A3) we have assumed Δf narrow enough, so that $\Delta f\Delta t\ll 1$, and therefore the narrow band noise current described by Eq. (A2) has a well-defined sinusoidal form. Furthermore, we have averaged over all possible phases of this waveform, to obtain Eq. (A3). Integrating Eq. (A3) over the whole frequency range yields

$$\langle\Delta q^2\rangle_{av}=\int df\langle\Delta q^2\rangle_{\Delta f}/\Delta f=(2kT/R)\Delta t. \tag{A4}$$

This has the desired form $\langle x^2\rangle_{av}=2Dt$, and gives us an effective diffusion coefficient, along the q coordinate, $D_q=kT/R$. The method used here can, incidentally, also be used to derive Eq. (4.2).

Are the resistive fluctuations adequately described by the simple diffusion term? The usual derivations[15] of the Fokker-Planck equation (3.12) point out the necessary condition: The higher moments $\langle\Delta q^n\rangle_{av}$ must go to zero more rapidly than Δt. Let us first evaluate these higher moments. The charge fluctuations in the resistor are produced by a great many independently moving carriers, subject to frequent scattering. In principle we can always choose a resistor where the carriers have particularly high scattering probabilities, and obtain the desired conductance by using a particularly large number of carriers. We are thus assured that, even in a very short time Δt, the capacitive charge fluctuation Δq produced by the resistor is the sum of a great many independent events. Hence, the capacitive charge fluctuations Δq must have a probability distribution of the form $(\Gamma/\pi)^{\frac{1}{2}}\exp[-\Gamma(\Delta q)^2]$. We know, however, from Eq. (A4) that the second moment $\langle\Delta q^2\rangle_{av}$ is linear in time. Hence Γ is proportional to $1/\Delta t$. This, however, implies that the higher moments, for example,

$$\int(\Delta q)^4(\Gamma/\pi)^{\frac{1}{2}}\exp[-\Gamma(\Delta q)^2]dq\sim 1/\Gamma^2\sim\Delta t^2, \tag{A5}$$

go to zero more rapidly than Δt.

TEMPORAL AND SPATIAL PATTERNS IN CHEMICAL SYSTEMS

Harry L. Swinney, W. Horsthemke, W.D. McCormick,
Z. Noszticzius, and W.Y. Tam
Center for Nonlinear Dynamics and the Department of Physics
University of Texas
Austin, Texas 78712 USA

1. INTRODUCTION

Laboratory experiments on nonequilibrium chemical reactions in continuous flow homogenous (that is, well-stirred) tank reactors (CSTRs) have revealed a variety of complex *temporal* states, including multi-peaked periodic oscillations and chaos [1-5]. Numerical simulations of mathematical models based on the chemical kinetics have yielded transition sequences similar to those observed in the experiments, and a comparison of theory and simulation has in some cases provided insight into the chemical kinetics [6-7].

This progress in the understanding of temporal dynamics has been achieved largely because CSTRs make it possible to maintain a reaction indefinitely in well-defined far-from-equilibrium state. In contrast, studies of *spatial* patterns have been conducted under transient conditions, where a pattern, formed usually on a filter paper or in a Petri dish, evolves continuously towards thermodynamic equilibrium. Recently our group has developed continuous flow *unstirred* reactors (CFURs) that now make it possible to study spatial patterns in systems maintained indefinitely in well-defined nonequilibrium states [8-9]. Thus the development and evolution of spatial patterns and the transitions between different patterns can be studied as a function of external control parameters, in the same way that temporal dynamical behavior has been studied in recent years.

We will describe some of the observations of complex temporal dynamics and spatial patterns.

2. COMPLEX TEMPORAL DYNAMICS AND CHAOS

One decade ago Schmitz et al. [10] reported the first observations of nonperiodic behavior in a well-controlled homogeneous chemical reaction maintained far from equilibrium. They were studying the Belousov-Zhabotinskii (BZ) reaction, which is the most thoroughly studied of the oscillating chemical reactions. The Schmitz et al. experiment was conducted in a CSTR of the type shown in Figure 1. Chemicals were fed continuously into the reactor, and for certain ranges in concentration and flow rate they found that the concentrations in the reactor oscillated erratically in time. At that time the diagnostic tools were not yet available to determine whether the observed nonperiodic behavior was *chaos*—the nonperiodic behavior found in some deterministic nonlinear systems—or whether the behavior arose from stochastic driving forces (e.g., fluctuations in the feed or stirring rate).

Subsequent experiments have demonstrated that some nonequilibrium chemical reactions do indeed exhibit deterministic chaotic behavior. Figure 2(a) shows a time series measured for the concentration of one the species in

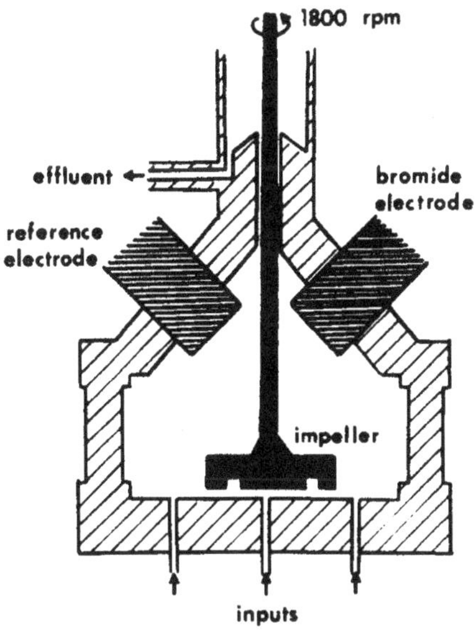

Fig. 1. A schematic diagram of the type of reactor (CSTR) used in studies of the dynamics of homogeneous nonequilibrium reactions. The cylindrical body is constructed of Plexiglas; the volume is 33 cm^3. (from ref. [3])

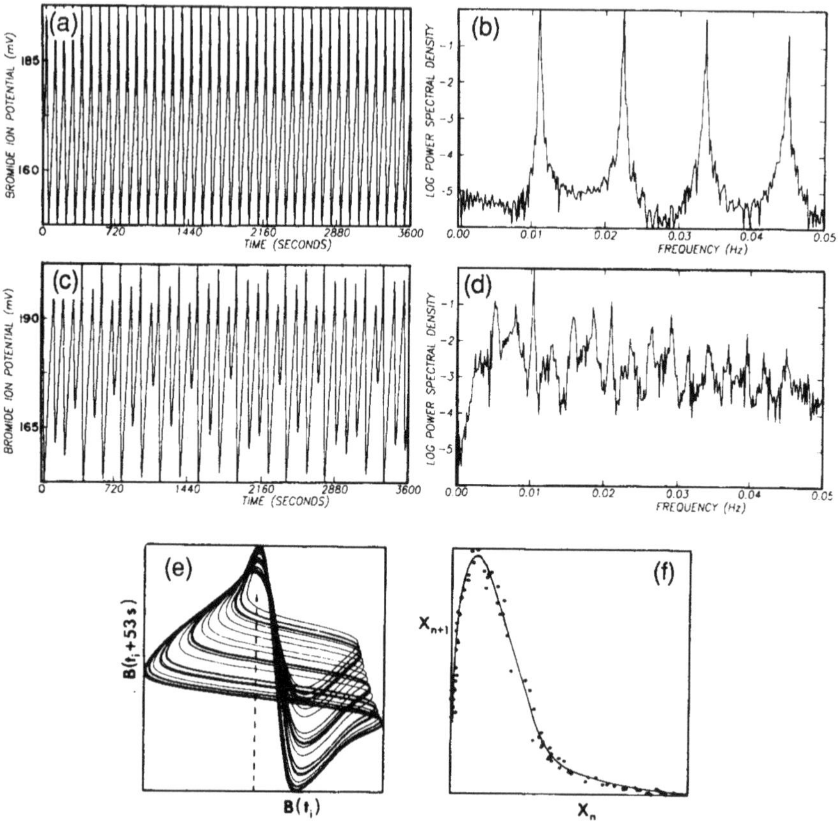

Fig. 2. (a) Time series for the bromide ion potential measured for a periodic state of the BZ reaction. (b) The power spectrum for the time series in (a); the spectrum consists of a single fundamental frequency and its harmonics. (c) Time series for the bromide ion potential measured for a nonperiodic state of the BZ reaction. (d) The power spectrum for the time series in (a); the broadband spectral noise is several orders of magnitude above the instrumental noise level ($< 10^{-6}$). (e) A two-dimensional projection of the strange attractor constructed by plotting B(t+T) versus B(t), where B(t) is the bromide ion concentration measured at time t, and T is a time delay (53 s in this case). (f) A one-dimensional map obtained by plotting successive minima in the time series as ordered pairs (X_n, X_{n+1}). The nonperiodic data all fall on a smooth single-valued curve; hence the system is deterministic. ((a)-(d) from [11], (e) and (f) from [12]).

the BZ reaction in a periodic state; the corresponding power spectrum is in (b). For a slightly different flow rate of the chemicals through the reactor, the concentration oscillates erratically instead of periodically (Fig. 2(c)) ; the power spectrum then contains broadband noise, so the system is definitely nonperiodic (Fig. 2(d)) . However, if successive minima in the nonperiodic time series, $X_1, X_2, \ldots, X_n, X_{n+1}, \ldots$ are plotted as X_{n+1} versus X_n, it is clear that the system is *deterministic*, not stochastic in its behavior: given X_n, the graph gives X_{n+1} (Fig. 2(f)). Analysis of these data by dynamical systems methods shows that the system has the properties expected of a chaotic system. In particular, errors grow exponentially fast, which means that long term prediction is impossible, even though the system is deterministic. A portrait of the system in phase space is the *strange attractor* shown in Fig. 2(e).

The transition from the periodic state in Fig. 2(a) to the chaotic state in Fig. 2(c) occurs through a period doubling sequence [13]: starting with the periodic state and changing the flow rate of the chemicals to move towards the region of chaos, a critical flow rate is reached beyond which every other oscillation has a slightly different amplitude—the period has doubled (see the second time series in Fig. 3). With further change in flow rate the period again doubles (the third time series in Fig. 3). The sequence of period doublings converges very rapidly; period 16 is observable but difficult to stabilize.

The behavior in the chaotic region beyond the accumulation point of the period doubling casade is quite complex. There are flow rate ranges (*windows*) in which the behavior is periodic with several oscillations per period. These periodic windows occur for all periods k, and each window has a distinctive signature (symbol sequence) describing the time order of relative oscillation amplitudes in one period. For one-dimensional maps the mathematical theory of the symbolic dynamics provides a unique classification of all periodic states for each period k. The theory also gives for all k the order in which the windows occur as a control parameter is varied—this order is called the *Universal Sequence* [14]. The experimentally observed sequence of states shown in Fig. 3 is in complete accord with the ordering and symbolic dynamics of the Universal Sequence. However, a recent experiment on the BZ reaction [3] has shown that small amounts of impurities in the chemicals

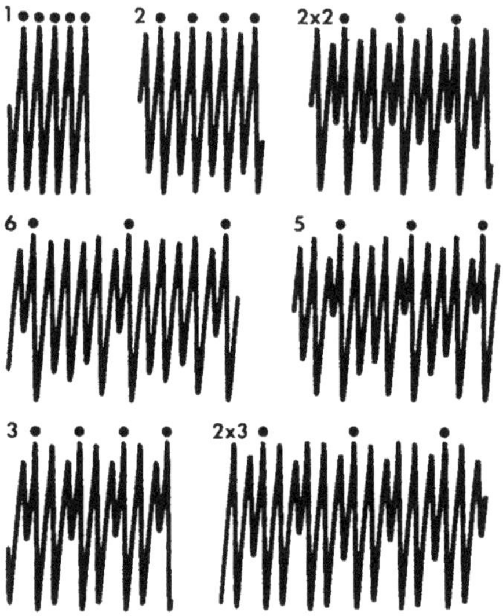

Fig. 3. Time series of the bromide ion concentration for several periodic states of the BZ reaction observed for different flow rates: the states have periods that are all multiples of a basic period— 1, 2 (period doubled), 2x2 (period doubled again), 6, 5, 3, and 2x3. There is an infinite number of states with higher periods between these states. (from [12])

can lead to a mixing of the order of the states of the Universal Sequence, even though the dynamics can still be described by a smooth one-dimensional map with a single maximum. This surprising sensitivity of the reaction dynamics to impurities in the reaction can be exploited to gain new insights into the mechanism of the reaction [15].

In another experiment on the BZ reaction, conducted with different chemical concentrations and a different catalyst, a sequence containing only periodic states was observed [4], and the number of distinct periodic states observed with varying flow rate was quite large. Transitions between states occurred with no intervening chaos and no hysteresis. The basic states were all of the form L^s, where L is the number of large amplitude oscillations and s is the number of small oscillations; see, e.g., the 2^3 and 2^4 states in Fig. 4. The intermediate states found at flow rates between those at which the basic states were observed were always concatenations of the basic states; see, e.g., the state $2^3 2^4$ in Fig. 4. Similar sequences have been well characterized for circle maps with a cubic inflection point with zero slope ("criticality"), but in experiments the path in parameter space would not in general be expected to correspond to the very special case of a map with zero slope. Thus it was puzzling that the BZ experiments and others (on convecting fluids, solid state devices, and the firing of neurons) yielded behavior so similar to that found for circle maps at criticality. Recently Barkley [7] solved the puzzle by showing that the periodic sequences can occur over a wide parameter range when the dynamics is described by circle maps with a slopea that approaches infinity. Analysis of the BZ data has yielded circle maps, and the maps do indeed have a large slope [4]. Thus, as in the example in the preceding paragraph, very complex dynamical behavior is found to be described by a surprisingly simple deterministic model.

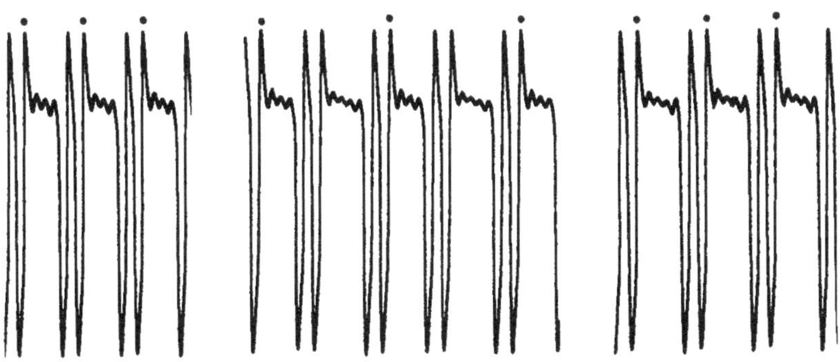

Fig. 4. Time series of the bromide ion concentration for three periodic states observed with increasing flow rate in an experiment on the BZ reaction. The dots above the time series are separated by one period. The 2^3 state on the left consists of two large amplitude oscillations followed by three small amplitude oscillations in each period; similarly, the 2^4 state on the right consists of two large and and four small amplitude oscillations. The middle state, $2^3 2^4$, is a concatenation of the states on the left and right. Only periodic states were observed in this experiment, while in the experiement that yielded the periodic states shown in Fig. 3, most of the observed states were chaotic. (from [4])

3. SUSTAINED SPATIAL PATTERNS

A continuous flow unstirred reactor (CFUR) designed by Noszticzius et al. [8] for the study of spatial pattern formation in a reaction-diffusion system consists of an annular inert gel that is in contact with well-stirred reservoirs of chemicals at the inner and outer rims. The polyacrylamide gel is 1 mm thick and has a 27 mm inner diameter and a 38 mm outer diameter. The reaction that takes place in the gel can be maintained indefinitely in a nonequilibrium state, and the concentration of the chemicals in the reservoirs at the inner and outer rims can be varied continuously to study transitions in the patterns. The gel surmounts two problems encountered in previous experiments on pattern formation: it prevents convection currents and inhibits bubble formation.

In one type of experiment the oxidizer [$KBrO_3$] of the BZ reaction was placed in the outer reservoir and malonic acid and ferroin in the inner reservoir [8,9]. The chemical front that then forms is initially axisymmetric, but for some range of reservoir concentrations it becomes unstable and azimuhal traveling waves form. Both clockwise and counterclockwise waves form at a pacemaker; these waves annihilate when they collide on the opposite side. However, with optical [9] or chemical [8)] it is possible to eliminate one or more waves; e.g., all the counterclockwise waves can be eliminated in this way, thus producing a "***chemical pinwheel***" with only clockwise waves, as shown in Fig. 5 (a counterclockwise pinwheel can be made similarly).

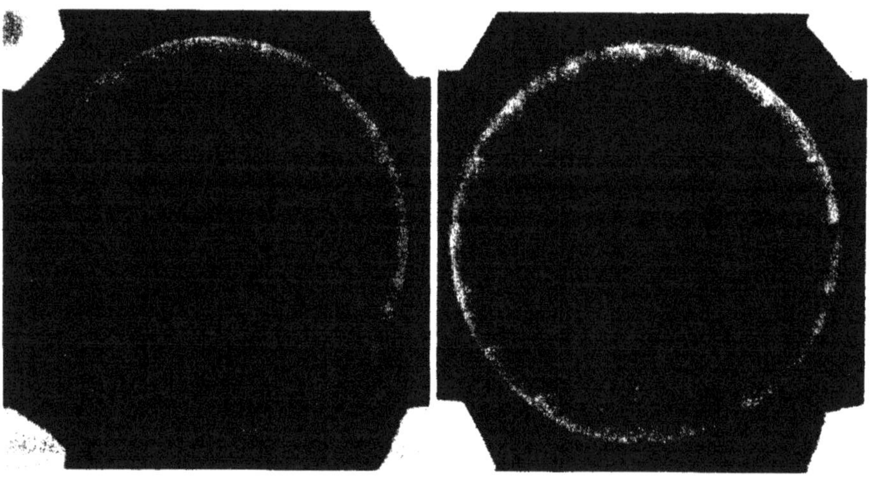

Fig. 5. Chemical pinwheels with 7 waves (left) and 17 waves (right). (from [8])

In addition to the annular gel reactor, we are developing several other geometries in which chemical patterns can form and be maintained in well-defined nonequilibrium states. These continuous flow unstirred reactors will make possible detailed studies of spatiotemporal patterns in chemical reactions.

ACKNOWLEDGMENTS

This work is supported by the Department of Energy Office of Basic Energy Sciences, the BP Venture Research Unit, and the Robert A. Welch Foundation.

REFERENCES

1. Temporal chaos is reviewed in F. Argoul, A. Arneodo, P. Richetti, J.C. Roux, and H.L. Swinney, Acc. Chem. Research **20** (Dec., 1987).
2. J. C. Roux, R.H. Simoyi, and H. L. Swinney, Physica **8D**, 257 (1983).
3. K.G. Coffman, W.D. McCormick, Z. Noszticzius, R.H. Simoyi, and H.L. Swinney, J. Chem. Phys. **86**, 119 (1987).
4. J. Maselko and H.L. Swinney, Physica Scripta **52**, 269 (1985); J. Chem. Phys. **85**, 6430 (1986); Phys. Lett. A **119**, 403 (1987).
5. F. Argoul, A. Arneodo, P. Richetti, and J.C. Roux, J. Chem. Phys. **86**, 3325 (1987).
6 P. Richetti, J.C. Roux, F. Argoul, and A. Arneodo, J. Chem. Phys. **86**, 3339 (1987).
7. D. Barkley, "Near critical behavior for one-parameter families of circle maps," submitted to Phys. Lett. A.
8. Z. Noszticzius, W. Horsthemke, W.D. McCormick, H.L. Swinney, and W.Y. Tam, Nature **329**, 619 (1987).
9. Z. Noszticzius, W. Horsthemke, W.D. McCormick, H.L. Swinney, and W.Y. Tam, "Sustained chemical waves in a Turing-Nicolis-Prigogine ring reactor," to appear in *Spatial Inhomogeneities and Transient Behavior in Chemical Kinetics*, edited by G. Nicolis and P. Gray (Manchester University Press, 1987)
10. R. A. Schmitz, K. R. Graziani, and J. L. Hudson, J. Chem. Phys. **67**, 3040 (1977).

11. J. S. Turner, J. C. Roux, W. C. McCormick, and H. L. Swinney, Phys. Lett. **85A**, 9 (1981).
12. R. H. Simoyi, A. Wolf, and H. L. Swinney, Phys. Rev. Lett. **49**, 245 (1982).
13. M. J. Feigenbaum, J. Stat. Phys. **19**, 25 (1978).
14. N. Metropolis, M.L. Stein, and P. R. Stein, J. Combinatorial Theory Ser. **A15**, 25 (1973).
15. Z. Noszticzius, W. D. McCormick, and H. L. Swinney, J. Phys. Chem. **91**, 5129 (1987).

Fractal Distributions Useful for Search Algorithms

Harold H. Szu

Naval Research Laboratory Code 5756, Washington D.C. 20375

ABSTRACT

Recent advances in nonconvex optimization problems are due to mathematical annealing techniques that mimic the classical solid state physics for finding the ground state, and are considerably faster than exhaustive global search techniques. The usefulness of random/fractal distributions of **Levy's** class for global minimum search algorithms is demonstrated to be **t/log(t)** times faster than conventional simulated annealing based on that of **Gaussian** class. We have applied it to a **phase transition double well potential**, and **multi-sensor location and tracking problem**.

1. Introduction

Most multi-dimensional optimization functions are nonconvex, meaning more than a minimum, and therefore expensive to compute using exhaustive global search techniques. A recently developed solution for this type of problem is a Monte Carlo technique of variable variances called simulated annealing, schematically shown in Fig. 1. The simulated annealing technique arrives at a globally optimum answer by simulating the annealing of a solid into its (globally) lowest energy state by first heating the solid to its melting point and then slowly cooling the material at a rate related to its heat transport speed.

Fig. 1 How does a monkey shake peanuts out of the bottom hole of a metal box of Rolling Hills? It shakes the box violently; soon learned as it gets tired, that a fast shaking following a gradual slow down yields the most fruitful result. This turns out to be universally true. It is intuitively clear in the molten-glass-solidification by a gradual cooling schedule. This is mathematically proved to be necessary and sufficient for Gaussian noises with a variable variance. A **fast convergence** is ultilizing **fractal noises.**

SERIAL MACHINE -- A PIN-BALL-IN-A-BOX

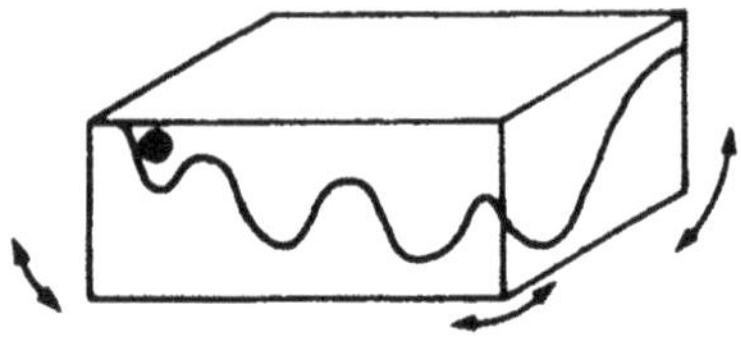

PARALLEL MACHINE -- MOLTEN-GLASS-SOLIDIFICATION

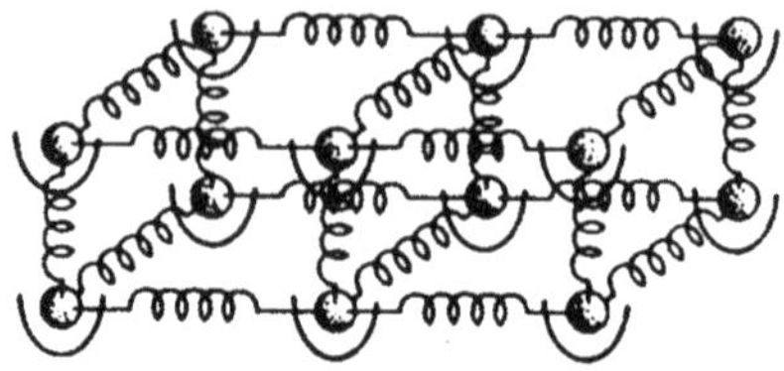

From a computer science point of view, there is no reason that a computer algorithm must have a physically realizable implementation. This paper discusses a mathematically realizable algorithm called Fast Simulated Annealing. This algorithm employs a semi-local search strategy using the (infinite variance) Cauchy probability density of Levy's to generate a random process of fractal dimensions that more often allows large steps in phase space than the strictly local step strategy using a Gaussian probability. The result of allowing more of these larger steps is to permit a faster cooling schedule, which is inverse linear in time, than the cooling schedule of the conventional local annealing algorithm, which is inverse in the logarithm of time. This **t/log t** speed increase is similar to the **N/log N** improvement between the discrete Fourier transform and the Cooley-Tukey Fast Fourier transform. This research should therefore have a major impact on the computational tractability of many nonconvex optimization problems. **Fig. 2 Convex versus Nonconvex means single minimum versus multiplr extrma**

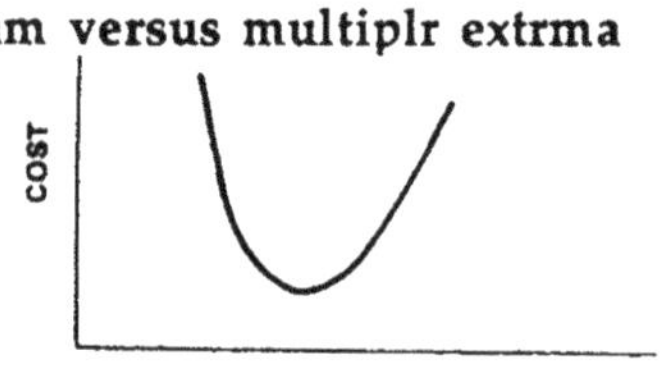

WIENER FILTER IN SPACE
ONE-STEP PRE-PROGRAMMED SOLUTION
OR STEEPEST DESCENT

KALMAN FILTER IN TIME
WITH ORTHOGONAL UPDATE

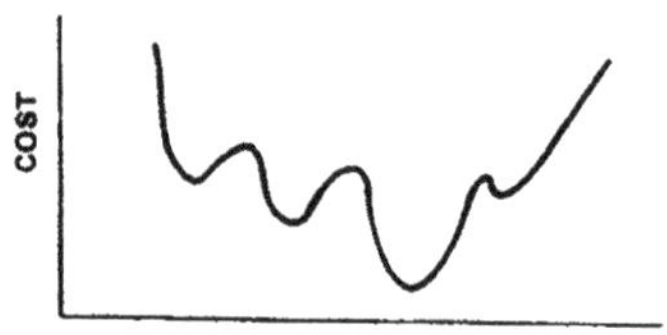

CONSTRAINED OPTIMIZATION
ALLOWING COST INCREASING
SIMULATED ANNEALING
REQUIRES SEARCH STRATEGY

(CAUCHY VS. BOLTZMANN)

$T(t) \cong \frac{1}{t}$ vs $\frac{1}{\mathrm{Log}\,(t)}$

An example of both methodologies is presented of the application of this technique to the problem of locating and tracking multiple, indistinguishable targets using noisy, bearing-only data with both missed detections and false alarms.

In a convex optimization problem, one can start at any point in the function space, measure the local gradient, and take a step in any direction which is lower in altitude than the current position. Repetition of this process will assure asymptotic convergence to the minimum (i.e., optimum) solution. In a nonconvex problem, the optimization function has multiple local minima, each with different depths, for which the optimum is defined to be the global minimum. The application of local gradient techniques to nonconvex optimization creates a problem where one is caught in a local minimum with no way of determining if the local minimum is also the desired global minimum. One solution to this dilemma is to permit steps whose magnitude and direction are dependent on the local gradient and an additive random process. Further, for the algorithm to converge, the magnitude of the random component of the step size must decrease in a statistically monotonic fashion. In the physical annealing process these steps can be equated with Brownian motion of a particle, traveling at statistical velocity V, over an inter-sample time Δt. There exists Levy's flights among local random walks as shown in Figure 3.

Fig. 3 Comparison between random walks: Gaussina (left) with Cauchy

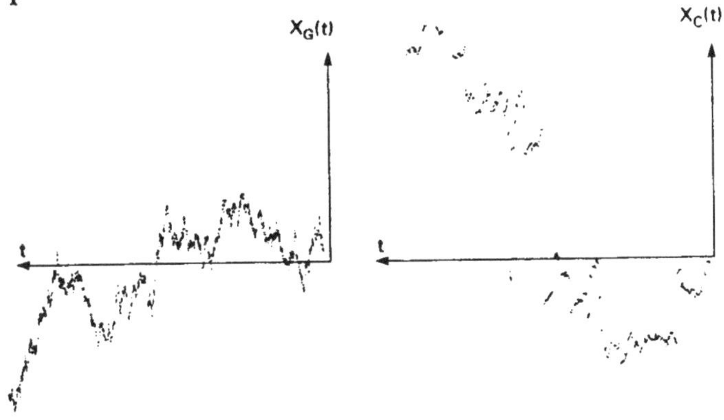

This behavior may be responsible for anomalous diffusions or growths as recently demonstrated by M.F. **Shlesinger et al** in turbulent fluids; B. **West**; and W. **Freeman** in biological systems (c.f. loc. cit.).

The expectation of V^2 is linearly related to the temperature of the particle. The simulated annealing community [1] [2] [3] therefore refers to the "temperature" of the random process and uses the term "cooling schedule" to refer to the algorithm for monotonically reducing the temperature. The cooling schedule is critical to the performance of the algorithm. For a given random process, cooling at too fast a rate will likely "freeze" in a non-global minimum. Cooling at a too slow rate, while reaching the desired minimum, is a waste of computational resources. The technical problem then becomes deriving the fastest cooling schedule that will guarantee convergence to the global minimum. With this understanding, this paper will use the term "cooling schedule" as synonymous with "fastest cooling schedule".

If the Gaussian random process is used, as in the **Conventional Simulated Annealing (CSA)**, it has been proved [4] that the cooling schedule T(t) must be inversely proportional to the logarithm of time t in order to guarantee convergence to the global minimum. This relatively slow convergence is due to the bounded variance of the Gaussian process which constrains the neighborhood of successive samples. This bounded variance random walk is called a local search strategy. If, on the other hand, one uses an infinite variance Cauchy random process, the cooling schedule has been proved [2] as in the **Fast Simulated Annealing (FSA)** to be sped up to be inversely proportional to time t. We call this new class of algorithms a semi-local search strategy because they permit occasionally long steps (the so-called **Lévy flights**) far from the neighborhood of the previous sample.

2. Comparison of Proofs of the Cooling Schedules for FSA and CSA

The **topology** of the state space plays an important role for any specific problem, in particular the **connectivity** of the state space. Given the identical state space, there are a number of similarities in the proofs of the cooling schedules for the **FSA** and **CSA** algorithms in D-dimensional vector spaces. For the convenience of comparison, the proofs will be demonstrated in parallel. In locating the minimum, one must start at some position or state in a D-dimensional space, evaluate the function at that state, and generate the next state vector. The **FSA** and **CSA** algorithms are different in that **FSA** uses a **Cauchy** distribution and **CSA** uses a Gaussian distribution in their respective state-generating functions. Both the **CSA** and **FSA** algorithms will use as its next state either the current state vector or the next state vector provided its incremental cost increase is less than the time dependent noise bound which is temperature (and time) dependent.

The **FSA** algorithm requires that ***state-generating*** be infinitely often in time *(i.o.t.)* whereas the **CSA** requires the ***state-visiting*** be ***i.o.t.***.

At some cooling temperature $T_c(t)$ at time t, let the state-generating probability of being within a specific neighborhood be lower bounded by g_t. Then the probability of not generating a state in that neighborhood is upper bounded by $1 - g_t$.

To insure a globally optimum solution for all temperatures, a state in an arbitrary neighborhood must be able to be generated *i.o.t.*, which however does not imply ergodicity which requires actual visits *i.o.t.*.

To prove that a specific cooling schedule maintains the state-generation *i.o.t.*, it is easier to prove the ***negation*** of the ***converse***, namely the ***impossibility*** **of *never*** generating a state in the neighborhood after an arbitrary time t_0.

Mathematically this is equivalent to stating that the infinite product of $\mathbf{1 - g_t}$ terms is zero. Taking the Taylor expansion of the logarithm of the product, one can alternatively prove that the sum of the $\mathbf{g_t}$ terms in infinite.

One can now verify cooling schedules in a **D-dimensional** neighborhood Δx_0 and arbitrary time t_0.

Among the various **Lévy** distributions (including **Cauchy, Holtzmach,** and **Gaussian**) there are two different classes, **local** (as in **CSA**) and **semi-local** (as in **FSA**). There exists an initial temperature T_0 and for $t > 0$, such that

CSA

$$T_a(t) = T_0/\log(t) \qquad (1A)$$

FSA

$$T_c(t) = T_0/t \qquad (1B)$$

$$g_t \approx \exp\left(\frac{|\Delta x_0|^2}{T_a(t)}\right) T_a(t)^{\,D/2} \qquad (2A)$$

$$g_t \approx \frac{T_c(t)}{\left(T_c^{\,2} - |\Delta x_0|^2\right)^{(D+1)/2}} \approx \frac{T_0}{t\,|\Delta x_0|^{D+1}} \qquad (2B)$$

$$\sum_{t=t_0}^{\infty} g_t \geq \exp(\log(t)) = \sum_{t=t_0}^{\infty} \frac{1}{t} = \infty. \quad (3A) \qquad \sum_{t=t_0}^{\infty} g_t \approx \frac{T_0}{\left|\Delta x_0\right|^{D+1}} \sum_{t=t_0}^{\infty} \frac{1}{t} = \infty. \quad (3B)$$

A 1-D example has been given in [5], that has so-to-speak simulated **Haken's** model of the complex phase transition dynamics by means of random walks in a double well potential. The Cauchy fractal distribution function generates a stable and convergent random walks to the state of x-axis of the global minimum phase(left side of Fig. 4). The **annealing algorithm** is simply that a new state of **lower** values of potential is automatically accepted; a new state of **higher** values is also acceptable in annealing algorithms (but **not so** in any other algorithms) if going uphill only means to be limited within the noise bounds, that will be gradually diminishing at every iteration steps, according to the cooling schedule. Thus, such a tunneling is allowed in annealing algorithms in order to escape the trapping of a local minimun. This mechanism permits the resolution of ambiguities associated with various possible local minima.

Fig. 4 Brownian Fractal (right side) and Cauchy Fractal (left side): Random Walks in a Double Well Potential

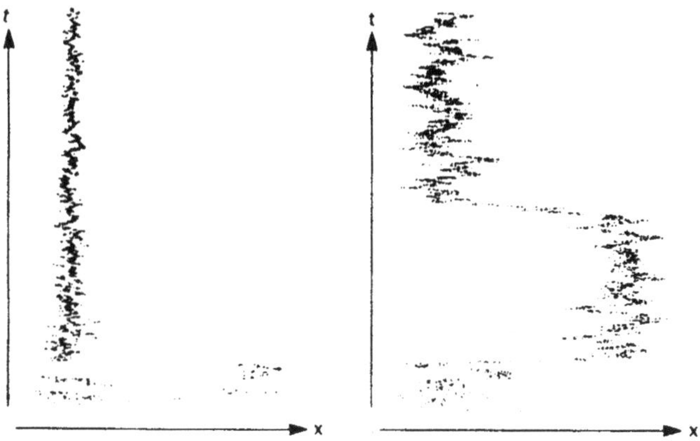

3. Generic Nonconvex Problem of Resolving Ambiguities in Unsupervised Learning

An example of a non-convex ambiguity resolution is the bearing fix problem. Given a known set of identical passive sensors at locations $\{(\xi_i, \eta_i), i = 1, .., n\}$, each sensor estimates to direction (bearing) to a set of indistinguisable targets at locations $\{(x_j, y_j), j = 1, ..., m\}$. The bearing from the ith sensor to the jth target is of the form $\theta_{ij} + \Delta\theta_{ij}$, where θ_{ij} are the true bearings and the parameters $\Delta\theta_{ij}$ are identically distributed random variables. The non-convex nature of this problem is demonstrated in Figure 5, where local minima (false targets 3, 4, and 5) appear at every intersection of two or more bearings.

Fig. 5 Target location from direction-finding data: ambiguity in multiple sensors and targets

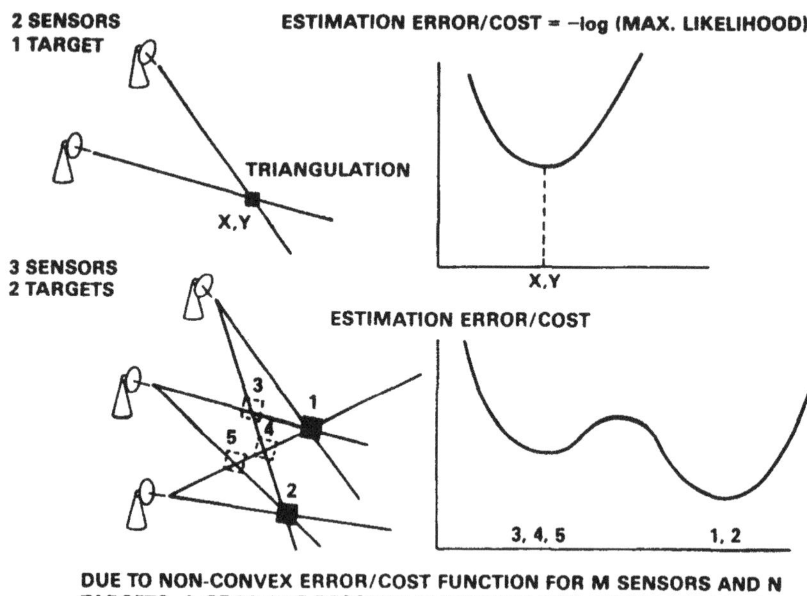

DUE TO NON-CONVEX ERROR/COST FUNCTION FOR M SENSORS AND N TARGETS, A GRADIENT DESCENT ALGORITHM WILL BE TRAPPED IN A LOCAL MINIMUM, PREDICTING FALSE TARGETS

The minima in the neighborhood of targets are generally lower but depend on the number of sensors and targets, and their relative geometry.

Mathematically, one can consider this process as deriving a set of vectors $\mathbf{x}_i$, from the origin to the estimated location of the ith target, given a set of line vectors $\mathbf{L}_j$ passing through the jth sensor with the measured bearing. The functional form of the conditional probability of $\mathbf{x}_i$ given $\mathbf{L}_j$ is identical to the conditional probability of $\mathbf{L}_j$ given $\mathbf{x}_i$. The latter problem is, given a set of positions, derive the course and speed of the various targets. The conditional probability is given by,

$$p(x_i \mid L_j) = (2\pi\sigma_{ij}^2)^{-1/2} \exp(-d^2(x_i,L_j)/2\sigma_{ij}^2) = p(L_j \mid x_i), \qquad (4)$$

where the distance $d(\mathbf{x}_i,\mathbf{L}_j)$ and its associated variance σ_{ij}^2 relate to the angular errors in the bearing fix problem and off-track distance errors in the tracking problem. The maximization of the appropriate maximum likelyhood is equivalent to the minimization of its exponent, namely the cost function is

$$C = -\log \prod_{\text{Lines}}^{n} \sum_{\text{Points}}^{m} p(L_j \mid x_i). \qquad (5)$$

Figure 6 is the results of a more real-world simulation. Here, there is some probability of wild bearing, which is flat distributed over 2π. Further, the probability of detection of a target by a sensor is limited. The five sensors are fixed at random locations on a circle and the five targets are randomly located in an inscribed square. The experimental parameters are standard deviation of bearing errors 2.5 degrees, probability of wild bearing .05, and probability of detection .95. The intermediate results in Figure 6 are after 100 iterations of the Cauchy annealing algorithm [5].

Fig. 6 Bearing-fix problem with missed detectors and wild bearings

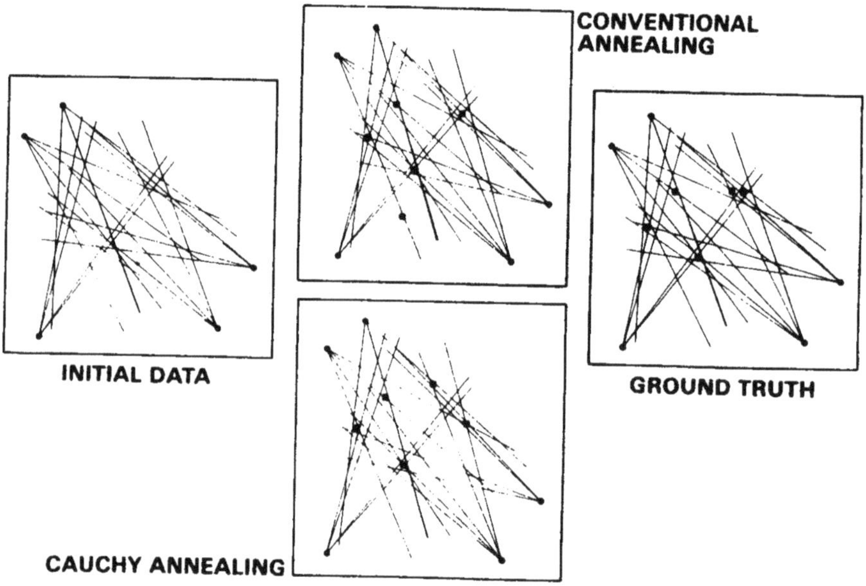

GROUND -TRUTH GENERATING FOR 5 SENSORS 5 TARGET BEARING-FIX ROBLEM:

ASSUMPTIONS:

A. WILD BEARING(FAR) 5 %

B. SENSOR DETECTION PROBABILITY 95 %

C. STANDARD DEVIATION OF BEARING ANGLES 2.5^0

INITIAL CONDITIONS:

1. 5 SENSORS LOCATED ON A CIRCLE
2. 5 TARGETS LOCATED IN A SQUARE
3. GENERATE BEARING ANGLE FOR EACH SENSOR AND TARGET

Figure 7 is an example of the tracking problem where the various locations are first associated with a set of straight lines and the direction of each line defines a course and the average velocity along the line determines the speed.

Fig 7 Track-correlation problem

VAX 780 1 MINUTE CPU

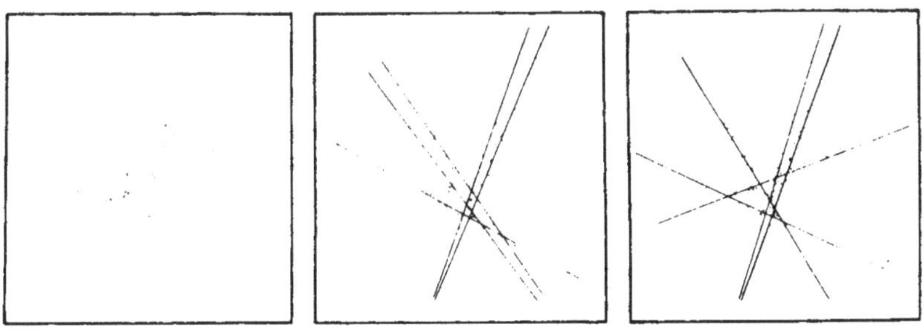

INPUT FIVE TARGETS HISTORY — OUTPUT TRACKS — GROUND TRUTH

In conclusion, we mention the importance of annealing mechanism that, perhaps as a tribute to the unified thinking of Professor **Haken**, the formation of the universe require the capture of neutrons and protons after the big bang birth, (or else a free space neutron would decay without the possibility of the atomic periodic table). Then the formation of molecules, and the biogical species, as it cools down find their deep minima in a disconnected phase spaces, schematically shown in Fig. 8.

Fig. 8 Self-Organization Adaptive Systems

MEASURE OF THE DEGREE OF SELF-ORGANIZING ADAPTIVE SYSTEMS

LOW DEGREE → HIGH DEGREE

MECHANICAL THROUGHOUT INPUT/OUTPUT	AB INITIO CALCULATION DESCENDING ALGORITHMS	INTERVENING ADAPTIVE SIMULATED ANNEALING	SELF-ADAPTIVE SUPERVISED LEARNING WITH HIDDEN UNITS	SELF-DETERMINED ADAPTIVE UNSUPERVISED LEARNING	LIFE SCIENCE GENETIC ADAPTIVE

LESS TIME → MORE TIME

MOST EXPERIMENTS (WIENER)	COMPUTER SIMULATIONS (KALMAN)	KIRKPATRICK et al (SZU, HARTLEY)	HINTON SEJNOWSKI	GROSSBERG HECHT-NELSON

EXAMPLES OF COST FUNCTIONS

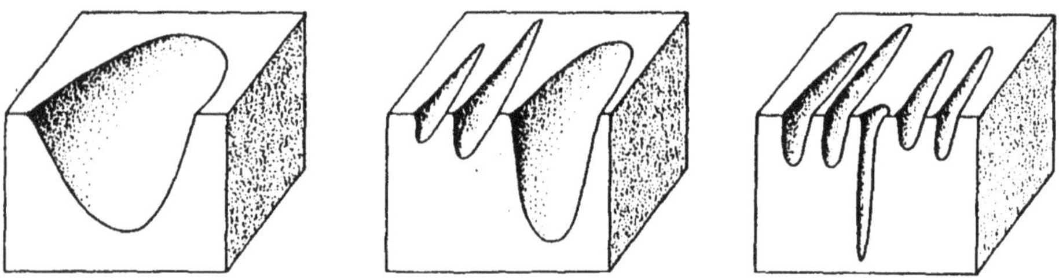

Acknowledgements

The author would like to thank Frank Polkinghorn for his help, and acknowledge the sponsorship of the ONT/NOSC TWAS program and the SDI/T/IS ONR office.

REFERENCES

1. H. H. Szu, and R. L. Hartley "*Nonconvex Optimization by Fast Simulated Annealing*", **Proceedings IEEE, Vol.75**, no.11, pp. 1538-1540, Nov. 1987

2. N. Metropolis, A. W. Rosenbluth, M. N. Rosenbluth, A. H. Teller, and E. Teller, "*Equation of State Calculation by Fast Computing Machines*", **J. Chem. Phys.** 21, 1087-1092, June, 1953.

3. S. Kirkpatrick, C. D. Gelatt, M. P. Vecchi, "*Optimization by Simulated Annealing*", **Science 220**, pp. 671-680, 13 May, 1983.

4. S. Geman and D. Geman, "*Stochastic Relaxation, Gibbs Distribution and the Bayesian Restoration in Images*", **IEEE Trans. Patt. Anan Mach. Int., PAMI-6** (No. 6), pp.721-741, Nov. 1984.

5. H. H. Szu and R. L. Hartley, "*Fast Simulated Annealing*", **Phys. Letters A, Vol.122**, nos.3,4, pp.157-162, Jan. 1987.

Part II

Pattern Generation in Neurobiological Systems

INVERTEBRATE CENTRAL PATTERN GENERATORS AS COMPUTATIONAL DATA BASES

Allen I. Selverston
Department of Biology, B-022
University of California, San Diego
La Jolla, CA 92093
(619) 534-2672

ABSTRACT

The lobster stomatogastric ganglion contains approximately thirty neurons which generates two distinct motor patterns, the pyloric and gastric. The basic patterns can be observed in isolated, deafferented preparations when two "upstream" ganglia, the commissurals, are left attached. The underlying mechanisms for the generation of the patterns is due to cellular bursting and network properties. Neuromodulatory substances, which are known to be present in lobsters, are capable of modifying the basic patterns by altering synaptic strengths and cellular properties such as bursting and plateauing. The changes in input/output properties of the individual neurons are able to alter the collective computational properties of the network. Strategies for modeling interesting small networks like the stomatogastric will be discussed.

Simple Neurons and Hebb Synapses

Computational strategies can play a critical role in giving direction and a theoretical framework to experimental neuroscience, but it requires some caution on both sides if the gains are to be truly lasting ones. The underlying assumption is that for both computational science and neuroscience, the brain can be reduced to its fundamental components and that the complex computations made in the central nervous system (CNS) result from the summation of elementary actions by the neurons and that these calculations can explain many psychophysical and cognitive phenomena. Can any of the new modeling algorithms tell us how the brain works? Can any but the most cellular phenomena (like the action potential) be described mathematically? Is there more to a modeling data base than simplified neurons and Hebbian synapses?

There have been enormous advances in biology at the cellular and molecular level. Within a short time it will probably be possible to express the genes for most membrane channels. We already have much of the structure for the ACh and other receptors and we know a great deal about how these two entities interact via second messenger systems. However there is much less empirical data about complex neural systems. Certainly to some this presents no problem; a "top-down" approach considering the brain only as a black box may have merit for psychological purposes. My own view is that a "bottom-up" approach is more powerful and that theoretical advances in the CNS cannot be made without an adequate data base. Computational approaches cannot gloss over this lack of empirical information. It should be noted that the enormous theoretical advances made in the physical sciences occurred because the properties of the elements and their interactions were simple compared to biological systems.

Physiological data from the mammalian CNS falls into only a few classes; single units recorded either intracellularly or extracellularly. The activities in these units are usually correlated with electrical stimulation of specific tracts, specific sensory inputs (particularly the visual system) or some motor output. More commonly, activity is recorded as populations such as EEGs or by using multiple electrode arrays placed directly on or in the brain. While this may be advantageous for some kinds of studies, it is clear that it provides no information about actual circuitry.

Invertebrate Model Systems

At this point in the evolution of computational approaches, the data bases which can be obtained from invertebrates, particularly from the area of rhythmic behavioral studies may be able to fill the computational data gap. Invertebrates offer many advantages to the experimentalist not seen in the vertebrates. The fundamental question of how these circuits relate to the human brain can be argued but in the end converges to pure faith. The biggest advantage of invertebrates are identifiable neurons. Invertebrate neurons are so large that once characterized, they can be reidentified in different animals. Additionally, intracellular soma recordings usually reveal subthreshold synaptic activity which means connections between cells can be determined. Nowhere has invertebrate "circuit chasing" been more successful than in the study of central pattern generators (CPGs), those ensembles of neurons capable of generating timing and pattern information for rhythmic behaviors entirely independently of peripheral reflex feedback loops. What we have learned from these circuits is that each neuron is a unique entity with its own set of static and dynamic conductances as well as a particular sensitivity to sensory input and neuromodulatory regulation. We have found that there are many different circuits used to produce quite similar behavior patterns (Selverston and Moulins, 1985). While the neurons and the circuits they form are each unique, they do not appear to be random. Instead there are common features to all CPG neurons and their circuit elements which might be considered as building blocks. These elements can be arranged in a variety of ways, just as a radio can be made up of different selections of electronic components as long as some general rules are followed. We can consider the lobster stomatogastric CPGs as examples of how this concept can be applied to real neural networks.

The Lobster Stomatogastric System

The stomatogastric ganglion contains approximately 30 neurons and is located in an artery just on top of the stomach. Unlike the stomachs of vertebrates, the lobster stomach is covered with discrete striated muscles which only contract when excited by bursts of impulses from the nervous system. Each muscle receives its own nerves and each contracts rhythmically to form two basic patterns. The faster or pyloric rhythm dilates and constricts the bottom of the stomach known as the pylorus. The slower pattern is associated with the gastric mill, three calcified ossicles located in the dorsal stomach which triturate the coarsely shredded food which has been stored in the cardiac sac. Fourteen neurons are involved in producing the three phase pyloric pattern and 12 are required for the four phase gastric. By recording from all pre- and postsynaptic elements in the ganglion, a neural circuit can be constructed (Fig. 1). Details of how this circuit was arrived at can be found in Selverston and Moulins (1987).

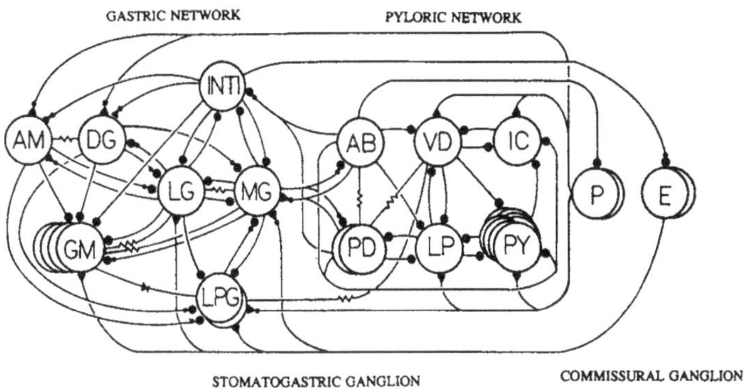

Fig. 1. The neural circuit underlying the gastric and pyloric central pattern generators. Black dots represent chemical inhibitory synapses. Triangles represent excitatory synapses and resistors signify electrotonic junctions. The E and P cells are in the commissural ganglia.

The general features of this circuit are that:

1. Most chemical synaptic connections are inhibitory but there is also extensive electrotonic coupling between neurons. Transmitter release is both graded and pulsatile.
2. A copy of the pattern (efference copy) is sent to upstream ganglia where it phasically inhibits the tonic firing of excitatory cells which, in turn, descend to the stomatogastric ganglion and supply excitatory drive to most of the neurons.
3. This descending excitation and additional descending inputs unmask bursting pacemaker and plateau potentials in many of the neurons.
4. Many neurons can be eliminated from each circuit without totally blocking the pattern, i.e. the pattern-generating properties are distributed throughout the synaptic network.
5. Each pattern is the result of the cooperative interactions between the cellular, synaptic and network properties.

While none of the well-described invertebrate CPGs uses the same mechanisms, they all use the same fundamental units.

Building Blocks for Oscillators

Mechanisms for the generation of cyclic burst patterns must be able to account for the production of timing and phasing information. This information derives from the combination of "building blocks" used.

We have found the most important cellular properties to include.

1. **Bursting pacemaker potentials:** These are most often expressed under appropriate neuromodulatory influence. Very few truly endogenous bursters exist.
2. **Plateau potentials:** Like bursting pacemaker potentials, these require synaptic drive and act to augment bursting in some neurons.
3. **Resting activity:** The spontaneous firing rate and regularity of neurons when synaptically isolated. Some neurons may just be silent.

The key synaptic properties include:

1. strength (efficacy) of the connection
2. accommodation or fatigue
3. delays
4. postinhibitory rebound -- especially recovery kinetics
5. summation and facilitation
6. sign and number of components
7. graded or pulsatile release - transfer function between cell voltage and spiking or transmitter release

The basic network configuration which is particularly important is reciprocal inhibition, often with weak parallel excitation.

Neuromodulation

What has become clear over the last five years is that all of the building blocks are not hard-wired or inflexible elements (except anatomically), but, under constantly changing chemical control. Both the pyloric and gastric rhythms must be thought of as a loosely organized system which can alter its structure with resultant alteration of motor output patterns.

The neuromodulators which have been found in the lobster include amines and peptides located in various parts of the nervous system, but many especially prominent in ganglia just "upstream" from the stomatogastric ganglion (Marder, 1987). Immunohistochemical studies have shown that the axons of these neurons enter the stomatogastric ganglion and branch profusely in the neuropile. Their release is local, not just across synaptic terminals, so that the CPG neurons are essentially flooded with the modulatory substance.

Unlike conventional synaptic transmission the modulators are not thought of as acting directly on receptors which have chemically sensitive channels, but instead at receptors which trigger an increase in several messengers with subsequent phosphorylation of channels. It is not yet clear how each cell responds to these substances, but it is possible that each chemical activates only a constellation of cells in each CPG and may affect individual neurons differently. Furthermore there is evidence that neuromodulators also affect synaptic transmission directly and may even alter the strength of electrical connections. At present it is possible to sort out all of the modifications, but not yet possible to resynthesize this disparate data into a whole pattern. We know what the modulator-induced pattern is when the circuit is intact and by use of a cell killing technique, we can isolate neurons and synapses and determine how the modulators affect them individually. We are not yet able to see if this unit data can produce the intact pattern and only a good model can assist us in this effort.

Any functional computer reconstruction must, at some point, be able to deal with the biological reality of circuits which are dynamic and easily "sculpted" into new funcitonal circuits. We do not know presently if similar forms of modulation occur in more complex circuits, but its occurrence is probably widespread given the thousands of peptides known to be present in mammalian brains.

An Interchange of Ideas

Computational neuroscience cannot succeed in unlocking general principles of brain function without empirical clues which will ground it in reality. Experimentalists on the other hand, are in desperate need of computational environments which can deal with limited numbers of neurons. As I alluded to earlier, simple ball and stick models do not suffice to explain how even a very simple system operates. Perhaps this isn't necessary and one might be able to make do with a qualitative understanding of the mechanism based on the available data. Adding the biological complexity to simulation algorithms which are presently available, may make them mathematically more difficult. But the flip side is that a good deal of the experimentally available data may be unimportant in terms of understanding mechanisms. One extremely useful contribution may be for modelers to delimit the kinds of data experimentalists should concern themselves with. At the very least, theorists should be able to suggest testable hypotheses.

Marder, E. (1987) Neurotransmitters and neuromodulators. In A.I. Selverston and M. Moulins (eds.) *The Crustacean Stomatogastric System*, Springer-Verlag, NY.

Selverston, A.I., and M. Moulins (1987) *The Crustacean Stomatogastric System* Springer-Verlag, NY.

STUDIES OF THE LAMPREY CENTRAL PATTERN GENERATOR FOR LOCOMOTION: A CLOSE RELATIONSHIP BETWEEN MODELING AND EXPERIMENTATION

Avis H. Cohen
Section of Neurobiology and Behavior
Cornell University
Ithaca, NY 14853

ABSTRACT

This paper describes central pattern generators, neural systems which generate temporal patterns of motor activity. The spinal pattern generator for locomotion of the lamprey is discussed at length. The description includes results of experiments which detail the frequency differences of the segmental oscillators along with the overall organization of their intersegmental coupling. These data are compared to predictions of dynamical systems models which have been developed to help understand the behavior of systems of coupled limit cycle oscillators. In an accompanying appendix, Nancy Kopell further outlines the relevant mathematics including recent results which help explain the behavior observed in the lamprey cord.

1. INTRODUCTION

Systems of coupled oscillators can be extremely difficult to understand under the best of circumstances. When the oscillators are uncharacterized except as being non-linear and the coupling is equally unidentified, the task is that much more difficult. The work of my colleagues and I over the years has been to unravel just such a system. The intersegmental coordinating system of the lamprey central pattern generator (CPG) has been the focus of the work. In this report I will outine what has been learned about this system indicating some of the interactions with the modeling, the insights provided back and forth and some predictions which have and have not been fulfilled. Especially interesting is new data from experiments not reported elsewhere which take on unexpected meaning in the light of recent mathematical work. The mathematics has appeared in various papers, beginning with Cohen et al. [1] and continuing most recently in Kopell and Ermentrout [2] which also discusses some other insights provided by the mathematics. An appendix by Kopell follows which deals with the most relevant recent results.

2. CENTRAL PATTERN GENERATORS

Central pattern generators are neuronal ensembles which generate patterns of neural activity without the need of sensory or higher order control. Examples are found throughout the animal kingdom wherever they have been sought. Well documented CPGs are known to exist all the way from leeches [3] and molluscs(cf. Mpitsos, this volume) up to mammals [4]. All vertebrates are most likely to be included in such a catalogue although the evidence for CPGs in humans is still indirect [5]. Not surprisingly, they are best understood at the cellular level in the simple nervous systems of invertebrates, nervous systems which survive and function when isolated from the organism. Most notable is the pattern generator for the stomach movements of lobsters and crabs, the stomatogastric ganglion(Selverston, this volume).

The existence of CPGs in mammals was not generally accepted until recently when Grillner and his associates [6] showed that cats spinalized, immobilized with curare, and devoid of sensory and descending input could produce virtually a complete locomotor motor pattern when injected with L-DOPA. This provided proof that the spinal cord had the complete neuronal apparatus required to generate the basic motor output to drive the limbs during locomotion. Sensory [7,8] and descending systems [9] have been shown to be crucially involved in making the movement adaptive and stable, but they are not required by the pattern generator or for the production of the motor pattern except for initiation. However, in normal functioning animals, motor pattern generation should be viewed as a fully integrated blend of all these systems working together.

3. LAMPREY SPINAL CORD AS A MODEL SYSTEM TO STUDY VERTEBRATE CPGS

Once the existence of CPGs had been established in higher vertebrates, the next task was to determine the mechanisms that underlay the pattern generation. How do neurons produce a temporally complex pattern in the absence of outside timing information? Given the vast number of cells likely to be involved the task of even identifying the neuronal elements of the pattern generator was daunting. To make the job more tractable, the lamprey spinal cord was chosen as a model system [10,11]. Although a bonafide vertebrate, the cord of the primitive lamprey is thin and has relatively few cells. Moreover, this fish-like animal's locomotor movements are uncomplicated since the aquatic environment it moves through is uniform and the animal's body is quite simple with no paired fins or other complex appendages. Finally, it was known that the cord of the lamprey could be isolated and kept alive and functional for days with relatively little effort [12].

It has now been well established that the isolated spinal cord of the lamprey can produce a stable pattern of activity when stimulated with bath applied compounds which act as neurotransmitters. The most effective have been the excitatory amino acids D-glutamate [11], N-methyl-D-aspartate(NMDA) or kainate [13]. These drugs can produce a stable periodic motor output pattern which has many aspects of the motor pattern seen during normal intact swimming [14]. The basic features that characterize the pattern include 1) strict alternation between the opposing motor outputs of a single segment (note R7-L7 in fig. 1); 2) the motor

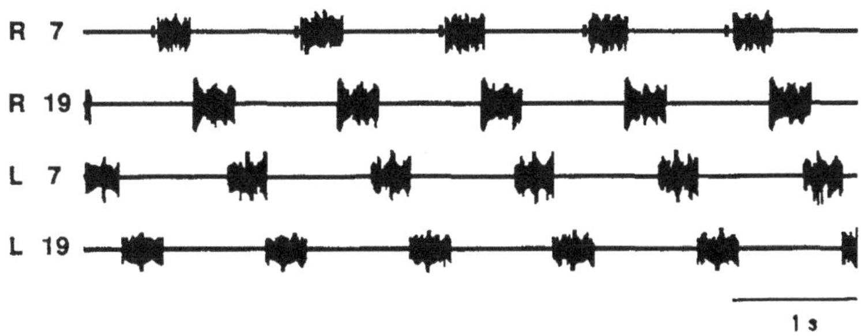

Figure 1: Diagram of motor output from isolated spinal cord of the lamprey elicited by bath application of D-glutamate. Letters and numbers refer to segment number in the isolated cord piece. (adapted from [11])

bursts of activity are always a constant proportion of the period, regardless of the period length. This proportion is often about 40%; 3) a constant phase delay between the bursts of any two segments. These phase delays have been described as being on the average 1% of the cycle per segment [11,14] along one side of the body with no printed report of systematic variations along the cord. The value of the segmental phases is independent of the swimming speed. Since there are about 100 segments, the 1% phase delays produce one wavelength along the body. These three features lead to the same hydrodynamically efficient traveling wave motion of swimming across all swimming speeds.

4. ORGANIZATION OF THE LOCOMOTOR CENTRAL PATTERN GENERATOR

A. The Segmental Oscillators

The overall organization of the CPG is believed to be that of a double chain of segmental neural oscillators coupled via inhibitory connections across the cord to give the observed alternation [15] and up and down the cord to produce the traveling wave [16]. The segmental oscillators are known only by inference. It is known, for example, that 4 segments can generate all aspects of the pattern [11] and 2 can generate alternating activity [17]. The oscillation itself could be a function of either membrane properties of individual neurons or the collective activity of networks of neurons lacking such properties. Within either framework the oscillators could be organized into discrete segmental structures or a diffuse network of cells strung out along the cord with little segmental demarcation. There could also be something intermediate between these two extremes, eg., there could be oscillatory cells all along the cord with denser concentrations segmentally. Relevant to this question is the fact that the drug which activates the rhythmic bursting, NMDA, is known to induce oscillations in many cell types [18]. The connections of these and any other oscillatory cells are known only preliminarily, but it has been suggested that they are organized into a network which even without the oscillatory membrane properties could give rise to an oscillation [17].

In most models of the lamprey locomotor CPG [1,19,20] the oscillators have been assumed to be <u>discrete</u> segmental structures. However, little has been done to determine the effect of this assumption. Recently, Kopell and Ermentrout have done preliminiary work in the direction of dealing with a more diffuse organization. They have looked at the behavior of segmental oscillators composed of oscillatory neurons which connect both upon each other and upon neighboring oscillators. If the coupling within a segment is comparable in strength to the coupling between segments, then the conceptual boundaries of the "segmental oscillators" are blurred, and yet the entire network can function like a chain of discrete oscillators.

The only other experimental information about the oscillators relates to the preferred frequencies of regional groups of oscillators. The oscillators are inactive when unstimulated, so one cannot speak of "natural" frequencies in this system, but it is possible to subject groups of 10 or more segments to identical conditions and compare their responses. The frequencies of the different groups of segments will typically differ when this is done. The first models for the CPG [1,20] predicted that the production of a traveling wave of motor activity required a gradient in the frequencies along the cord. Indeed, when cords were cut apart it was found that the groups of segments did differ in their frequencies. The first experiments used cords cut into only 2 or 3 pieces. The more rostral segments were generally faster. Thus, there appeared to be a gradient which most often descended along the cord. Occasionally, the frequencies were faster in the caudal segments (ascending gradient) and in some they were nearly identical [16].

These observations led to a problem, however, because although a gradient was required for the traveling wave, the phase lags among the segments were non-uniform along the cord. In the early models the differences in the phase lags could be made quite small with particular assumptions, but they could not be eliminated, and none had been reported [1].

More recently, the original cord cutting experiments were redone to obtain a more precise picture of the pattern of frequencies along the cord. In these experiments 50 segment cord pieces were cut into 4 parts and the frequencies monitored under identical conditions. It was found that the frequencies did differ, but the differences did not form a gradient; they were not typically monotonic in their changes, although they could be (fig. 2).

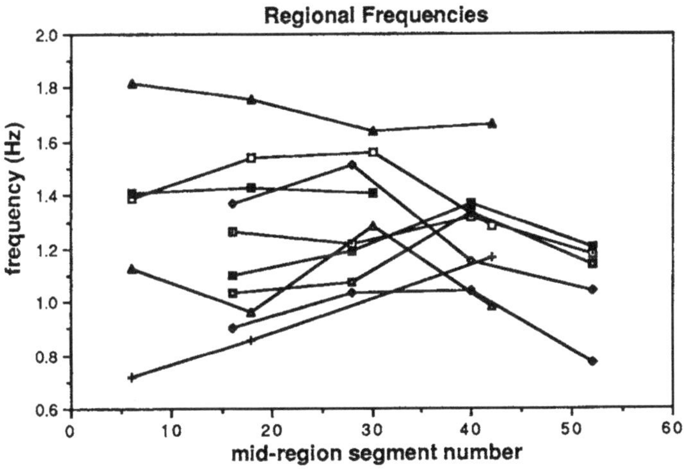

Figure 2: Diagram of frequency differences in the locomotor CPG along 50 segment pieces of lamprey spinal cord after cords were cut into 4 equal segment pieces. Mid-piece segment number is plotted versus the frequency of bursting when all pieces are held at the same concentration of D-glutamate. Some pieces were taken directly behind the gills and others were taken 10 segments caudal to that point. (see phase lags along the intact cord of the same pieces.)

Therefore, there do exist clear differences in the frequencies of the oscillators, but the pattern of differences does not form any consistent picture. However, the clearest lesson of these experiments is that one must be cautious in one's optimism when model predictions appear to be fulfilled. In spite of the pitfalls, intuitions derived from the role of the frequency differences in the models have led to a number of successful experiments. These are outlined in Cohen [21].

B. The intersegmental Coordinating System

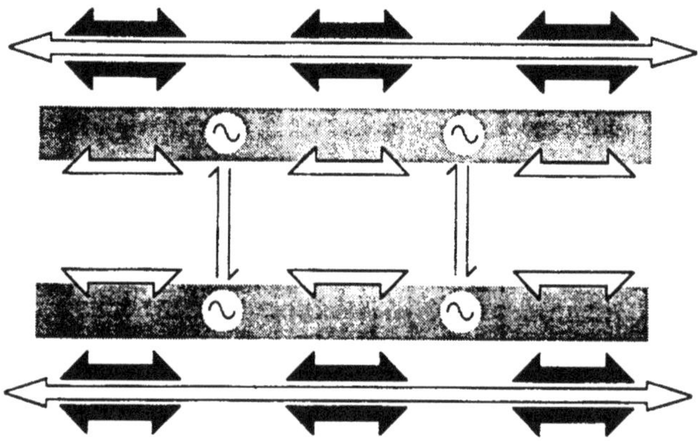

Figure 3. Diagram of the overall organization of the lamprey CPG for locomotion. The circles represent segmental oscillators which may or may not be discrete structures(see discussion in 4A). The arrows represent subdivisions of the coordinating system. There are two short components, one in the lateral tracts (outer arrows) and one in the medial area (inner arrows) and one long component in the lateral tracts (outer long arrows). The actual lengths of the short subdivisions is unclear and may differ considerably.

The lamprey spinal cord lends itself nicely to detailed organizational studies of the CPG. The cord is thin and transparent and discrete surgical lesions can easily be made to interrupt visually recognizable fiber tracts and morphologically distinct regions. On this basis the cord can be divided into three parts, the cellular gray region and the paired lateral and medial tracts.

When lesions and various other techniques were used to characterize the coordinating capacity of each of these regions, it was found that the lateral tracts contain two components of the coordinating system, one with short fibers and one with long fibers (fig.3). The long fibers are at least 10 segments long, but their precise length is unknown at this time. The length of the short fibers is unknown except by exclusion -- they are less than 10 segments long. Each subsystem can maintain coordination in isolation, but the short fibers are more robust and less susceptible to damage caused by small lesions than are the long fibers. The medial tracts and the cellular gray regions each by itself is incapable of coordination. However, when the medial tracts are combined with the medial edge of the cellular gray region, they form a short fiber subsystem which can maintain coordination alone, but is quite prone to disruption by small lesions. Thus, the coordinating system appears to have three sub-divisions, two short and one long. Each has the potency to maintain coordinated bursting stably by itself; together they form a highly stable and distributed system.

5. MODELING STUDIES OF THE COORDINATING SYSTEM

The early models [1] assumed that the CPG was composed of a double chain of coupled non-linear (limit cycle) oscillators. The coupling was simple diffusive coupling, with the specific models being chosen for their simplicity and computational tractability. Although they provided some welcomed insights at the time, as described above, one of the drawbacks of the diffusive coupling models was that the traveling wave resulted from a frequency gradient along the chain of oscillators which we now know need not exist (fig. 2).

Recently, Kopell and her colleague Ermentrout have developed a new theoretical framework in which the phase delays emerge from a different mechanism and do not require frequency differences to produce traveling waves [2,19]. Their models can be generalized to incorporate the earlier models, thereby permitting many interesting mechanistic comparisons to be made. As in the earlier models the coupling has been primarily nearest-neighbor, but the coupling is what Kopell and Ermentrout call "synaptic" coupling, a type which more accurately mirrors the neural connections with chemical synapses. The diffusive coupling better represents electrical synapses which do exist, but are not as common as chemical synapses at which a neurotransmitter mediates responses. For a full discussion of their models, the reader should see [2,19].

An important fact that has emerged from the mathemtical treatments and that has seemed to hold across the various models is that with frequency differences along the chain it is difficult to obtain uniform phase delays among the oscillators. Because of the robustness of this observation and the indisputable fact that there are typically differences in the oscillator frequencies of the lamprey locomotor CPG, it was important to know if the phase delays really were uniform as many of us had been assuming.

There existed a strictly mechanical problem in obtaining this information. The segmental motor outputs are close together (1mm apart) and highly numerous in the preparations that are of interest for intersegmental coordination (typically 50-60 per side). To avoid the necessity of positioning many electrodes along the cord to measure the segment by segment phase delays, I devised the following protocol:

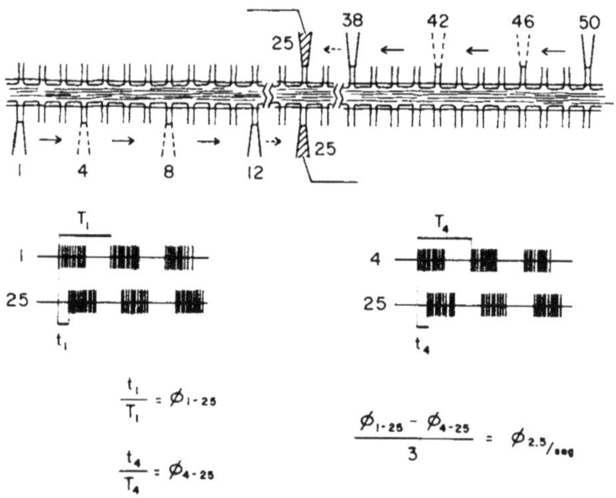

Figure 4. Diagram of protocol for measuring phase delays between nearly successive segments. See text for details.

Fifty segment cord pieces were dissected out and six electrodes positioned three on each side of the cord (fig. 4). The two middle electrodes were kept as reference electrodes opposite each other at roughly the same cord segment. Two electrodes were rostral and two caudal to the reference electrodes. These were moved in three segment increments, recording 25 cycles at each position. The phase delays between the moving and reference electrodes was computed (where phase delay is defined as the delay from the more rostral to the more caudal segment, divided by the period of the more rostral segment). The values for the successive positions was then subtracted and divided by the actual number of segments moved. This gave the value of the per segment phase delay over the interval traveled by the moving electrode. For each experiment (n=6) the per segment phase values were plotted against the mid-segment for the interval over which they were computed (fig. 5).

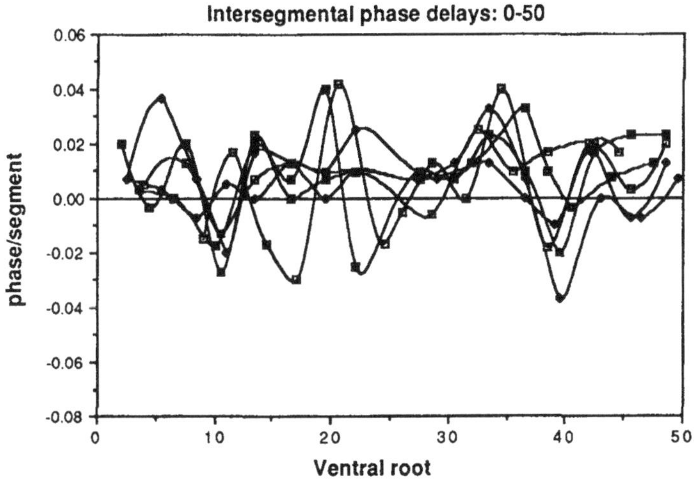

Figure 5: Intersegmental phase delays from segment to segment along a 50 segment cord piece. (see fig. 4 for method of computing phases) The cord pieces are from those whose frequencies are illustrated in fig. 2.

Looking at figure 5, it is clear that the phase delays are not in fact uniform along the cord piece. The data are quite variable, but consistently there are two pronounced dips in the phase values, both dips near 10 segments from their respective ends of the cord. Near the dips are often rises in the phase values. The dips and rises could be "edge effects" or they could be a function of some peculiarity in the anatomy of the region I chose to examine. To test which of these possibilities was more likely, I examined new pieces dissected from fresh animals from the same batch. These pieces were taken exactly ten segments caudally from the position of the original group. If the dips and rises were a function of the cord anatomy, then the rostral dip should be eliminated and the caudal dip should now be 20 segments from the end . However, what was observed were two dips, again roughly ten segments from the end (fig. 6). The precise segment number of the rostral dip is a bit less in these segments, but the dips are nowhere near where they would be if they were strictly related to the anatomy of the cord. They appear to be true edge effects.

As will be discussed in the Appendix by Kopell, edge effects such as those seen in the experiments occur as a consequence of having an organization in which each oscillator is connected to several or even many of its neighbors, instead of just one. Indeed, in the preliminary simulations with

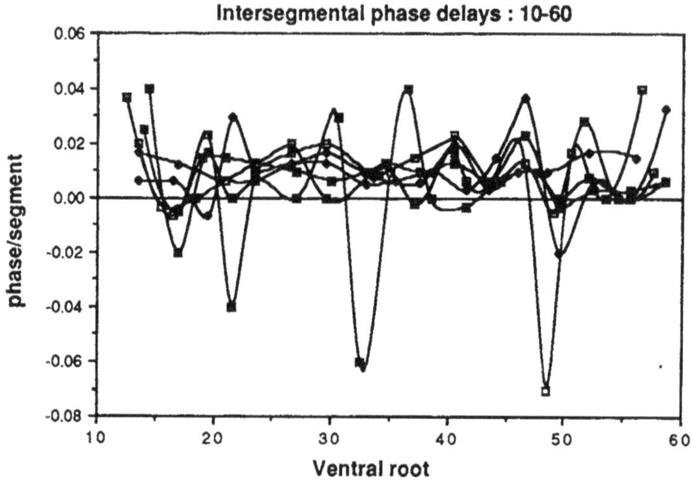

Figure 6: Intersegmental phase delays in cord pieces taken from animals 10 segments caudal to those in fig. 5. Everything else was the same. The frequencies for these cord pieces are also plotted in fig. 2.

such models, the dips came at the segment corresponding to the number of segments connected on each side to any given segment. This organization has the further property that the multiple coupling helps to buffer the phase lags against differences in local frequency of the sort that was found in the experiments described above. That is, for models with only nearest neighbor coupling, such frequency differences generally lead to travelling waves of activity with phase lags nonconstant all along the cord, not just near the edges. Thus, we might think of the multiple coupling as a system designed to produce particular phase lags, even in the presence of accidental frequency differences; the edge effects are then artifacts of such a design caused by removing the cord from the animal.

6. CONCLUSION

Central pattern generators are well established, but not well understood. This is especially true of those found in vertebrate nervous systems. In this paper, I have presented our attempts to unravel the functional organization of the lamprey CPG for locomotion. The effort has included a strong interaction between experimental and theoretical treatments. The interaction has been clearly two-way, with the theory enlightening the experiments and the results suggesting new forms for the equations. And, in the process of developing the models to explain the locomotor CPG, new understanding of systems of coupled biological oscillators has emerged, making this an extremely fruitful interaction, indeed.

Acknowledgements
The author would like to thank M.Baker, M.Nelson and P.Brodfuehrer for assistance in the production of the figures, and N.Kopell for helpful comments. The work presented here was supported by NIH grant no. 16803 (to AHC) and AFOSR contract no. F49620-87-C-0013 to Northeastern University.

REFERENCES

[1] Cohen, A.H., Holmes, P.J. and Rand, R.H., The nature of the coupling between segmental oscillators of the lamprey spinal generator for locomotion; a mathematical model. J. Math Biol., 13,345-369(1982).

[2] Kopell, N. and Ermentrout, G.B., Coupled oscillators and the design of central pattern generators, Math. Biosci., in press.

[3] Friesen, W. O., Poon, M. & Stent, G.S., Neuronal control of swimming in a medicinal leech IV. Identification of a network of oscillatory inerneurons. J. exp. Biol., 75:25-43(l978).

[4] Grillner, S., Control of locomotion in bipeds, tetrapods and fish. In V. B. Brooks(Ed), Handbook of Physiology, Section 1: The Nervous System, Vol. II, Part 2, Amer. Physiol. Soc, Bethesda, pp. 1179-1236(1981).

[5] Forssberg, H. Ontogeny of human locomotor control. I. Infant stepping, supported locomotion and transition to independent locomotion. Exp. Brain Res., 57:480-493(1985).

[6] Grillner, S. Zangger, P. How detailed is the central pattern generator for locomotion? Brain Res. 88:367-37l(l975).

[7] Rossignol, S., Lund, J. P. and Drew, T. The Role of Sensory Inputs in Regulating Patterns of Rhythmical Movements in Higher Vertebrates. In: Neural Control of Rhythmic Movements in Vertebrates. Eds. A. H. Cohen, S. Rossignol, and S. Grillner, John Wiley & Sons, Inc., New York pp.201-283(l988).

[8] Wallén, P. Spinal mechanisms controlling locomotion in dogfish and lamprey. Acta Physiol. Suppl. 503, pp. 1-45(l982).

[9] Gelfand, I. M., Orlovsky, G.N., and Shik, M.L. Locomotion and Scratching in Tetrapods. IN: Neural Control of Rhythmic Movements in Vertebrates. Eds. A. H. Cohen, S. Rossignol and S. Grillner. John Wiley and Sons, Inc. New York, pp.167-199(l988).

[10] Grillner, S., McClellan, A., Sigvardt, K., Wallén, P., Wilén, M., Activation of NMDA-receptors elicits "fictive locomotion" in lamprey spinal cord in vitro, Acta Physiol. Scand., 113:549-551(1981).

[11] Cohen, A. H. and Wallén, P., The neuronal correlate of locomotion in fish: "Fictive swimming" induced in an *in vitro* preparation of the lamprey spinal cord, Exp. Brain Res., 41:11-18(1980).

[12] Rovainen, C.M. Neurobiology of lampreys, Physiol. Rev., 59:1007-1077(1979).

[13] Brodin,L., Grillner,S., and Rovainen,C.M., N-methyl-D-aspartate, kainate and quisqualate receptors and the generation of fictive locomotion in the lamprey spinal cord, Brain Res., 325:302-306(1985).

[14] Wallén, P., and Williams, T. L., Fictive locomotion in the lamprey spinal cord in vitro compared with swimming in the intact and spinal animal, J. Physiol.(London), 347:225-239(1984).

[15] Cohen, A. H. & Harris-Warrick, R. M. Strychnine eliminates alternating motor output during fictive locomotion in the lamprey. Brain Res. 293:164-167(l984) .

[16] Cohen, A. H. The structure and function of the intersegmental coordinating system in the lamprey spinal cord. J. comp. Physiol., J. Comp. Physiol.,A, 160:181-193(1987).

[17] Grillner, S., Wallén, P. Dale, N., Brodin, L. Buchanan, J. and Hill, R. Transmitters, membrane properties and network circuitry in the control of locomotion in the lamprey. TINS 10:34-31(l987).

[18] Wallén, P. and Grillner, S. N-Methyl-D-Aspartate Redceptor-Induced, Inherent Oscillatory Activity in Neurons Active During Fictive Locomotion in the Lamprey. J. Neurosci. 7:2745-2755(l987).

[19] Kopell,N., Toward a theory of modeling central pattern generators. IN: Neural Control of Rhythmic Movement in Vertebrates, Eds. A.H.Cohen, S.Rossignol, and S.Grillner. John Wiley & Sons, Inc., New York, pp. 369-413(1988).

[20] Rand, R. H., Cohen, A. H. and Holmes, P. J. Systems of Coupled Oscillators as Models of CPG's. IN: Neural Control of Rhythmic Movements in Vertebrates. Eds. A. H. Cohen, S. Rossignol, and S. Grillner. John Wiley & Sons, Inc., New York, pp. 33-367(l988) .

[21] Cohen, A. H. Effects of oscillator frequency on phase-locking in the lamprey central pattern generator, J. Neurosci. Meth., (1987)21:113-125.

APPENDIX: Chains of oscillators and the effects of multiple coupling.

N. Kopell
Department of Mathematics
Boston University
Boston, MA 02215

As discussed above, the lamprey spinal cord can be roughly modelled as a double chain of oscillators, coupled both across the cord and along the cord. In normal locomotion, the contralateral oscillators operate in antiphase, which can easily be accounted for by the inhibitory connections across the cord. Thus, the modelling efforts are directed to a better understanding of how the organization of the network allows it to produce the appropriate phase lags between segments. For this problem, one can model the cord as a single chain of oscillators, with the local contralateral networks lumped together into "segmental oscillators". The work to be discussed is joint with G.B. Ermentrout.

Since the available information does not permit a detailed description of the local oscillators, our strategy was to use mathematics to sort out which features of the observed behavior are consequences simply of the existence of local oscillators and the linear geometry, and which require more restrictions on the oscillators or on their coupling; the restrictions then provide some focus for the biological investigations. It is not within the scope of this appendix to describe all of that work (see e.g. [1–3].) Instead, we shall briefly describe some features of the framework of the investigation, and our present conclusions about the possible functions of some of the anatomical organization of the cord.

In the mathematics, the segmental oscillators are almost completely unspecified, except to require that they each be describable by an ordinary differential equation having a stable periodic solution. (This requires a nonlinear equation.) The state variables of these equations can be vectors of arbitrary size, including any variables, such as membrane potentials, firing rates, channel openings, etc., that may be relevant. The oscillators at different positions along the chain need not be the same (for example, they need not have the same frequency), but the changes must be gradual along the chain. The large amount of generality about the oscillators reflects the fact that the behavior of the system of coupled oscillators is relatively insensitive to the details of the oscillators.

By contrast, some of the details of the coupling are quite important. We shall write out the form of the equations in a very general way, and then discuss the restrictions. The full equations are of the form

$$(1) \qquad X_k' = F_k(X_k) + \sum_{i=1}^{m} G_i^+(X_{k+i}, X_k) + \sum_{i=1}^{m} G_i^-(X_{k-i}, X_k).$$

In equations (1), X_k is a vector of variables for the kth oscillator. $X_k' = F_k(X_k)$ represents the equation of the oscillator at the kth position, where $1 \leq k \leq N$. G_i^+ represents the effect of the $(k+i)^{th}$ oscillator on oscillator k; i.e., the sum involving $\{G_i^+\}$ represent the affects of the ascending coupling. Similarly, the sum involving the $\{G_i^-\}$ represents the affects of the descending coupling, which is allowed to be completely different from the ascending coupling. The integer m denotes the number of neighbors on each side of the kth oscillator that is connected to the latter. These equations can be generalized to allow the strengths of coupling to vary along the cord [2]; for simplicity, we do not do so here.

Even in this generality, it is possible to investigate equations (1) analytically, provided that we make a few hypotheses on the coupling functions $G_i^\pm$. The first hypothesis is that all the state variables of the oscillators are completely determined by the phases of the oscillators; this is always true provided that the coupling is not so strong as to overcome the tendency of the limit cycle oscillators to maintain their wave–form and amplitudes [4].

The second is less intuitive: it is assumed that if the coupling terms in (1) are replaced by their averages over a cycle of the kth oscillation, then the resulting equations have the same phaselocking behavior as (1). If the coupling is "weak", it follows from so–called "averaging theory" that this hypothesis can be made without loss of generality. For stronger coupling, the hypothesis is not in general valid, but we argue in [2] and [5] that unless the circuit is designed so that the hypothesis is correct, the system is vulnerable to a catastrophic cessation of the rhythm. Furthermore, we suggest a design to accomplish this that is consistent with the known physiology.

In the presence of these hypotheses, equations (1) reduce to equations having a much lower dimension. These have the form

$$\theta_k' = \omega_k + \sum_{i=1}^{m} H_i^+(\theta_{k+1} - \theta_k) + \sum_{i=1}^{m} H_i^-(\theta_{k-1} - \theta_k). \tag{2}$$

Here θ_k is the phase of the kth oscillator, which has frequency ω_k. H^+ and H^- are some scalar, 2π–periodic functions, such as $A \sin(\cdot) + B \cos(\cdot)$. The ability to reduce to phase equations comes from the first hypothesis; the fact that the coupling terms in (2) depend only on the difference of the phases is a consequence of the second.

As mentioned in the main part of the paper, the existence of the travelling wave of electrical activity is not, in this framework, primarily a result of frequency differences along the cord. (Such differences can produce waves of activity, but these waves have the wrong quantitative properties.) Instead, it follows from the hypothesis of "synaptic coupling" [1–3], which says, roughly, that the coupling is such that, on the average over a cycle, the oscillators influence the frequency of one another even when they are in the same state. (The paradigm is an oscillatory neuron or neural subnetwork which influences other local oscillators via excitatory or inhibitory neurotransmitters. Technically, the hypothesis is that $H^{\pm}(0) \neq 0$.) When the coupling is "synaptic" in the above sense, a chain of identical oscillators, coupled only to its nearest neighbors, does not phase lock in synchrony; instead, there is a travelling wave of activity with nonzero constant phase lags $\theta_k - \theta_{k+1}$ (except possibly right near one or both of the boundaries). If, in addition, the oscillators have frequencies that vary along the cord, the phase lags also vary.

It is more biologically correct to regard the chain of oscillators as coupled to many neighbors, rather than to just nearest neighbors. This more complicated description also helps to account for what would otherwise be a puzzle: frequency differences of significant size have been found in the cord by Cohen, but there are no measurable variations in the phase lags of a form that could be attributed to the frequency differences. Mathematical analysis and numerical simulation of equations (2), compared to the nearest neighbor version ($m = 1$), show that a multiply coupled chain has a more constant phase lag in the presence of variation in frequency at positions along the cord [6]; for a chain that has a relatively large m, such as $m = 10$, the frequency variations of the range seen experimentally might have negligeable effect on the lags. Furthermore, the multiple coupling provides a mechanism for producing, in a highly stable manner, very small phase lags between neighboring segments [6]; this is significant, since the phase lag along the lamprey cord is, on the average, only 1% per segment.

Because the chain of oscillators is finite, the oscillators at the ends experience different interactions from those in the middle, even if all the oscillators and all of the coupling is uniform along the chain. Indeed, it is this difference that creates the travelling wave of activity when the coupling is synaptic [1–3]. (If the ends are identified to form a circle of oscillators, the effect of synaptic coupling vanishes, and synchrony is then a possible solution.) When the chain is multiply coupled, not only the first and last oscillators, but the first m and last m oscillators experience different interactions from the others. This is the source of the "edge effect" that was discussed in the main part of the paper. This effect exists over the range of the multiple coupling.

The mathematical analysis of the edge effect, like the experimental results, is still preliminary. There are many possible choices for the frequencies and the coupling functions H^+ and H^-, and the properties of the solutions can be very diverse [3]; we concentrated on those choices we believed to be most relevant to the experiments described above. To remove the influence of the frequencies, we used $\{\omega_k\}$ constant. Another effect we wished to remove was the "boundary layer" near one or the other end. This boundary layer can exist even when there is nearest neighbor coupling; unlike the edge effect we have been considering, the boundary layer (if it exists at all) extends over a region of size $(1/N)$, independent of the number of coupled neighbors. This boundary layer can be essentially eliminated if we require of H^+ and H^- that they each have the following property: if the other were eliminated, the remaining coupling term would produce waves going in the descending direction. (If H^+ and H^- are the same up to a scalar multiple, then the waves that they would each produce alone would be in opposite directions. Thus, this condition requires significant anisotropy in the coupling.) This restriction on the coupling functions is also motivated by data from forcing experiments [7]; preliminary analysis suggests that that this data is consistent only with coupling functions that have that property [2,8]

Numerical simulations were done under the above restrictions. The solutions that were obtained were qualitatively similar to somewhat smoothed versions of the phase lag data of Cohen, with $m = 10$. That is, near the left end, the solution decreased from a positive value as the index increased up to m; there was a minimum around m, and then a rapidly damped oscillation to a constant value. A similar description applied to the right hand end. Simulations were also done with coupling in only one direction. Similar results were obtained, with the edge effect at only one end (left end for forcing from the left, and right end for forcing from the right.) The edge effect was essentially independent of the size of the chain, provided that the chain has at least $2m$ oscillators; thus, the simulations suggest the prediction that as the lamprey cord is gradually cut down in size, and the phase lags measured, then the maximal dips will continue to be at a constant distance from each of the ends. The simulations differed from the data in one important qualitative way that is still not well understood: In the data, the phase lags appear to dip below $\varphi = 0$, thus locally changing the direction of the wave. In the simulations to date, the phase lags, as defined above, remained positive. We are currently investigating some possible sources of such negative lags.

REFERENCES

[1] Kopell and Ermentrout, "Symmetry and phaselocking in chains of weakly coupled oscillators", Comm. in Pure and Appl. Math., 39 (1986), 623–660.

[2] N. Kopell and G.B. Ermentrout, "Coupled oscillators and the design of central pattern generators", Math. Biosciences, in press, 1988.

[3] N. Kopell, "Toward a theory of modelling central pattern generators", in The Neural Control of Rhythmic Movements in Vertebrates, A.H. Cohen, S. Rossignol and S. Grillner, eds., J. Wiley and Sons, N.Y., in press, 1987.

[4] G.B. Ermetrout and N. Kopell, "Frequency plateaus in chains of weakly coupled oscillators, I", SIAM J. Math. Anal. 15 (1984), 215–237.

[5] G.B. Ermentrout and N. Kopell, "Oscillator death in systems of coupled neural oscillators", in preparation.

[6] N. Kopell, G.B. Ermentrout, W. Zhang, "Multiple coupling in chains of oscillators", in preparation.

[7] A.D. McClellan and K.A. Sigvardt, "Features of entrainment of spinal pattern generators for locomotor activity in the lamprey spinal cord", J. Neuroscience, in press.

[8] G.B. Ermentrout and N. Kopell, "Phaselocking in forced chains of oscillators", in preparation.

VARIABILITY AND CHAOS: NEUROINTEGRATIVE PRINCIPLES IN SELF-ORGANIZATION OF MOTOR PATTERNS[1]

G. J. Mpitsos, H. C. Creech, C. S. Cohan[2], and M. Mendelson[3]
Oregon State University
M.O. Hatfield Marine Science Center
Newport, OR 97365
USA

ABSTRACT

In this chapter we discuss the possibility that variability may be a central feature of self-organizing processes. We suggest that variability may be inherently part of the mechanisms by which adaptive "neurocircuits" emerge, and contrast such functional neurocircuits against definitions involving anatomical or dynamical structures which the self-organizational definition both contains and supercedes. The experimental work focuses on an invertebrate animal, the sea slug, *Pleurobranchaea californica*, which has a rich behavioral repertoire of buccal/oral behaviors, and a relatively simple nervous system containing identifiable neurons. We present evidence from work on a set of 20 neurons, which we refer to as BCNs (buccal-cerebral neurons), that communicate between the buccal ganglion and cerebral ganglion. These neurons are crucial for generating all buccal/oral behaviors, and provide an advantageous source of experimental material for inquiring into the self-organization of group activity. Variability in the activity of the BCNs, and in the motoneurons that they drive, is attributable to low-dimensional chaos, as shown by: 1) autocorrelation functions; 2) correlation analysis of phase portrait dimensions; 3) calculation of Lyapunov exponents; and 4) the structure of 1D maps of Poincaré sections. These findings indicate that some variability may arise from the same mechanisms that generate the patterned activity: i.e., that the observed variations are not noise that is superimposed on the code underlying a behavior, but rather that they constitute the code itself. We discuss the findings with respect to the role of sensory feedback in the production of adaptive behavior of animals as they interact with complex and often unpredictable environments, and we suggest that chaotic neural activity provides a means for the nervous system to create new informational space rendering animals more stably adaptable in such changing environments.

INTRODUCTION

Self-organization represents the ability of groups of individuals to act "cooperatively" [7,12,16,17,20,53]. These groups consist of relatively autonomous individuals, each acting nonlinearly and usually having information of what only part of the group is doing. Some individuals may have more persuasive powers than others, but given the information concerning a single individual, an observer could not determine the behavior nor future action of the group.

Hidden in the above account is the role that variability may have in establishing cooperative behavior. In neural systems, computer simulation studies have shown that a network of interconnected elements can produce different patterns of activity, depending of the

---

[1]This work was supported by grant AFOSR-86-0076.

[2]Present address: Department of anatomy, SUNY, Buffalo, NY 14226.

[3]Present address: Department of Family Medicine, University of Washington, Seattle, WA 98195, and Tacoma Family Medicine.

parametric state of the network [18,37,44]. In the same way, recent neurophysiological studies on simple systems indicate that the same population of identifiable neurons can produce different patterns of activity, depending on differences in sensory inputs or neurohumoral factors [11,13,31,42,47]. Because of the high degree of parallel interconnections among neurons in most nervous systems, it has become obvious that it is not possible to predict how the system will work dynamically by simply looking at the architecture of the neurocircuit; i.e., to determine the functional characteristics of any given neuron or of a group of neurons, it is necessary to examine the system during its temporal expression.

However, while the attribute of being parallel implies dynamical, it does not necessarily account for self-organization adequately. A functional neurocircuit arising from parallel networks may require dynamics in order to appear, but if the temporal characteristics and final result of the dynamics are always the same, the definition of self-organization has not progressed much beyond the original one in which the information for a motor pattern or a behavior resides in the neuroanatomy itself. This is not to say that the framework of anatomical, physiological, and neurohumoral factors are not essential. But while changes in these frameworks, such as synaptic modifications arising from associative learning or changes in hormone levels, may lead to different behaviors, they do not necessarily speak directly to the issue of self-organization. The central issue, we believe, is to account for the process by which both similar and different behaviors emerge from the same underlying framework, and how this framework can produce the same behavior in different ways. Thus, at least conceptually, self-organization seems to be tied to variability.

Observations of extensive variations in behavioral and neural responses have led us to suggest that variability may be an inseparable part of the mechanisms generating motor patterns rather than representing extraneous noise [38,39]. This is to say that variation is not noise riding on the neural code, but rather that the variation resides within the code itself. One of the simplest examples of this relationship between code and variation is the logistic equation, $X_{n+1} = A(1-X_n)X_n$, where each successive value X_{n+1} is generated from the previous value X_n [32]. For values of the constant A just below 4, the equation gives unpredictable results: although there is no noise in the equation, given a solution to the equation for a particular value of X, the probability of predicting the solution at some future value quickly decreases. This sensitivity to initial conditions is a characteristic feature of chaotic systems; the systems follow deterministic laws such as the logistic equation, but the behavior of the systems is not predictable into the future.

This inseparable connection between signal and noise has led us to inquire into the possibility that variability in our experimental system may also be chaotic. Our approach has been to take advantage of the technical amenities of invertebrate animals in which it is possible to identify specific neurons or small groups of neurons repeatedly from one animal to another [41]. In order to understand how adaptive behavior arises in whole animals, such groups of neurons should have a central role in generating behaviors, but the number of neurons in the group should be small enough to permit a quantitative analysis of all of their responses. In broad terms, the experimental problem may best be viewed as an inquiry into the process by which an individual affects the functioning group, and, in turn, how the group affects the individual. In the present paper we begin to address this problem by examining the temporal characteristics of activity in individual neurons during the production of repetitive activity arising from coordinated interactions in the group of neurons to which the neuron in question belongs.

THE EXPERIMENTAL SYSTEM

The sea slug *Pleurobranchaea californica*, a marine gastropod mollusc, produces many different buccal/oral behaviors that outwardly seem to have similar repetitive movements. Some of the most interesting of these behaviors consist of: several components of feeding, such as biting, ingestion, and swallowing movements; an obvious, active form of regurgitation; a food-rejection behavior by which the animal sequentially expels unwanted substances from the mouth in a fashion resembling a reverse of the bite-ingestion phases of feeding; and self- and interanimal gill grooming. An example of a feeding response appears in Fig. 1.

The nervous system of this animal consists of four major ganglia and several minor ones. Of the major structures, there are two pedal ganglia, one for each of half of the foot; one buccal ganglion which drives the opening and closing of the jaws and controls the movements of the radula, a structure which is analogous to a tongue; and a cerebral ganglion (brain) which innervates the mouth, lips, and the anterior regions of the head. Connectives (cables of nerve axons) convey information between specific ganglia.

In *Pleurobranchaea*, as in many other invertebrate animals and in some vertebrates, many neurons can be visually or physiologically identified either individually or in small groups [8,13,24,41,42,45,47]. Despite the relative simplicity of these animals, hundreds or thousands of neurons may become active in the production of a given behavior. However, although there are approximately 10,000 neurons in the nervous system of *Pleurobranchaea*, a peculiarity in its neuroanatomy (and probably in the nervous systems of many of its cousins) significantly reduces the number of neurons one needs to consider in order to inquire into self-organization. For proper coordination to occur between the movements of the jaws and mouth in *Pleurobranchaea*, it is necessary that the brain receive information about activity in the buccal ganglion (Fig. 2). It so happens that there is only one group of 15-20 neurons in each buccal hemiganglion that can perform this information-carrying capacity. We refer to them as BCNs (buccal-cerebral neurons), and they, along with one giant neuron, are the only neurons that send axons from the buccal ganglion to the brain.

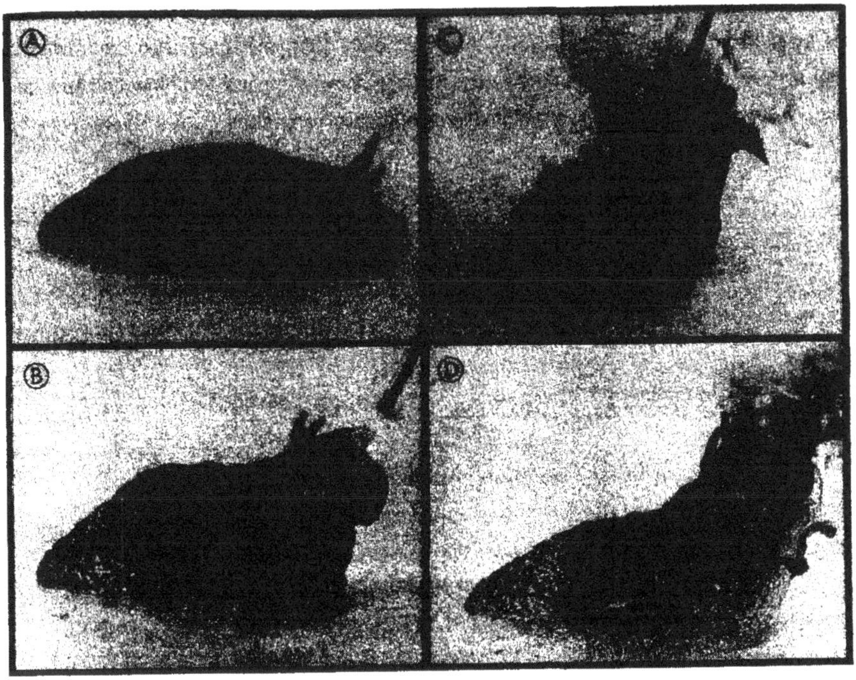

Figure 1: Photographs of the carnivorous mollusc *Pleurobranchaea* (A) locomoting from left to right. (B-D) increasing responses to food descending from a small tube at the upper right. Note progressive extension of the proboscis and opening of the mouth in (C). Response in (D) occurred within a fraction of a second from a starting posture as in (A). Cowcatcher-like structure lying over the proboscis is a sensory oral veil for detecting chemical and mechanical stimuli. Pair of upwardly directed objects are rhinophores which perform chemosensory functions. The unpaired gill is on the right side; best seen in (C) and (D). (from Mpitsos, Collins, and McClellan [40]).

It is relatively easy to record intracellularly from individual neurons, and extracellularly from the nerve bundles that communicate between the various ganglion. Because of the anatomical separation of the ganglia, it is possible either to examine local effects arising from activity within a group of neurons or within a single ganglion by severing the connectives, or to examine the interactions between groups of neurons by leaving the connectives intact. Such work has shown that the BCNs have a multiplicity of functions, but their most important feature is that the functional attributes of a given neuron, or of the group as a whole, emerge from the interrelationship or context of firing among all the coactive neurons [39]. Moreover, the firing of individual neurons and of the group is quite variable, as are the behaviors that these neurons help to form and regulate [38]. The BCNs intercommunicate among themselves, converge and diverge onto interneurons and motoneurons in the brain and buccal ganglion, they receive feedback from by the motoneurons that they drive [39], and they feed back onto the interneurons that drive them [14]. As such findings show, the neuroanatomic architecture of *Pleurobranchaea* is designed for a high degree of parallel processing.

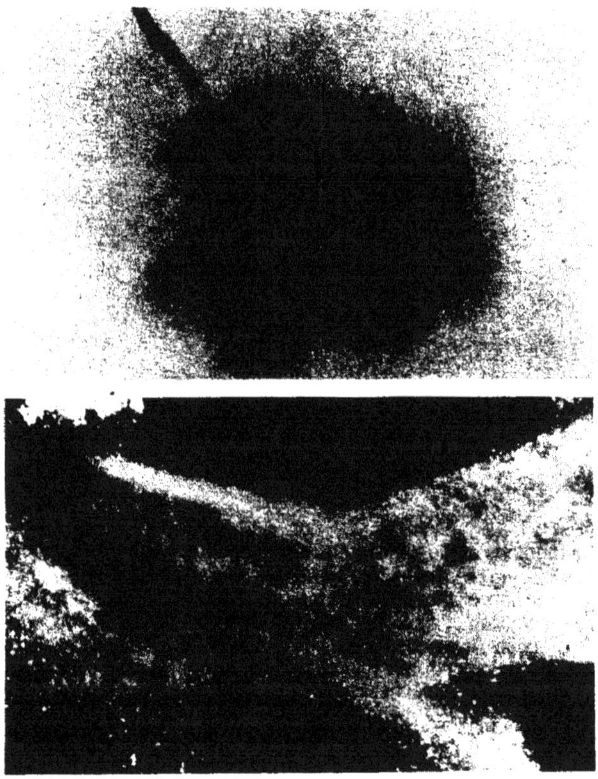

Figure 2: Brain (top panel; dorsal view); buccal ganglion (bottom panel; ventral view). Small dark objects are neurons that have been filled with a black precipitate (cobalt sulfide) for better visualization. Cells in the brain are motoneurons that innervate the mouth and lips. Cells in the buccal ganglion are BCNs, many of which contact brain motoneurons either monosynaptically or polysynaptically. Not shown is that each buccal hemiganglion connects with the corresponding half of the brain by means of a long cerebral-buccal nerve connective (CBC). Magnification: brain (22X); buccal ganglion (32X).

As in most invertebrate animals, the *Pleurobranchaea* nervous system functions well after being removed from the rest of the animal. Such isolated nervous systems can generate two characteristic types of motor patterns relating to different behaviors, one relating to the swallowing phase of feeding, and the other relating to the active phase of regurgitation (see Mpitsos and Cohan [38,39] for a critical discussion). In the following sections we examine first the motor pattern relating to feeding and then discuss an example relating to regurgitation.

COOPERATIVITY AND VARIABILITY IN SELF-ORGANIZING PROCESSES

Among their various functions, the BCNs generate the cyclical rhythm for opening and closing the jaws. This rhythmic activity is shown in Fig. 3 as sequences of bursts in buccal ganglion nerve root R3 which contains motoneurons that activate muscles for closing the jaws,

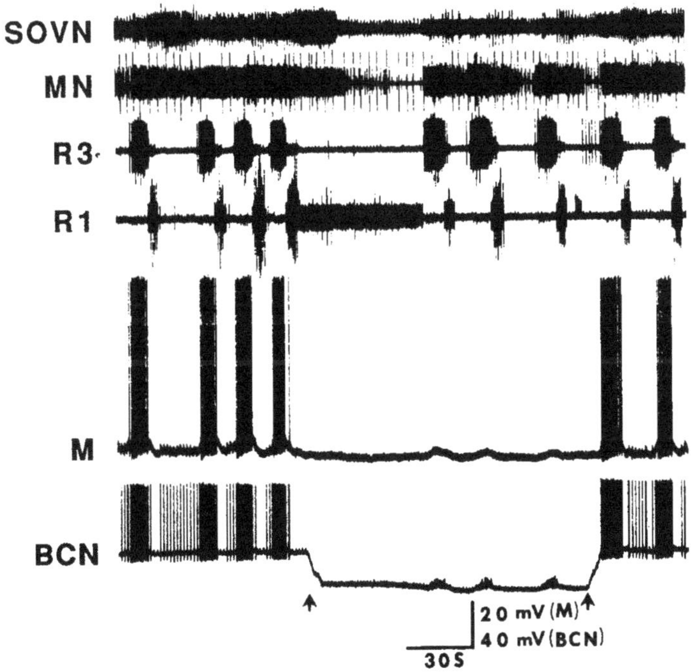

Figure 3: Functional "cooperativity" following perturbation of patterned activity. Top four traces are extracellular recordings from nerve roots, each containing activity of many motoneuron axons. Bottom two traces are intracellular recordings from a motoneuron (M) and a BCN. Removal of the BCN by hyperpolarizing it (between arrows) caused cessation of all activity, but the activity in the rest of nervous system eventually recovered despite the fact that the hyperpolarization forced the BCN to remain quiescent. Activity here, as in all physiological records shown below, was elicited by tonic electrical stimulation of a sensory root of the buccal ganglion. SOVN: brain nerve root innervating the oral veil and anterior mouth. MN: brain nerve

root innervating the mouth; contains axon of M. R3: buccal ganglion nerve root containing motoneurons for closing jaws. R1: buccal ganglion root containing motoneurons for opening jaws. (From Mpitsos and Cohan [39]).

and in root R1 which contains motoneurons that activate muscles for opening the jaws. As a group, the BCNs have heterogeneous effects on the generation of patterned activity. Figure 3 shows a BCN that initially had a strong effect. Removal of this neuron from the circuit (by passing hyperpolarizing current through the intracellular recording electrode) immediately stopped all cyclical activity in motoneuron (M), in buccal roots R3 an R1, and in brain nerve roots SOVN and MN. After about 60 sec, the pattern reestablished itself, despite the fact that the BCN was still hyperpolarized. With each cycle in the reestblished pattern the membrane potential of the BCN and MN exhibited small excitatory synaptic bumps originating from parallel inputs from the BCNs and other newly activated neurons. Release of the BCN from hyperpolarization (second arrow in Fig. 3) also released the motoneuron but had little effect on the pattern of activity in the rest of the nervous system.

Another example of changeable function in single neurons appears in Fig. 4 which shows the effect of the history of activity in the nervous system on subsequent activity in single neurons. Sections A and B of Fig. 4 each contain a motor pattern relating to the swallowing phase of feeding. However, in Fig. 4A the BCN and motoneuron fired most actively in phase with the R1 portion of the buccal cycle, whereas in Fig. 4B they fired in phase with the R3 portion. The only difference between these two recording situations was that in Fig. 4A the activity was recorded soon after the nervous system had generated a regurgitation motor pattern, whereas in Fig. 4B the activity was recorded after the nervous system had been generating the pattern relating to swallowing behavior.

The recordings in Figs. 3 and 4 illustrate two important interrelated features of cooperativity and self-organization: 1) Patterns of responses can be established in different ways, and 2) individual neurons can exhibit different functional properties within similar patterns. In Fig. 3 the BCN initially had a strong effect in pattern generation, but then it lost this effect when other BCNs reinstated the pattern. In Fig. 4 the BCN and motoneuron remained active throughout, but completely changed their phase of activity as a consequence of the preceding activity.

Both of these examples are extreme cases of variability. Closer inspection of the traces in Figs. 3 and 4 shows that there were many instances of relatively small variations. For example, the number of action potentials in the bursts of R1 in Fig. 3, and the number of action potentials in the BCN and motoneuron in Fig. 4, differed from one burst to another. We believe that the more extreme forms of variability arise from the same nonlinear dynamics that produce the small variations. Some of these smaller variations are examined quantitatively in the following section.

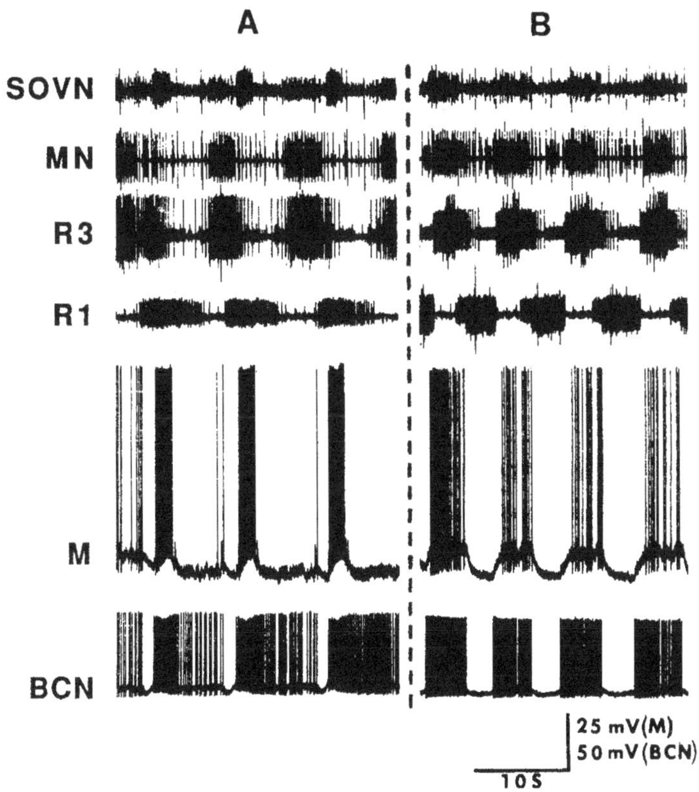

Figure 4: History of activity in the nervous system affects the way in which activity among neurons self-organizes to produce similar motor patterns. Same captions as in Fig. 3; different M and BCN. Note that the BCN and M were most active during the R1 phase of the buccal cycle in (A) whereas they were most active during the R3 phase in (B). (From Mpitsos and Cohan [39]).

DYNAMICS OF FIRING IN SINGLE NEURONS DURING PATTERNED ACTIVITY

Interpolation of Time Series

The time series of intracellular responses of BCNs and motoneurons (as in Figs. 3 and 4) were first recorded on FM tape. Selected portions were then played back and digitized at 1K Hz for computer analysis. The rate of digitization was adequate for analysis of the 10 to 15 msec action potential durations that are typical of molluscan neurons. As in Figs. 3 and 4, the activity described below was activated by electrically stimulating the stomatogastric nerve which innervates the esophagus and carries chemosensory information to the buccal ganglion. The stimuli were presented at 1 Hz, and at an electrical current strength near the threshold for generating the bite-swallow motor pattern [see also 34-36,39].

To illustrate the analysis, we selected the activity of a brain motoneuron which received converging inputs from several BCNs. The motor pattern was about 168 sec long and contained 548 action potentials. Bouts of feeding in whole animals varies from several seconds to several minutes, and, therefore, the selected record was within the range of typical behavior in whole animals. At digitization rates of 1K Hz, the number of points in the 168 sec time series was too long for analysis on a typical laboratory microcomputer. In order to reduce the amount of data, we used the sequence of unequal intervals occurring between action potentials instead of the equally spaced samples in the digitized series. Longer patterns have been recorded from other neurons, but we chose the present ones so as to examine situations that physiologists may typically encounter in experimentation: i.e., the length of the data was representative of adaptive responses in whole animals; the data required compression, but, once compressed, the series was relatively short and contained few cycles of structured activity.

Analysis of time series composed of such unequally spaced spike intervals in BCNs and motoneurons in *Pleurobranchaea* have yielded evidence for low-dimensional choas, as has been described for spontaneous activity in cat cortex [46]. We believe, however, that it is physiologically justifiable to convert the unequal time series into an equally spaced one by

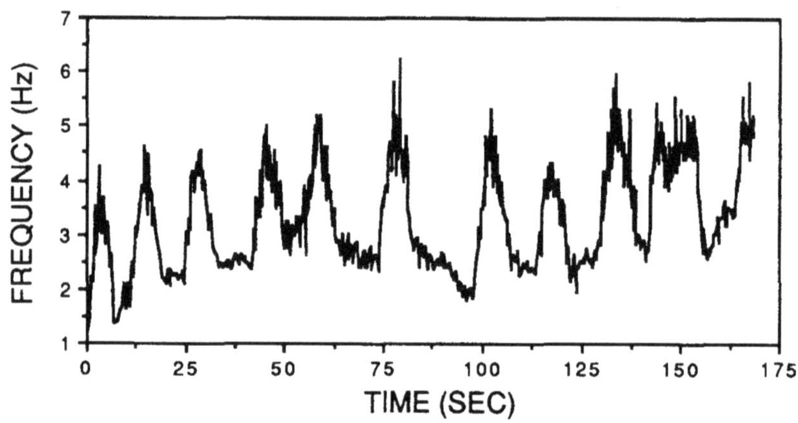

Figure 5: Firing frequency of a brain motoneuron during a 168 sec bout of the bite-swallow motor pattern. Frequency represents the reciprocal of the time between a spike in the motoneuron and the preceding spike. Data shown here is uninterpolated. For subsequent analyses, a variety of interpolation rates were used. Major criterion for effective interpolation was to reproduce all aspects of the shape in the uninterpolated series.

interpolation. First, it is a relatively safe assumption to make that the frequency of action potentials arriving at a synapse carries the code of the information in the input neuron's activity. Figure 5 shows the frequency series for the uninterpolated data obtained from the motoneuron described in the preceding paragraph. This series has three components: a slow trend in which the baseline frequency slowly increased throughout the series; 11 consecutive bursts of activity; and a high-frequency component which is superimposed on both the slow trend and the repeating bursts. Second, we propose that the shape or the envelope of such a frequency series, including initially all the components, contains the essential elements for reconstructing the information that one neuron conveys to another. Third, we propose that it is permissible to sample the series at equally spaced intervals as long as the interpolation rates retain the characteristics of the uninterpolated frequency function. This transformation to equally-spaced intervals allowed for the application of more analytical methods in examining the activity of single neurons than are available for unequally spaced intervals.

Because interpolation distributes the density of points evenly throughout the series, too low a rate underrepresented the high-frequency bursts. However, once we obtained good congruence between the shapes of the uninterpolated and interpolated series, the analyzed results remained stationary over a wide range of interpolation rates. In the present example, similar results were obtained using interpolation rates between 0.05 and 0.25 sec.

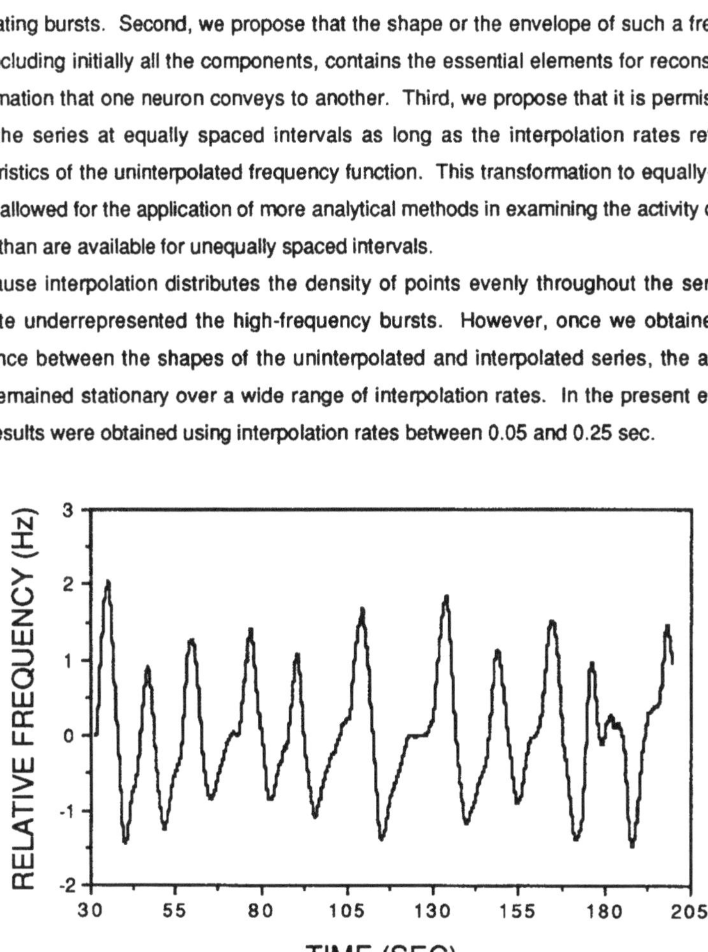

Figure 6: Same patterned activity as in Fig. 5, except that the activity was linearly interpolated at 0.10 sec intervals and passed through a filter having a high pass of 0.05 Hz, and a low-pass of 0.15 Hz.

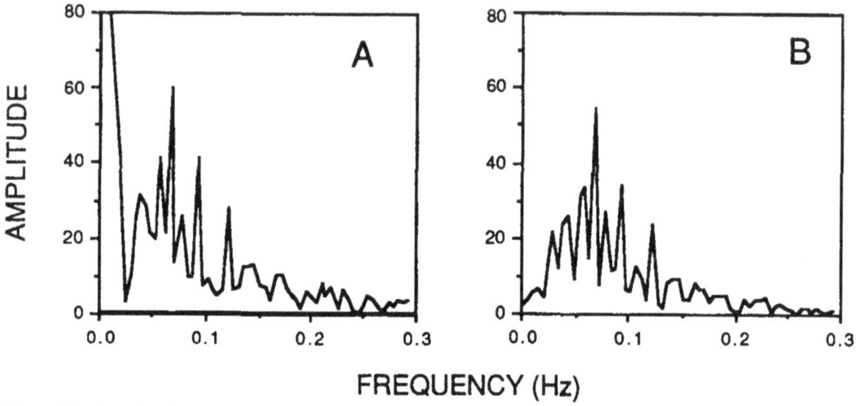

Figure 7: Amplitude spectra. (A) interpolated, unfiltered data. (B) interpolated and filtered data using a band pass of 0.05 to 0.15 Hz.

The amplitude spectra of the unfiltered and filtered series are shown in Fig. 7. Interestingly, the peaks are near but not exact multiples of one another. Examination of the data using different digitization and interpolation rates showed that the characteristics of the filtered and unfiltered spectra arose from the time series of the neuronal response rather than from the manipulations performed on the data. Moreover, when the pattern of activity spontaneously switched to another pattern representing the activity underlying regurgitation, the spectrum also changed, but, again, many of the peaks appeared to be close multiples of one another. Such spontaneous shifts in the pattern of neural activity are particularly interesting, and we shall return to them later in this chapter.

Autocorrelation Function

We examined autocorrelation functions for two reasons: 1) To obtain an indication of whether the activity was sensitive to initial conditions, i.e., whether the predictability of the activity decreased with increasing lags into its evolution. 2) To determine the appropriate lag for use in dimensional analysis.

The autocorrelation functions of the unfiltered time series slowly declined toward zero, reflecting the slow trend in the baseline frequency of Fig. 5, whereas the autocorrelation function of the filtered series quickly dropped to zero. Figure 8 shows that the autocorrelation function of the filtered series in Fig. 7B first crossed zero at a lag of 3.7 sec, and within 60 sec (approximately three cycles in the data of Fig. 6) the autocorrelation declined toward a more stable value near zero. Given the fact that the times series was repetitive and of relatively short duration, it is not surprising that harmonics appear at 75 and 150 sec, but the overall autocorrelation shows a consistent decline.

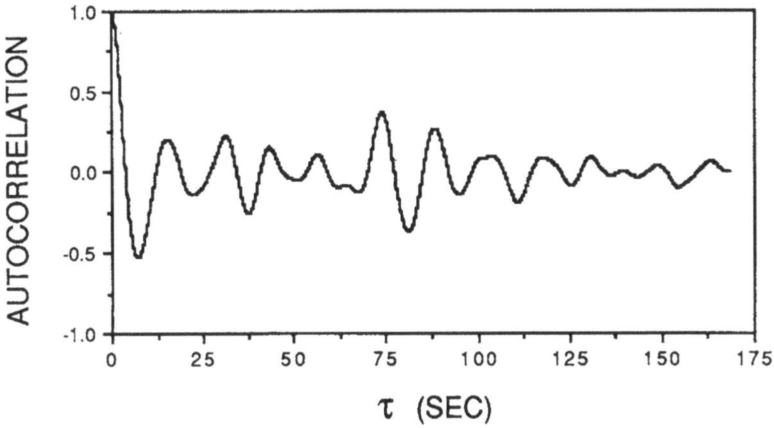

Figure 8: Normalized autocorrelation function of activity shown in Fig. 6.

Phase Portrait

From a time series of a single variable representing the combined effects of several processes, it is possible to obtain an indication of the dynamics of the activity and of the number of variables governing it. Following the work of Packard, Crutchfield, Farmer, and Shaw [43] and of Takens [54], we defined a multidimensional space by mapping the original frequency series f(t) along the x axis, then mapping the same data on the y axis after shifting it by a lag (τ), and on the z axis after a shift of 2τ. In d-dimensional space the procedure follows $f(t+(d-1)\tau)$, where τ is an integer multiple of the sampling interval. The evolution of the time series through successive points in multidimensional space defines the trajectory of the activity, and the overall evolution defines the phase portrait. A 3-D view of the first six cycles in the time series of Fig. 6 is shown in Fig. 9. The trajectories begin at the upper right at the asterisk and progress clockwise and outward until reaching the $f(t+2\tau)$ axis, then they swing in toward the origin and turn upward and outward again. The phase portrait resembles a bent coil. Even with so few cycles it is possible to see some divergence in the interrelationship of the trajectories, particularly at the sharp corner near the $f(t+2\tau)$ axis. To construct this phase portrait and to conduct the analyses presented in subsequent sections, we used $\tau = 3.7$ sec, or multiples of it, as determined from the first lag that generated a zero autocorrelation (Fig. 8). Karl Eugen Graf (Department of Clinical and Physiological Psychology, Tübingen) has obtained good congruence of results on EEGs using this method of selecting τ and one based on mutual information theory (personal communication).

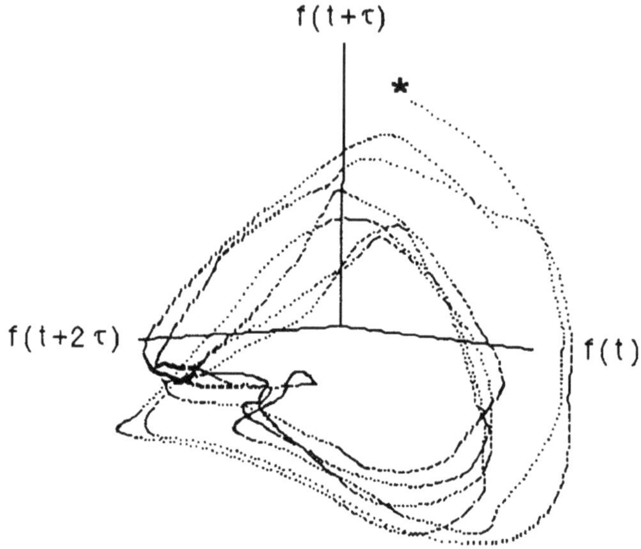

Figure 9: Phase portrait of activity shown in Fig. 6; for clarity, only the first six cycles are shown. The original series, f(t), was mapped on the x-axis, and then successively shifted by a lag τ on the y-axis and by 2 τ on the z-axis. The τ (3.7 sec) was selected from the first zero crossing of the autocorrelation function.

Correlation Dimension

Dimensional analyses provide an indication of two important features of time series: the number of dimensions (d) that govern the phase space, and the topological dimension (n) of the phase portrait, which denotes the dynamical nature of the time series. Following Grassberger and Procaccia [15], we defined an integral autocorrelation function:

$$C(r) = \frac{1}{N^2} \sum_{\substack{i,j=1 \\ i \neq j}}^{N} \Theta\,(r - |X_i - X_j|) \ . \qquad (1)$$

This algorithm selects a point X_i and measures the number of points X_j lying within a prescribed range of distances r until all N-1 points of the series are counted. The function θ excludes points lying outside each r. The process is then repeated by selecting another point as X_i for all N points. For small r, C(r) scales as r^n. Thus, n may be obtained from the slopes of plots of log C(r) = n log (r). The calculations are made for a series of embedding dimensions d: For random noise n scales linearly with increasing d; i.e., noise fills all space as dimensions are added, but in deterministic processes, the value of n levels off (saturates) after a certain d. Integer n values usually indicate predictable activity tending asymptotically toward some limit, whereas noninteger (fractal) values indicate chaotic activity [30].

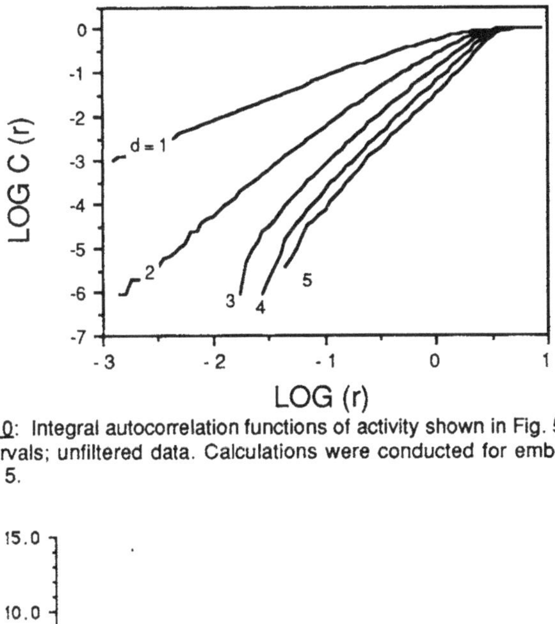

Figure 10: Integral autocorrelation functions of activity shown in Fig. 5 after interpolation at 0.1 sec intervals; unfiltered data. Calculations were conducted for embedding dimension d = 1 through 5.

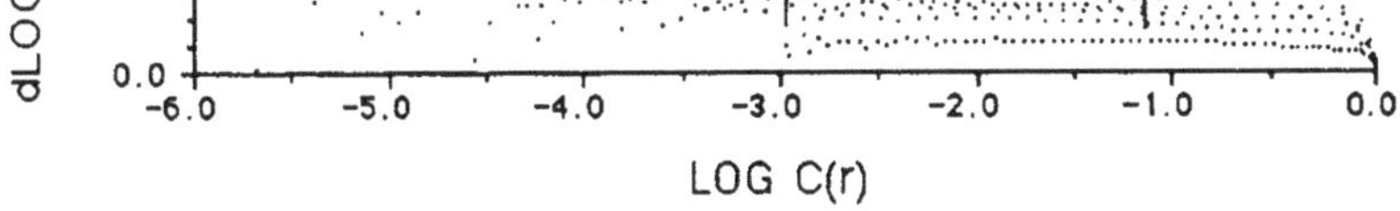

Figure 11: Determination of regions along curves of Fig. 10 having linear and saturating slopes.

The slopes along the plots of log C(r) versus log (r) curves often show variability (Fig. 10). In order to estimate the appropriate regions at which the slopes are linear and saturate toward a common value, we followed the example of Rapp et al. [46] and calculated the slope of the curves between successive points. These slopes were then plotted against log C(r) in Fig. 11. In the indicated range of Fig. 11, the points for d = 1,2 are easily distinguishable at the bottom near the log C(r) axis, but those for d = 3,4,5 intermingle. The indicated regions along the log C(r) dimension were then used to calculate the slopes n along the same regions for each d in

Fig. 10. Figure 12 shows the results of plotting n against d for the unfiltered activity of the motoneuron. Also shown are the results for the filtered data of Fig. 6, and for random noise having the same number of points, mean, and standard deviation as in the motoneuron series. For random noise, n scaled linearly and equally with increasing d. The unfiltered data saturated at a fractal dimension above 2 at d = 4, whereas the filtered date saturated at a fractal dimension below 2 at d = 2. The error levels were quite low, being smaller than the points in the graph. Using selective filters, we found that it was the high-frequency components that produced the difference between the filtered and unfiltered saturation levels in Fig. 12.

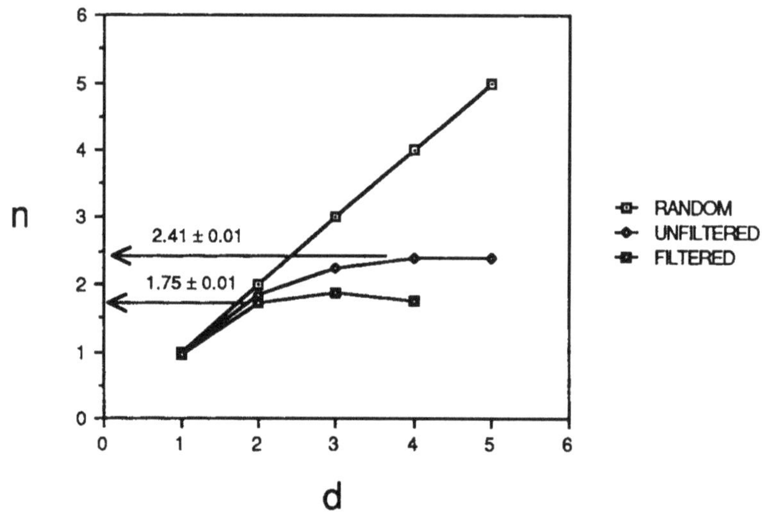

Figure 12: Saturation of dimension n with increasing embedding dimension d. Unfiltered data saturated at d = 4, whereas the filtered data saturated at d = 2. Random noise did not saturate. Standard error for each calculation was smaller than the size of the symbol in the illustration. Curves are drawn through points in order to aid visualization. See Figs. 10 and 11.

The saturation values of n were quite stationary when examined with different scaling factors. Similar saturation levels were obtained with interpolation rates of 0.25 and 0.05 sec as with the 0.1 sec rate used above. In addition, as shown in Fig.13 for the filtered and 0.1 interpolated data, the saturation levels were relatively unaffected by changing τ in the calculations of C(r). The average orbital period, as calculated from Poincaré sections of the phase portrait (discussed below), was 15 sec, and, therefore, the series of τ shown in Fig. 13 properly covered a large range of the orbit time.

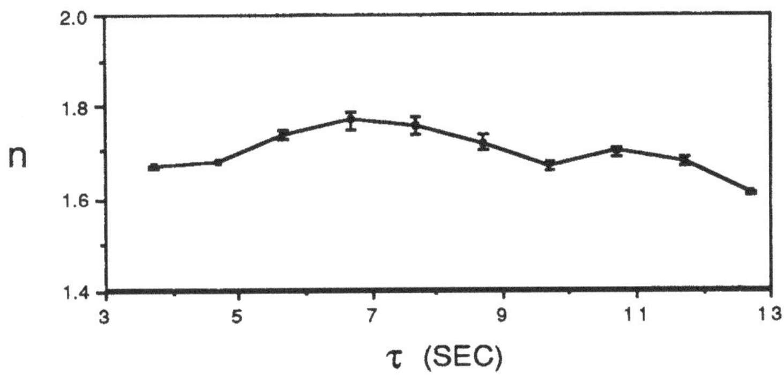

Figure 13: Topological dimension n remained stationary with increasing τ; d also remained constant, but is not shown here. Calculations were conducted on the filtered series shown in Fig. 6.

Lyapunov Exponents

Lyapunov exponents provide a means for determining the topological growth of the attractor as the time series evolves. We used the computer programs of Alan Wolf (personal communication) and Wolf, Swift, Swinney, and Vastano [55] to calculate the first Lyapunov exponent. Conceptually, the algorithm defines a d-dimensional sphere. Each dimension has a Lyapunov exponent (λ) indicating the rate of growth of the attractor in that dimension. The sum of the exponents in a dissipative system is negative. Chaotic attractors have at least one positive exponent denoting exponential divergence of nearby trajectories; i.e., stretching occurs on the attractor in certain direction(s) of phase space (positive λ), while contraction occurs in other directions (negative λ). In systems having positive λ, folding must occur in order to keep the diverging trajectories within a bounded surface.

From Wolf et al. [55], the algorithm for determining λ_1 is:

$$\lambda_1 = \frac{1}{t_M - t_0} \sum_{k=1}^{M} \log_2 \frac{L'(t_k)}{L(t_{k-1})} . \qquad (2)$$

The total time for evolving through data consisting of M equally spaced points is t_M, and t_0 is the initial time. $L(t_{k-1})$ is the distance separating a point on a fiducial trajectory of the attractor and a nearest point in phase space at time t_{k-1}. $L'(t_k)$ is a new distance between the fiducial point and another nearest replacement point after the trajectory has evolved a prescribed number of steps to time t_k. As a simple example, one might start by determining the distance between two points lying in adjacent trajectories in Fig. 10. After evolving a number of time steps along a trajectory, the distance between the points may have increased. After a series of similar iterations involving increasing replacement distances, the above equation yields a positive change in λ_1 in bits per sec.

Two scaling factors critically influence the calculations of λ_1: the number of time steps in the evolution through the attractor before making the calculations, and the maximum distance through the range of the data in searching for the replacement points at each calculation. To determine the appropriate range for each of these scaling factors, we calculated λ_1 for a series of evolution times and maximum replacement distances: First, λ_1 was calculated for a given evolution time (multiple of the 0.1 sampling interval) and a series of maximum allowable replacement distances (percent of the range of the data). For each pair of evolution time and replacement distance, the calculations generated relatively stationary λ_1 values as the iterations progressed through the time series (Fig. 14). Second, the stationary λ_1 obtained for the series

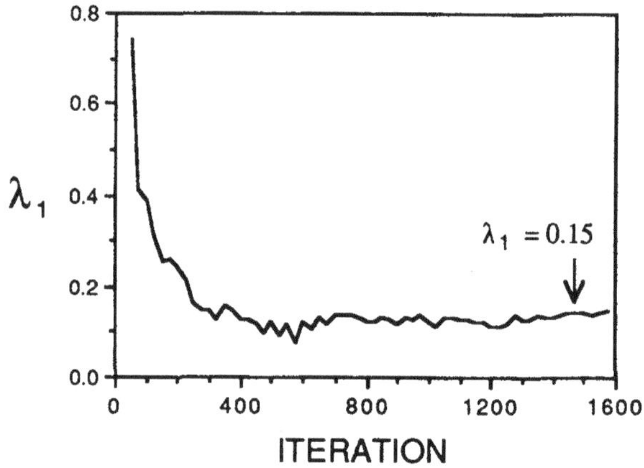

Figure 14: First Lyapunov exponent (in bits/sec). Calculations were made at fixed intervals (multiples of the sampling interval) in the evolution through the attractor. Note that the exponent reached a positive, stationary value. The embedding space was constructed using $\tau = 3.7$ sec and $d = 3$.

of calculations with each evolve step was plotted against a range of maximum allowable replacement distances (Fig. 15). The results show that replacement distance L'_{max} was a more critical factor than evolution time EVOLVE. For small replacement distances, different evolution

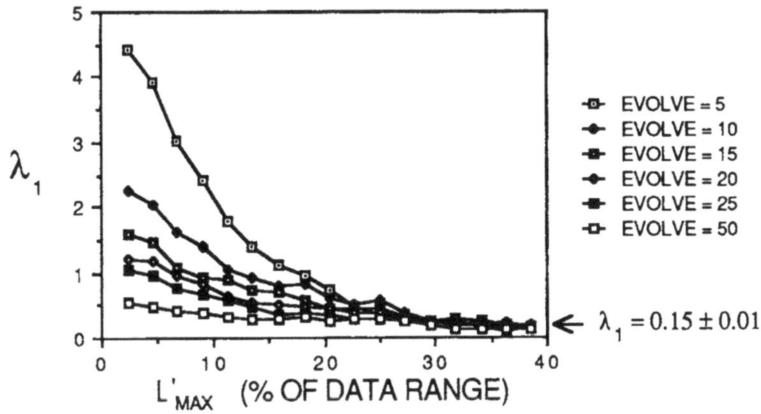

Figure 15: The first Lyapunov exponent plotted as a function of EVOLVE (multiples of the sampling interval) and L'_{max}, (the maximum prescribed scaling through the range of the data before making each calculation). Embedding space here was constructed using τ =3.7 sec and d = 3.

steps generated different λ_1, but for replacement distances of about 30% of the data range, all evolution steps generated similar λ_1. Using an EVOLVE of 20 and a maximum replacement distance of 30% of the data range, we also determined that λ_1 was reasonably stationary over a wide range of τ (Fig. 16).

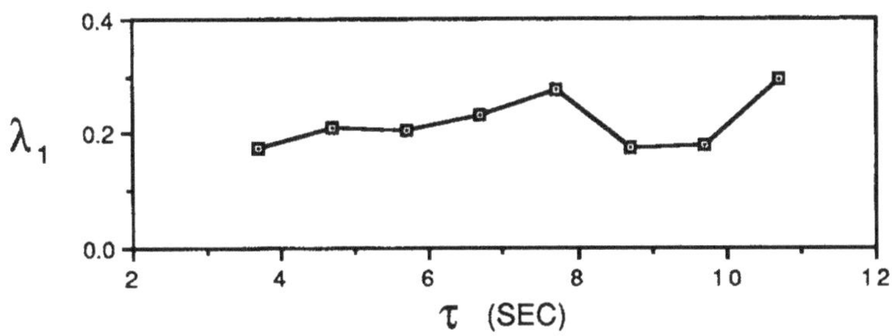

Figure 16: First Lyapunov exponent remained relatively stationary when the calculations were made in embedding space constructed with different lags τ.

1D Map

The trajectories representing the activity of the motoneuron that we have been analyzing have an unpredictable yet deterministic interrelationship with one another. To demonstrate this, we first obtained a Poincaré cross section of the flow of the trajectories through phase space, and then constructed a 1D map of the relative position of the trajectories within the Poincaré section, much as Roux, Simoyi, and Swinney [49] have described in their analysis of the Belousov-Zhabotinskii reaction.

The horizontal line in Fig. 17A shows the level at which we made a cut in the 2-D phase portrait. The one-way crossings of the trajectories through this line yielded the average orbital period and the interrelationship of one trajectory to another. Each trajectory crossed the line at some parameter value and with an average period of 15 sec. We then normalized the parameter positions of the crossings and plotted them in Fig. 17B as a 1D map showing the relative position of a particular crossing (t_{n+1}) with respect to the previous crossing (t_n). The number next to each point in Fig. 17B shows which ordered-pair of Poincaré crossings the point represents.

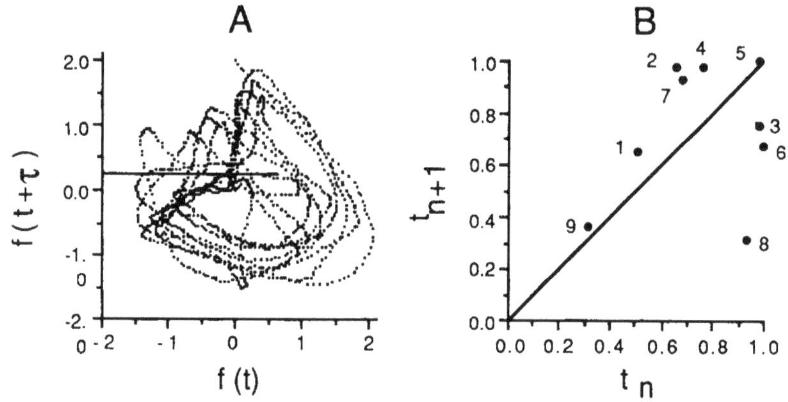

Figure 17: (A) Poincaré section was taken at the horizontal line through a 2-D phase portrait. (B) 1D map constructed by plotting the relative position of ordered pairs of trajectories passing in the same direction through the horizontal line. The numbers next to each point show the succession of ordered-pairs.

We can extract two important features from the 1D map. First, the points are not randomly scattered. In fact, they appear to fall on a relatively smooth curve having a positive slope above the 1:1 line and a very steep negative slope below the 1:1 line. Second, although there are not enough Poincaré crossings to define the exact slope of a smooth curve as it crosses the 1:1 line, the steepness of the fall-off below the line strongly suggests that the slope is greater than -1. Both of these findings are consistent with the findings we presented in the previous sections indicating that the activity of the motoneuron was chaotic: the nonrandom positions of the points show that the activity was governed by a deterministic process, and the steep fall-off shows that the activity was not a limit cycle, but rather that it was unpredictable (nonperiodic). The shape of the series of points in Fig. 17B is similar to the 1D map in Chay's [9] computer studies in which she found that spiking activity of simulation neurons bifrucated from periodic activity into chaos through period doubling. One interesting difference, is that while both Chay's simulation study and our study dealt with single neurons, the simulations dealt solely with the generation of chaos in the interspike intervals, whereas our analyses dealt with the cycle of high-frequency bursts that that comprised the motor pattern. Moreover, because the motoneuron was driven by BCNs, which generate the rhythmic pattern of activity [39], it seems likely that our results apply not only to chaos in a single motoneuron neurons but also to chaos reflecting the integrated activity of the group of coactive neurons. Examination of the activity in individual BCNs has demonstrated similar evidence of low-dimensional chaos as in the motoneurons.

MOTOR PATTERN HETEROGENEITY

In the above discussion we have examined variability occurring within a relatively homogeneous motor pattern. However, even in response to constant stimulation, motor patterns can be structurally nonhomogeneous. An example of this is shown in Fig. 18. The first

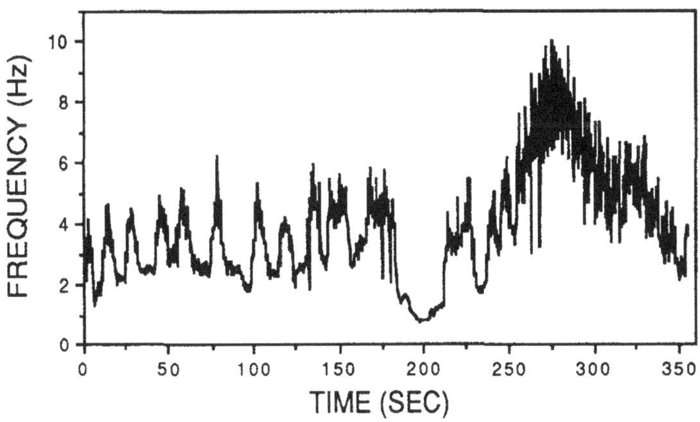

Figure 18: Heterogeneity of firing in a brain motoneuron. First half of the record is the same as in Fig. 5. Second half or the record, representing regurgitation behavior, arose spontaneously. The activating stimulus was the same throughout and consisted of brief (1 msec) electrical pulses that were applied at a rate of 1 Hz to the stomatogastric nerve, an afferent root of the buccal ganglion.

200 sec of this activity is the same as shown in Fig. 5. After that point, the repeating bursts increased rapidly and overlapped into a peak of high-frequency activity. Subsequently, as the overall frequency decreased, the individual bouts of high-frequency firing again reappeared.

During instances of alterations in the pattern of firing, some neurons become more active while others become more quiescent. The new motor pattern usually has a shorter duration than the one from which it emerges, and often has characteristics that resemble the motor pattern relating to the active phase of regurgitation [34,35,38]. Figure 19 shows a segment of a motor pattern involving three neurons that can be reidentified in successive preparations: the

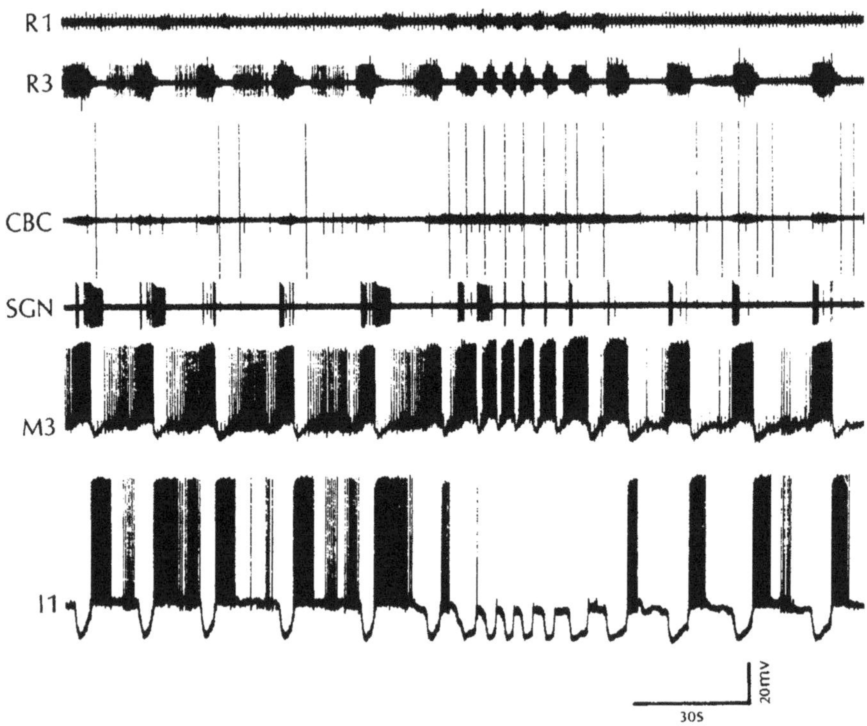

Figure 19. Heterogeneity of firing in many neurons. Note the spontaneous shift of activity, first, from a low-frequency pattern, which relates to bite-swallow behavior, to a higher-frequency pattern, which relates to regurgitation behavior, and then the return to the first. M3 is a reidentifiable buccal ganglion motoneuron, and I1 is a reidentifiable interneuron. SGN is the stomatogastric nerve which innervates the salivary gland, and CBC is the cerebral-buccal connective. Small-amplitude bursts in the CBC were primarily from the BCNs; large spike was from the metacerebral giant neuron in the brain. Tonic electrical pulses were applied to the SGN on the opposite side of the buccal ganglion. Other captions are the same as in Fig. 3.

metacerebral giant neuron (located in the brain) whose axon spike appears in the CBC trace of Fig. 19; a buccal ganglion motoneuron (trace M3), and a buccal ganglion interneuron (trace I1). The illustrated segment was taken from a 10 min recording, during which the pattern switched approximately every 90 sec. As in the case observed in recordings taken from relatively intact animals [38], some episodes of motor pattern switching were often composed of blends of the two extremes shown in Figs. 18 and 19. As we shall discuss, such motor pattern variations pose difficult though interesting problems in the application of dynamical theories to behavior.

DISCUSSION

In order to focus explicitly on variability arising from processes within the central nervous system, we used nervous systems that had been completely removed from sensory inputs and motor effectors. Such isolated nervous systems have been characterized previously [e.g.,14,34-36,39]. In this chapter we have analyzed the activity of a motoneuron which received inputs from several BCNs that are responsible for generating and controlling rhythmic motor output underlying different behaviors. The neurophysiological records that we chose to analyze had characteristics that posed problems, as presented above, that neurobiologists often face in experimental situations. Despite these problems, the results appear to be relatively robust. In further work, we have analyzed the activity of other motoneurons and BCNs, some of whose activity was considerably longer then the present example, and have obtained similar results.

Chaotic Attractors in the Generation of Motor Patterns

Evidence for presence of chaotic attractors. The definition of attractors has been widely discussed [e.g.,1,12,15,30,53]. In reference to neural networks, they may be considered simply as energy states that form and constrain the integrated activity within a limited parameter space defined by the phase portrait. The conclusion that the attractors in *Pleurobranchaea* may be chaotic follows from the findings that: 1) Autocorrelation functions quickly fall to zero. 2) Calculation of correlation dimensions shows that the activity is governed by fractal topological structures that are embedded in low-dimensional state space. 3) Trajectories in the phase portraits appear to diverge from one another. This finding is supported by calculations showing that the first Lyapunov exponent is positive which indicates that the trajectories diverge exponential, and that there must be folding of the attractor in order to keep the trajectories within a bounded surface. 4) 1D maps taken from Poincaré sections show that the interrelated positions of the trajectories are not random, but, rather, appear to follow some relatively simple function whose structure is indicative of nonperiodic behavior. These findings indicate that the pattern-generating mechanisms in *Pleurobranchaea* themselves generate unpredictable though deterministic activity in the same way that mathematical relationships such as the logistic equation [32] or the Rössler attractor [48] generate chaotic or unpredictable activity.

Behavior-specific chaotic attractors. Previous analyses of repetitive behavior or oscillatory brain function have fruitfully characterized the dynamics of repetitive behavior, or of the underlying neural activity, as limit cycles [21-23,25,45]. In rabbit olfactory EEGs, basal chaotic activity provides a route for limit cycles to emerge that represent particular odors [12,53]. Similarly, "critical fluctuations" have been proposed as a means by which phase transitions occur in human hand and finger movements [22]. Chaotic activity may also provide a route for pattern switching in *Pleurobranchaea*. However, the predominance of evidence for chaos in our records suggests that chaotic attractors themselves may be behavior-specific. Although chaos has been proposed to occur in a variety of biological systems [3-5,12,27,33,50,53], to our knowledge the work reported here constitutes the first evidence of chaos in patterned activity that is relatable to adaptive behavior in whole animals.

Nonstationary Motor Patterns

Intrinsic variation and adaptive stability. By being sensitive to initial conditions, a signal exhibiting intrinsic variation, in contrast to a nonvarying signal or one on which extrinsic noise is superimposed, can carry new information into the future. The intrinsic variations that we have ascribed above to chaos, represent in effect the attempt of the nervous system to generate new informational space which, we propose, provides for stability in the animal's adaption to a varying and often unpredictable environment.

In the natural environment, animals are often confronted by many stimuli simultaneously requiring a "behavioral choice" to respond selectively [10] or to exhibit a blend [6,28,29,38]. Considered from a classical perspective, there may be reflex interactions between separate neurocircuits relating to the different behaviors [26]. However, even given evidence of such reflex interactions, a view of the "neurocircuit" as arising from self-organizing cooperativity among groups of neurons, as defined in the Introduction, may give some insight into the spontaneous pattern switching and blending that we have presented above in reference to Figs. 18 and 19.

Consider the following example: inverted animals exhibit slower righting behavior when food is present than when there is no food [10]. In order to right themselves, animals usually have to twist the anterior portion of the foot, but they also direct the foot toward the direction of the food stimulus (Fig. 1) and may even try to "grasp" it. Thus, when animals are inverted and presented food, the neurons governing the movement of the anterior regions of the foot may receive instructions simultaneously for two different behaviors. Such influxes of mixed sensory instructions may lead to the production of hybrid or blended behaviors, and thereby cause an increase in the execution time (or complete suppression) of one or both behaviors, not so much by reflex inhibition of one behavior to another, but more because of a lack of sufficient "consensus" of information arising among the converging sensory inputs as they compete for activation of target neurons. Because the buccal-oral system of *Pleurobranchaea* has a greater range of multifunctional capabilities than the foot, and because the BCNs are key elements in the generation and control of all of these behaviors [39], it is not surprising that blends of motor patterns should be observed in the activity of the BCNs and in the neurons with which they interact.

Adaptive attractors and blending. The present findings lead to two interrelated questions that pose difficult experimental problems: 1) If neural output is so variable and unpredictable, what determines the appropriate response? 2) Can the attractor concept account for blending. In response to the first question, we propose that there may be no appropriate, centrally programmed motor pattern or genetically established neurocircuit for a behavior [38,39]. Rather, the self-organizing process can lead to many types of motor patterns, some of which may be nonadaptive for the particular conditions surrounding the animal. The correct motor pattern and the effective "neurocircuit" arise from continual dialectic interactions between the animal and the environment. Variability occurring naturally in the *Pleurobranchaea* nervous system gives these emerging "neurocircuits" a fluid quality. The interaction of the animal with the environment further accentuates such fluidity. It is necessary, therefore, to consider the combined effects of two sources of influence, one intrinsic and the other extrinsic, in the process by which functional neurocircuits emerge during the production of adaptive behavior.

In response to the second question, the adavantage of state space analyses and attractor formulations is that many quantitative features of dynamical processes may be obtained without having to determine the equations of motion. However, the apparent fluidity of neural activity poses difficulties in applying dynamical theories. For example, nonhomogeneous attractors that have topological features representing different types of motor patterns may be appropriate when the patterns represent different aspects of the same behavior, but are more difficult to invoke in order to account for the appearance of different behaviors. It is probably more useful to consider motor pattern switching (Figs. 18 and 19) as bifrucations from an attractor to another, with each attractor representing a different behavior or motor pattern. Since the entire phase space represents the possible dynamical combinations of the group of coactive neurons, it is not unreasonable that intermediate areas between the attractors, or the basins of individual attractors, could have mixtures of several motor patterns. An interesting possibility is that the attractors themselves may not be stationary and that together the different attractors are part of a larger, "hyper" attractor having its own dynamics.

Unlike known mathematical and physical attractors that have stationary qualities, neural systems are subject to many temporal changes that alter the underlying framework from which the patterns emerge. There may be long-term effects relating to learning or hormone levels, but short-term effects are probably more common. Examples of short-term effects were shown in Figs. 3 and 4 regarding the effect of history on the generation of motor patters. Given a quiescent nervous system or a recent startup of a particular motor pattern, the activity may appear as shown for the first sections of Fig. 18 and 19. As the activity progresses, the underlying context of group activity may change because individual neurons are affected by the preceding activity in which they take part. As each neuron changes, its parametric effect on the group can also change. Depending on the extent of such parametric changes, the activity of the group may show reversible temporal fluctuations and graded shifts into new types of activity.

CONCLUSION

We have attempted to illustrate in this chapter that functional "neurocircuits" leading to adaptive behavior have a fluid quality. In order for concepts such as attractors to be useful they must be able to account for these fluid qualities. While these constructs, which have arisen primarily from studies in the dynamics of idealized physical and mathematical models, have been useful in the study of variability in our experimental system, the experimental system in turn seems to pose questions for further development of dynamical theory. Although we have examined the possibility that there may be chaos in neural processing, our central question has not so much to do with chaos *per se*, but rather with the role of variation and of the types of variations that become involved in the emergence of self-organization. Moreover, inasmuch as nervous systems (as most biological systems) are distributed, parallel, variable, self-organizing, and dialectical, it is essential, as discussed with regard to the neural basis of learning [40,41], to develop a conceptual language for addressing these issues more effectively than present theory allows. Inasmuch as our experimental system is not unique in its essential characteristics, comparative analyses of such relatively "simple" systems may help to uncover general principles underlying the self-organization of activity in cooperative groups.

ACKNOWLEDGEMENTS

We thank Al Albano, Robert Burton, John Edstrom, Paul Rapp, William Shaffer, Alan Wolf, and David Zopf for helpful discussions. We are especially grateful to Alan Wolf for letting us use his computer programs to calculate Lyapunov exponents, and to Lavern Weber for the generous use of facilities at the Hatfield Marine Science Center.

REFERENCES

1. Abraham, R. H., and C. D. Shaw (1983) Dynamics--The Geometry of Behavior, Parts 1-4, Aerial Press, Santa Cruz.

2. Ayers, J., G. Carpenter, S. Currie, and J. Kinch (1983) Which behavior does the lamprey central motor program mediate? Science 221: 1312-1315.

3. Aihara, K., and G. Matsumoto (1986) Chaotic oscillations and bifrucations in squid giant axons. In Chaos, A. V. Holden,ed., pp. 257-269, Princeton University Press, Princeton.

4. Babloyantz, A., and A. Destexhe (1986) Low-dimensional chaos in an instance of epilepsy. Proc. Natl. Acad. Sci. USA 83: 3513-3517.

5. Babloyantz, A, J. M. Salazar, and C. Nicolis (1985) Evidence of chaotic dynamics of brain activity during the sleep cycle. Phys. Lett. 111A: 152-155.

6. Bellman, K. L. (1979) The conflict behavior of the lizard, *Sceloporus occidentalis*, and its implication for the organization of motor behavior. PhD Dissertation. La Jolla, CA: University of California, San Diego.

7. Bullock, T. H. (1980) Reassessment of neural connectivity and its specification. In Information Processing in the Nervous System, H. M. Pinsker and W. D. Willis, eds., pp. 199-220, Jr. Raven Press, New York.

8. Carew, T. J., and C. L. Sahley (1986) Invertebrate learning and memory: From behavior to molecules. Ann. Rev. Neurosci. 9:435-487

9. Chay, T. R. (1985) Chaos in a three-variable model of an excitable cell. Physica 16D: 233-242.

10. Davis, W. J., G. J. Mpitsos, and M. Pinneo (1974) The behavioral hierarchy of the mollusk *Pleurobranchaea*. I. The dominant position of the feeding behavior. J. Comp. Physiol. 90: 207-224.

11. DiDominico, R., and R. C. Eaton (1985) Seven principles for command and the neural causation of behavior. Brain Behav. Evol. In press.

12. Freeman, W. J, and C. A. Skarda (1985) A perspective on brain theory: Nonlinear dynamics of neural masses. Brain Res. Rev. 10: 147-175.

13. Getting, P. A., and M. S. Dekin (1985) *Tritonia* swimming: A model system for integration within rhythmic motor systems. In Model Neural Networks and Behavior. A. I. Selverston, ed., pp. 3-20, Plenum Press, New York.

14. Gillette, R., M. Kovac, and W. J. Davis (1978) Command neurons in *Pleurobranchaea* receive synaptic feedback from the motor network they excite. Science 199: 798-801.

15. Grassberger, P., and I. Procaccia (1983) Characterization of strange attractors. Phys. Rev. Lett. 50: 346-349.

16. Grossberg, S. (1980). Studies of Mind and Brain. Reidel Publishing Co. Boston.

17. Haken, H. (1983). Synergetics: An Introduction. Nonequilibrium Phase Transitions and Self-Organization in Physics, Chemistry, and Biology. 3rd Edn., Springer-Verlag, New York.

18. Harmon, L. (1964) Neuromimes: Action of a reciprocally inhibitory pair. Science 146: 1323-1325.

19. John, E. R. (1972). Switchboard versus statistical theories of learning and memory. Science 177: 850-864.

20. Katchalsky, A., V. Rowland, and R. Blumenthal (1974) Dynamic patterns of brain cell assemblies. Neurosci. Res. Prog. Bull. 12: 3-187.

21. Kelso, J. A. S., E. L. Saltzman, and B. Tuller (1986a) The dynamical perspective on speech production: Data and theory. J. Phonet. 14: 29-59.

22. Kelso, J. A. S., Scholz, J. P., and G. Schöner (1986) Nonequilibrium phase transitions in coordinated biological motion: Critical fluctuations. Phys. Lett. 118A: 279-284.

23. Kelso, J. A. S., and B. Tuller (1986) A dynamical basis for action systems. In Handbook of Cognitive Neuroscience, M.S. Gazzaniga, ed.,pp. 321-356, Plenum Press, New York.

24. Kennedy, D., and W. J. Davis (1977) Organization of invertebrate motor systems. In Handbook of Physiology, Vol. 2: Neurophysiology, 2nd Edition, E. R. Kandel, ed., pp. 1023-1087, American Physiological Society, Bethesda.

25. Kopell, N. (1987) Toward a theory of modelling central pattern generators. In Neural Control of Rhythmic Movements, A. H. Cohen, S. Rossignol, and S. Grillner, eds., John Wiley and Sons, Inc., New York. In press.

26. Kovac, M. P., and W. J. Davis (1980) Reciprocal inhibition between feeding and withdrawal behaviors in *Pleurobranchaea*. J. Comp. Physiol. 139: 77-86.

27. Mackey, M. C., and L. Glass (1977) Oscillation and chaos in physiological control systems. Science 197: 287-289.

28. Leonard, J. L., and K. Lukowiak (1985) The forms of spontaneous and evoked gill movements in *Aplysia*. Soc. Neurosci. Abstr. 11: 191.7.

29. Leonard, J. L., J. Edstrom, and K. Lukowiak (1987) A re-examination of the "gill withdrawal reflex" of *Aplysia californica* Cooper (Gastropoda; Opisthobranchia). Preprint.

30. Mandelbrot, B. B. (1985) The Fractal Geometry of Nature, Freeman & Co., New York.

31. Marder, E. and S. L. Hooper (1985) Neurotransmitter modulation of the stomatogastric ganglion of decapod crustaceans. In Model Neural Networks and Behavior, A. I. Selverston, ed., pp. 319-338, Plenum Press, New York.

32. May, R. M. (1976) Simple mathematical models with complicated dynamics. Nature 261: 459-467.

33. Markus, M., D. Kuschmitz, and B. Hess (1985) Properties of strange attractors in yeast glycolysis. Biophys. Chem. 22: 95-105.

34. McClellan, A. D. (1980) Feeding and regurgitation in *Pleurobranchaea californica*: Multibehavioral organization of pattern generation and higher order control. Ph.D. thesis. Case Western Reserve University, Cleveland, Ohio

35. McClellan, A. D. (1982) Movements and motor patterns of the buccal mass of *Pleurobranchaea* during feeding, regurgitation, and rejection. J. Exp. Biol. 98: 195-211.

36. McClellan, A. D. (1982) Re-examination of presumed feeding motor activity in the isolated nervous system of *Pleurobranchaea*. J. Exp. Biol. 98: 212-228.

37. McDougall, W. (1903) The nature of inhibitory processes within the nervous system. Brain 26:153-193.

38. Mpitsos, G. J., and C. S. Cohan (1986) Comparison of differential Pavlovian conditioning in whole animals and physiological preparations of *Pleurobranchaea*: Implications of motor pattern variability. J. Neurobiol. 17: 499-516.

39. Mpitsos, G. J., and C. S. Cohan (1986) Convergence in a distributed motor system: Parallel processing and self-organization. J. Neurobiol. 17: 517-545.

40. Mpitsos, G. J., S. D. Collins, and A. D. McClellan (1978) Learning: A model system for physiological studies. Science 199: 497-506.

41. Mpitsos, G. J., and K. Lukowiak (1985) Learning in gastropod molluscs. In The Mollusca, A. O. D. Willows, ed., pp. 95-267, Academic Press, New York.

42. Nolen, T. G., and R. R. Hoy (1984) Initiation of behavior of single neurons: The role of behavioral context. Science 226: 992-994.

43. Packard, N. H., J. P. Crutchfield, J. D. Farmer, and R. S. Shaw (1980) Geometry from a time series. Phys. Rev. Lett. 45: 712-716.

44. Perkel, D., and B. Mulloney (1974) Motor pattern production in reciprocally inhibitory neurons exhibiting inhibitory rebound. Science 185: 181-183.

45. Rand H., A. H. Cohen, and P. J. Holmes (1987) Systems of coupled oscillators as models of CPG's. In Neural Control of Movements in Vertebrates, A. H. Cohen, S. Rossignol, and S. Grillner, eds., Wiley and Sons, New York. In press.

46. Rapp, P. E., I. D. Zimmerman, A. M. Albano, G. C. Deguzman, and N. N. Greenbaun(1985) Dynamics of spontaneous Neural activity in the simian motor cortex: The dimension of chaotic neurons. Phys. Lett. 110A: 335-338.

47. Ritzmann, R. E., M. L. Tobias, and C. R. Fourtner (1980) Flight activity initiated by giant interneurons of the cockroach: Evidence for bifunctional trigger interneurons. Science 210:443-445.

48. Rössler, O. E. (1979) An equation for continuous chaos. Phys. Lett. 57A: 397-398.

49. Roux, J.-C., R. H. Simoyi, and H. L. Swinney (1983) Observation of a strange attractor. Physica 8D: 257-266.

50. Schaffer, W. M., S. Ellner, and M. Kot (1986) Effects of noise on some dynamical models in ecology. J. Math. Biol. 24: 479-523.

51. Schöner, G., H. Haken, and J. A. S. Kelso (1986). A stochastic theory of phase transitions in human hand movement. Biol. Cybern. 53: 1-11.

52. Shaw, R. (1986) The Dripping Faucet. Aerial Press, Santa Cruz.

53. Skarda, C. A., and W. J. Freeman (1987) How brains make chaos in order to make sense of the world. Behav. Brain Sci. 10: 161-195

54. Takens, F. (1981) Detecting strange attractors in turbulence. In Lecture Notes In Mathematics, Vol. 898. D. A. Rand and L. S. Young, eds., pp. 366-381, Springer-Verlag, Berlin.

55. Wolf, A., J. B. Swift, H. L. Swinney, and J. A. Vastano (1985). Determining Lyapunov exponents from a time series. Physica 16D: 285-317.

Calculation of correlation dimensions from experimental data: Progress and problems

P. E. Rapp
Department of Physiology and Biochemistry
The Medical College of Pennsylvania
3300 Henry Ave.
Philadelphia, PA 19129

A. M. Albano
Department of Physics
Bryn Mawr College
Bryn Mawr, PA 19010

A. I. Mees
Department of Mathematics
University of Western Australia
Nedlands, W.A. 6009
Australia

ABSTRACT

Dynamical analysis, particularly estimation of correlation dimension, has been applied to a variety of signal analysis problems including the analysis of electroencephalographic signals. Three experimental design questions must be addressed. What is an appropriate epoch length? What is an appropriate sample interval and what is an appropriate embedding dimension? We are able to demonstrate that in practice embedding dimension and sample interval should not be considered separately. Rather, it is sufficient to determine an appropriate value for the data window.

However, the result presented here do not establish a simple criterion for selecting window length. It is shown that autocorrelation functions and singular value spectra do not provide reliable means of estimating an appropriate window.

I. Introduction

It is frequently observed that while our technical capacity for acquiring observational data is increasing exponentially, our capacity to extract meaningful information from that data is increasing only linearly. This is particularly true in the neurosciences. One possible avenue of resolution of the data acquisition/interpretation dilemma is dynamical systems theory and the dynamical characterization of these signals. A central component of this mathematical technology is the estimation of correlation dimension using the Grassberger-Procaccia algorithm [1,2]. The dimension of a signal is a quantitative characterization of its geometric properties. Informally, it is a measure of the inherent complexity of the signal. Systems whose behavior is random have an unbounded dimension and chaotic systems have finite fractional dimensions. There are several mathematically interrelated, but variant, definitions of dimension. While the Hausdorff dimension is the best characterized mathematically, the correlation dimension has similar mathematical properties and is more readily applied to experimental data. In our studies, the absolute value of the correlation dimension is not of central importance. At present, it is more important to determine if changes in the dimension are correlated with changes in physiological activity.

Correlation dimension measurements have been made from neuronal spike train data [3,4] and from human electroencephalograms. In a study with healthy, adult subjects [5] preliminary results indicated that the correlation dimension of the EEG signal recorded at site O_2 increased on the initiation of cognitive activity. Results published by Layne, Mayer-Kress and Holzfuss [6,7] indicated that the induction of moderate fluroxene anesthesia results in an increase in dimension. Babloyantz, Salazar and Nicolis [8] found that the dimension estimate for Sleep Stage 2 is greater than that obtained for Sleep Stage 4. Babloyantz's subsequent results [9] suggested that the dimension of an EEG signal displays an abrupt decrease at the initiation of an absence seizure, and that the dimension of the beta wave is greater than the alpha [10].

While the applications of dynamical characterization to electrophysiological signals is promising, the magnitude of the remaining technical problems should not be underestimated . These problems have been addressed in general terms elsewhere [11]. In Sections II and III of this paper we focus on a specific issue. In order to establish the context of that discussion. It is necessary to present, at least in outline, the structure of the Grassberger-Procaccia algorithm.

A central problem encountered in the experimental investigation of complex systems is the investigator's inability to measure all of the system's dynamical variables. Indeed, it is often impossible to even identify all of the variables. Suppose that M values of a single variable are measured, v_1, v_2, ... v_M. The question that must be addressed is how can we infer something about the dynamical structure of the generator from a single channel time series? (Generalizations of multiple channel recordings are straightforward and will not be discussed here.) The process begins by using the original time series to create a set of points in an N-dimensional embedding space according to the rule:

$$X_1 = (v_1, v_2, ... v_N)$$

$$X_2 = (v_2, v_3, \ldots v_{N+1})$$

.

.

$$X_K = (v_K, v_{K+1}, \ldots v_M)$$

where K is the number of N-dimensional points that can be created from M data points, K=M-N+1. (Alternative, equally valid, embedding rules can be used.) If certain restrictions are meet, then it is possible to infer much about the structure of the generator from the structure of {X}. These requirements include selection of a sufficiently large embedding dimension, a sufficiently large data set, and the absence of noise and transient behavior in the data. Specifically, Takens [12] used the Whitney embedding theorem to demonstrate that in the limit of an infinite amount of noise free data, an embedding exists between the original dynamical system and {X} whenever N is sufficiently large. When and embedding exists between these sets, they have the same dimension.

Using this result Grassberger and Procaccia [1,2] estimated the correlation dimension for {X}. Calculation of the correlation integral is the first step in the process.

$$C_N(r) = (1/N_p)\sum_{i=1}^{K}\sum_{j=i+1}^{K} H(r-|X_i-X_j|)$$

where H is the Heaviside function, N_p is the number of distinct pairs of points in set {X}, N_p=K(K-1)/2, K is the number of points in set {X}, N is the embedding dimension and M is the number of measurements. Grassberger and Procaccia demonstrated that for M and

N sufficiently large and r sufficiently small, the slope of the logarithm of the correlation integral calculated as a function of ln(r) will have a linear region. This linear region is called the scaling region. The value of the slope in the scaling region is the correlation dimension D_c. In practice, $C_N(r)$ is calculated and $d\ln C_N(r)/d\ln(r)$ is then plotted against ln(r). The region of constant slope, the plateau, corresponds to a level region in the derivative function which is used to establish the existence and range of the scaling region. However, the value of D_c is estimated from the correlation integral, not from its derivative.

While a statement of the procedure is straightforward, its implementation is not as simple as this description might suggest. A naive application of the method can produce spurious dimension estimates. For example, in any experiment, care must be made in the selection of T_s, the sample time. This is the time interval between v_i measurement and v_{i+1}. Additionally, the choice of embedding dimension, N, has not been made explicit. Selection of these two parameters is an important element of experimental design. This selection is addressed in the next two sections.

II. Restatement of the problem: Windows

What criterion can be used to assess the appropriateness of a choice of T_s and N? We propose, provisionally, to use the length of the scaling region as a selection criterion and suppose that good choices of T_s and N are those that result in a large scaling region.

This is not a complete specification of the problem. Scaling region length can be a misleading criterion. One must also consider the duration of recording, epoch length, $T_E=(M-1)T_s$. It is known [13] that over-sampling a signal, small T_s and small T_E, can result in a large plateau and spuriously low estimates of dimension. A more precise statement of the criterion would be: for a given epoch length T_E, the best choices of T_s and N are those choices that result in the largest scaling region. This leaves the choice of T_E unresolved.

An example of the effect of varying N and T_s on scaling region length is shown in Figure 1. The data are generated by integrating the Rossler system [14].

$$dx/dt = -y-z$$

$$dy/dt = x+ay$$

$$dz/dt = b-cz+xz$$

for parameter values a=.2, b=.4 and c=5.7. The differential equation was solved for T_s=.05 and was first integrated forward for 1000 iterations to produce an initial point close to the attractor. In order to facilitate comparisons in data sampling protocols, it is necessary to introduce two additional elements of nomenclature: lag and window. The lag, L, is the number of sampling intervals between successive components of a point in {X}.

$$X_1 = (v_1, v_{1+L}, v_{1+2L}, \ldots, v_{1+(N-1)L})$$

Thus for the case L=3 and N=4, the first point in {X} would be:

$$x_1 = (v_1, v_4, v_7, v_{10}).$$

Window, W, is the number of sampling intervals spanned by a point in set {X}; W=(N-1)L. Strictly speaking, the time spanned by a point in {X} is $WT_s=(N-1)LT_s$. However, for comparisons within a given data set, that is, whenever T_s is fixed, it is sufficient to use W and not WT_s. Examination of Figure 1 shows that variation in N and L can have significant effects on plateau length. In each case the points in {X} were spread out uniformly over the entire data set of 10,000 points so that epoch length, T_E, is the same in each case. It is seen that an inappropriate choice of N and L can result in the failure to resolve a measurable linear scaling region.

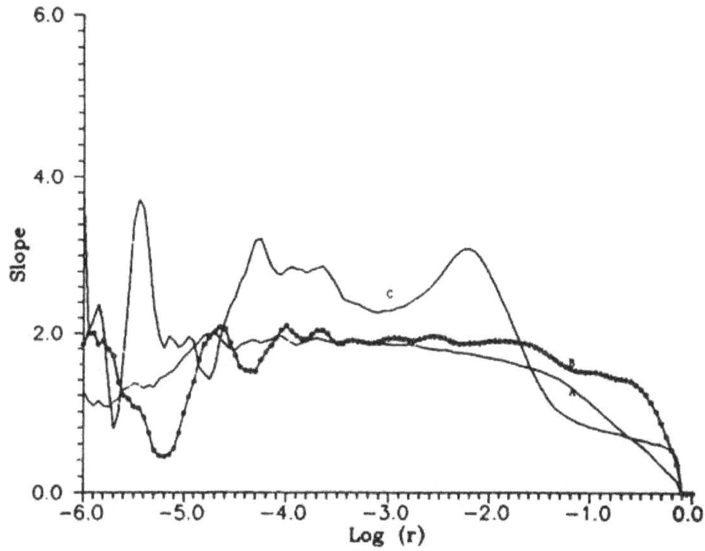

Figure 1:
Plot of slope=$d\ln C_N(r)/d\ln(r)$ versus ln(r) for data generated by the Rossler equations for a=.2, b=.4, c=5.7 and T_s=.05. One thousand points in set {X} were used for each calculation. Curve a: N=9, L=3, W=24; Curve b: N=7, L=11, W=66; Curve c: N=9, L=12, W=96.

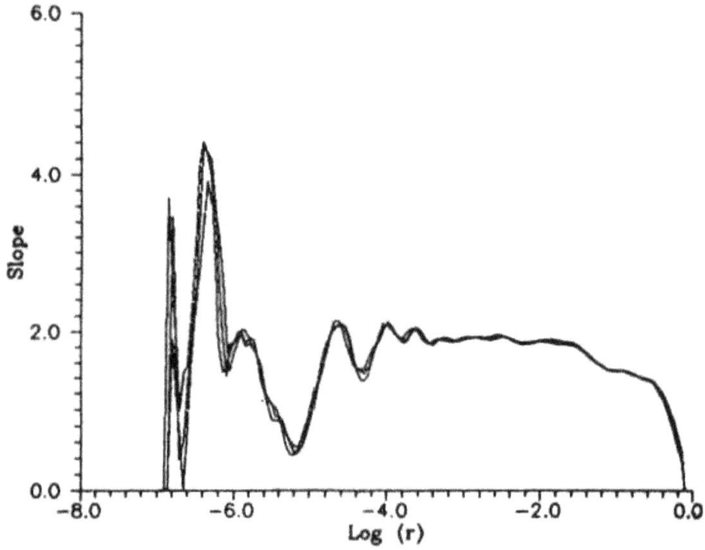

Figure 2:
Plot of Slope=$d\ln C_N(r)/d\ln(r)$ versus $\ln(r)$ for the same data used in Figure 1. In this instance the three curves correspond to (N,L,W)=(7,11,66), (9,8,64), (6,13,65).

The calculations shown in Figure 2 suggest that the significant determinant of scaling region length is not L and N specified independently, but rather the value of W. The curves in Figure 2 correspond to (N,L,W)=(7,11,66), (9,8,64) and (6,13,65). The curves coincide. Similar results were obtained with data from the Lorenz attractor and the 3-torus. These empirical results suggest that, once epoch length is fixed, window rather than N and L separately is the primary determinant of scaling region length. Upon reflection this is as it should be since the window specifies the component of the trajectory covered by any element of {X}. Further numerical experimentation and thought revealed a qualification to the principle "window specifies scaling region length." The embedding dimension N must be greater than some lower bound. This lower bound is determined by the dimension of the attractor. Following Taken's theorem, one would require $N \geq 2d+1$, where d is the dimension of the attractor. However, once this restriction on a lower bound on N is satisfied, all available evidence suggests that the results summarized in Figure 2 is valid. Thus the original problem of selecting N and L has been replaced by the problem of constructing an appropriate criterion for selecting W.

III. Criteria for selecting an appropriate window

Can properties of the time series' autocorrelation function give an indication of the appropriate embedding window? The calculations summarized in Figure 1 suggest that of the three values of window examined, W=66 is the better window length. The autocorrelation function for the Rossler attractor was calculated using 8192 points. The first zero crossing occurred at tau=60. The first minimum occurred at tau=31, and the autocorrelation time was tau=23. (The autocorrelation time is defined as the time required for the signal to drop to 1/e of its maximum value.) These calculations indicate that the first minimum of the autocorrelation function might be an acceptable criterion for selecting window length.

Unfortunately, similar numerical experiments with the Lorenz attractor and the 3-torus failed to reproduce this correlation, and thus indicate that the first minimum of the autocorrelation function is not a reliable predictor of optimal window length. Additionally, there are difficulties associated with calculating the first minimum and the first zero of an autocorrelation function. Both of these values are sensitive to the number of data points used in the autocorrelation function calculation. The Rossler attractor data are well behaved but problems were observed when estimating the properties of the Lorenz attractor's autocorrelation function. As the number of data points used in the autocorrelation calculation change, an inflection point may become a local minimum. Similarly, the location of the first zero crossing can change with the size of the time series. Large data sets are required to produce stable values for the first minimum and the first zero crossing. The autocorrelation time is a much more stable number and can be estimated reliably with limited data. This is precisely why autocorrelation times appear so prominently in the signal processing literature. In summary, autocorrelation functions do not appear to provide prediction of appropriate window length.

The second possibility investigated was the singular value spectrum. We begin the discussion with a statement of the singular value decomposition theorem [15]. A real rxc matrix A, $r \geq c$, has the unique decomposition

$$A_{rxc} = V_{rxc} S_{cxc} U^{T}_{cxc}$$

where the columns of V and orthonormal; U is an orthogonal matrix and S is a diagonal matrix with nonzero elements that follow the ordering

$$s_1 \geq s_2 \geq \ldots\ldots \geq s_c \geq 0$$

the s_j's are the singular values. If A has rank $j<c$, then

$$s_1 \geq \ldots \geq s_j > s_{j+1} = \ldots = s_c = 0$$

In practice, singular values are not exactly zero, but there may be an abrupt step-like decrease in their numerical values of the form

$$s_1 \sim s_2 \sim s_3 \ldots \sim s_k >> s_{k+1} \geq \ldots \geq s_c$$

As will be described presently, when a spectrum of this type occurs, the singular value decomposition can be used as a noise reduction technique.

This general result can be used in dynamical analysis by applying the singular value decomposition to the embedding matrix that is formed by using the K points of {X} to form a KxN embedding matrix

$$E_N = \begin{bmatrix} v_1 & v_2 \ldots\ldots & v_N \\ v_2 & v_3 \ldots\ldots & v_{N+1} \\ \cdot & & \\ v_K & v_{K+1} & v_N \end{bmatrix}$$

The matrix E_N can be decomposed to

$$E_N = VSU^T$$

Recall that the rows of E_N are the points in {X} that are used to calculate correlation integrals and the correlation dimension. Albano, et al, [16] proposed using the singular value decomposition to reduce the noise in experimental data. Matrix U is an orthogonal transformation. Therefore, using rows of matrix $E_N U$ instead of E_N

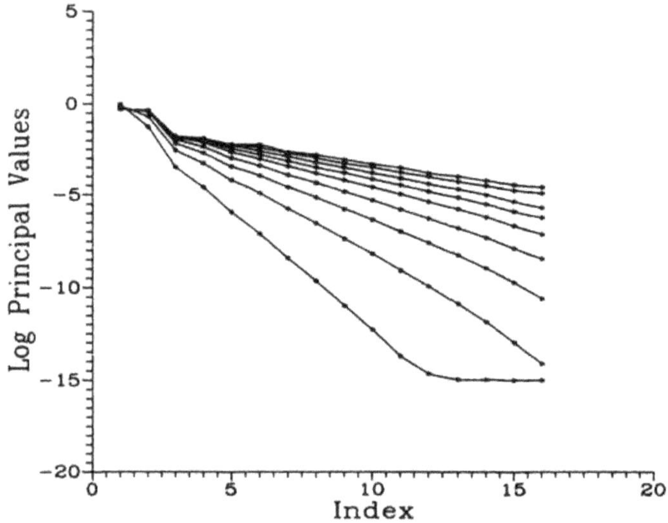

Figure 3:
Normalized singular value spectrum for data generated by the Rossler equations. Plot shows $\log_{10}(s_j^2/\Sigma s_i^2)$ as a function of index j. The nine spectra correspond (from bottom to top) lag=1, 2, 3, 4, 5, 6, ... 9. Embedding dimension is N=16 in each case.

will not effect the dimension of the set of points. However, if there is an abrupt drop in singular values, then the last N-k rows of the matrix can be discarded and a reduction in noise results.

This noise reduction technique has a long history in engineering and will be effective only if there is a marked decrease in magnitude of the singular values with increasing index. This requirement for noise reduction could therefore be used as a criterion for selecting a window for embedding dynamical data. By this criterion, a successful window is one which results in a singular value spectrum with a marked step-like decrease in the singular value spectrum. Figure 3 shows the singular value spectrum for embedded Rossler data for windows 15, 30, 45, ... 135. An examination of Figure 3 indicates that the qualitative shapes of the spectra do not offer a basis for choosing the window. Qualitatively similar results were obtained with the Lorenz attractor and the 3-torus.

IV. Conclusions.

The conclusions presented in this contribution are largely negative. The results suggest that both autocorrelation functions and singular value spectra do not provide reliable guides for window selection. Fraser and Swinney [17] have suggested that the mutual information function may succeed. This possibility is presently being investigated.

The hardest problem in this analysis has not been addressed. This is the selection of appropriate epoch length, T_E. Because uncertainty in T_E selection is a consequence of nonstationarity of the signal, this problem is not subject to a simple resolution. We have suggested elsewhere [11] that an analog of the Wigner transformation used in classical spectral analysis could be applied to dimension spectra.

Although the presently available results are sufficiently promising to encourage continued research into the application of dynamical analysis to electroencephalographic signals, it is also clear that many pressing technical questions are unanswered.

However, these questions are being asked with increasing precision and this alone encourages optimism.

Acknowledgements:

PER would like to acknowledge support from NIH Grant NS19716, a visiting faculty appointment at the University of Western Australia and the hospitality of Kingswood and St. George's College, University of Western Australia. We also wish to acknowledge with gratitude the generous support of the Medical College of Pennsylvania College Computer Center.

References:

1. Grassberger,P. and Procaccia,I. Characterization of strange attractors. Phys. Rev. Lett. 50, 346-349 (1983).

2. Grassberger,P. and Procaccia,I. Estimation of the Kolmogorov entropy from a chaotic signal. Phys. Rev. 28A, 2591-2593 (1983).

3. Mpitsos,G.J., Creech,H.C., Cohan,C.S. and Mendelson,M. Low-dimensional chaos in neurons during generation of rhythmic motor patterns. preprint. Hatfield Marine Science Center.

4. Rapp,P.E., Zimmerman,I.D., Albano,A.M., deGuzman,G.C. and Greenbaun,N.N. Dynamics of spontaneous neural activity in the simian motor cortex: The dimension of chaotic neurons. Phys. Lett. 110A, 335-338 (1985).

5. Rapp,P.E., Zimmerman,I.D., Albano,A.M. deGuzman,G.C., Greenbaun,N.N. and Bashore,T.R. Experimental studies of chaotic neural dynamics: Cellular activity and electroencephalographic signals. In: Nonlinear Oscillations in Chemistry and Biology. H.G.Othmer, ed. 175-205, Springer Verlag, NY (1985).

6. Layne,S.P., Mayer-Kress,G. and Holfuss,J. Problems associated with dimensional analysis of electroencephalographic data. In: Dimensions and Entropies in Chaotic Systems. G. Mayer-Kress, ed. pp. 246-256. Springer-Verlag, NY (1986).

7. Mayer-Kress,G. and Holzfuss,J. Analysis of the human electroencephalogram with methods from nonlinear dynamics. In: Temporal Disorder in Human Oscillatory Systems. L.Rensing, U. an der Heiden, and M.C.Mackey, eds. pp. 57-68, Springer-Verlag, NY (1987).

8. Babloyantz,A., Salazar,J.M. and Nicolis,C. Evidence of chaotic dynamics of brain activity during the sleep cycle. Phys. Lett. 111A, 152-156 (1985).

9. Babloyantz,A. Low-dimensional chaos in an instance of epilepsy. Proc. natn. Acad. Sci. U.S.A. 83, 3513-3517 (1986).

10. Babloyantz,A. and Destexhe,A. Strange attractors in the human cortex. In: Temporal Disorder and Human Oscillatory Systems. L.Rensing, U. an der Heiden and M.C.Mackey, eds. pp. 48-56, Springer Verlag, NY (1987).

11. Albano,A.M., Mees,A.I., deGuzman,G.C. and Rapp,P.E. Data requirements for reliable estimation of correlation dimension. In: Chaotic Biological Systems. H.Degn, A.V.Holden and L.F.Olson, eds. pp. 207-220. Plenum Press, NY (1987).

12. Takens,F. Detecting strange attractors in turbulence. In: Dynamical Systems and Turbulence. Lecture Notes in Mathematics. Volume 898. D.A.Rand and L.S.Young, eds. pp. 365-381, Springer-Verlag, NY (1980).

13. Theiler,J. Spurious dimensions from correlation algorithms applied to limited time series data. Phys. Rev. 34A, 2427-2433.

14. Rossler,O.E. An equation for continuous chaos. Phys. Lett. 57A, 397-398 (1976).

15. Golub,G.H. and van Loan,C.F. Matrix Computations. Johns Hopkins University Press, Baltimore (1983).

16. Albano,A.M., Smilowitz,L., Rapp,P.E., deGuzman,G.C. and Bashore,T.R. Dimension calculations in a minimal embedding space: Low dimensional attractors for human electroencephalograms. In: Physics of Phase Space. Y.S.Kim and W.W.Zachary, eds. pp. 15-17, Springer-Verlag, NY (1987).

17. Fraser,A.M. and Swinney,H.L. Independent coordinates for strange attractors from mutual information. Physical Rev. 33A, 1134-1140 (1986).

Part III

Chaos, Coding and Fractals

Self-similar Basin Boundary in an Invertible System (Folded-Towel Map)

O.E. ROSSLER[1], J.L. HUDSON[2], M. KLEIN[1] and R. WAIS[3]

[1]Instiute for Physical and Theoretical Chemistry
University of Tübingen
7400 Tübingen, F.R.G.

[2]Department of Chemical Engineering
University of Virginia
Charlottesville, VA 22091
U.S.A.

[3]Computing Center
University of Hamburg
2000 Hamburg,
F.R.G.

ABSTRACT

An invertible 3-variable difference equation - a folded-towel map with but two quadratic terms and a constant Jacobi determinant - is investigated numerically. For a large range of parameters, a self-similar basin boundary is found. Such a limiting set was apparently never found before in a differentiable system of generic type. The new complexity is bound to be a characteristic of "most" realistic 4-variable continuous systems - like dripping faucets, car engines, chemical oscillators and neural networks.

INTRODUCTION

The "new science" of chaos was recently reviewed by Gleick [1]. He treated all nonlinear phenomena - including the beautiful results of Julia and Mandelbrot on complex-analytic noninvertible maps - on essentially the same footing. Can fractals be subsumed under the word chaos, or is it rather the other way around ?

Self-similar basin boundaries exist so far only in noninvertible 2-D maps that are either complex-analytic [2] or piece-wise continuous [3], or general nonlinear [4,5]. There is also an example of the existence type (non-explicit) due to Harrison [6] of an invertible map which is 2-D. It is less than 3 times differentiable. More importantly, it is apparently non-generic, that is, special (rare) points in parameter space are needed to generate the behavior. An explicit example of the latter type (fractalization of torus) is due to Kaneko [7].

No generic example with this kind of behavior - a basin boundary that looks self-similar when cut in a certain way - appears to be known. In [4], it was argued that such behavior is generic in 3-D invertible maps which as a certain parameter approaches zero become a non-invertible 2-D example. An equation for which this can be demonstrated analytically for a subregion of state space was proposed in [8].

In the following, a first explicit example will be presented.

THE SIMPLEST FOLDED-TOWEL MAP

Consider the following 3-variable difference equation:

$$
\begin{aligned}
x_{n+1} &= y_n \\
y_{n+1} &= a\, y_n(1 - y_n) - b\, z_n \\
z_{n+1} &= a\, z_n(1 - z_n) - b\, x_n \; .
\end{aligned}
\qquad (1)
$$

This map in the limit $b \longrightarrow 0$ decomposes into two uncoupled logistic equations (in y and z). It has a constant Jacobi determinant (contraction factor), $\det(J) = b^2$. It is the simplest global diffeomorphism with two positive Lyapunov characteristic exponents - two independent directions of exponential divergence between neighboring initial points. Such maps were called folded-towel maps (hyper horseshoes) in [9]. A map that is equally simple as Eq.(1) was first indicated by Stefański [10]. In fact, Eq.(1) is obtained from Stefański's example by transposing two right-hand sides and inverting one sign.

NUMERICAL RESULTS

If b is allowed to approach zero, the two uncoupled logistic maps explode individually if a becomes larger than 4. Nevertheless the two different attractors at minus infinity that then exist make for a non-trivial (self-similar) basin boundary, see Fig. 1. This picture ("iterated Swiss flag") was already given in [4]. It "survives" the transition toward invertibility and a higher dimension by virtue of the fact that the self-similar separatrix plotted (the boundary between black and white) has a transversal manifold in the third dimension by Smale's [11] theory, see [4, 8].

For a larger value of b, for example the picture of Fig. 2 is obtained. The qualitative structure and the self-similarity are still there. A different regime is shown in Fig. 3. The types of fractal to be found in Eq.(1) are still uncharted. Since it is possible to insert an arbitrary function of y and z as an additional term on the right (third line) of Eq.(1), without affecting the constant Jacobi determinant, the variety to expect becomes even larger. We are positive that non-differentiable self-similar structures will be among those found in the long run. So far we were unable to pinpoint a clear-cut example, though.

DISCUSSION

Neural systems behave qualitatively like other dynamical systems, even when investigated _in vivo_ [12]. The qualitative dynamical behavior of nonlinear systems is therefore of interest also in the field of neurodynamics. At the time being, the qualitative behavior of 3-D dynamical

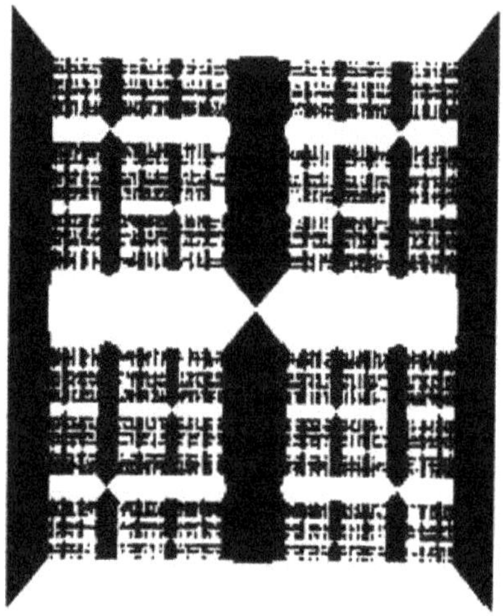

Fig. 1. Self-similar basin boundary (intertwined basins) in Eq.(1), with $b = 0.001$ and $a = 4.1$. Initial points that lead to an attractor at $y \longrightarrow -\infty$ are plotted in black, the others in white. Numerical thresholds chosen: absolute value of y or z larger than 500, $y < z$ (additional criterion for the present picture); maximum iteration number: 1000 per point. Axes: - 0.1 to 1.1 for y (abscissa), the same for z (ordinate). Initial value of x: 0.01.

Fig. 2 (next two pages). Fig. 2a: Analogous picture as in Fig. 1, but with $b = 0.1$ and $a = 4.0$. Conventions as in Fig. 1. Fig. 2b: Inset of Fig. 2a, magnified. Axes: 0.0865 to 0.22 for y, 0.00575 to 0.182 for z. Further magnifications (not shown) have the same qualitative appearance.

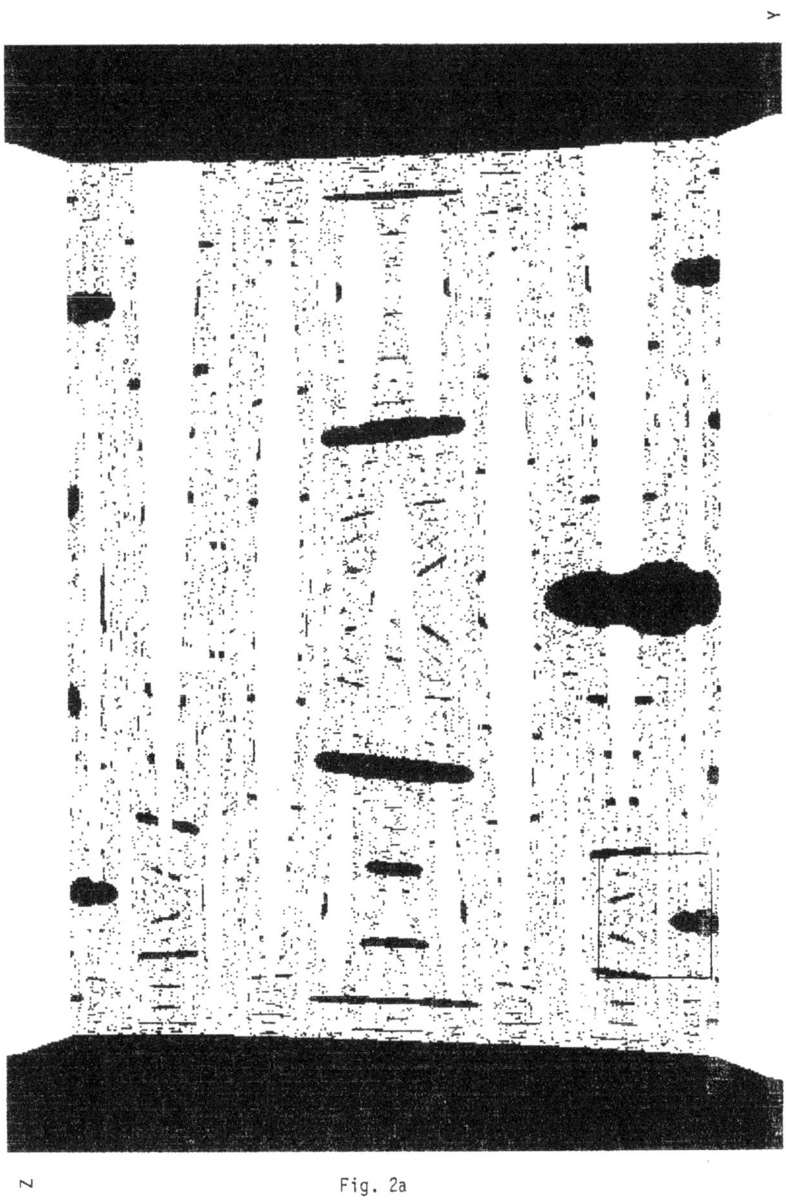

Fig. 2a

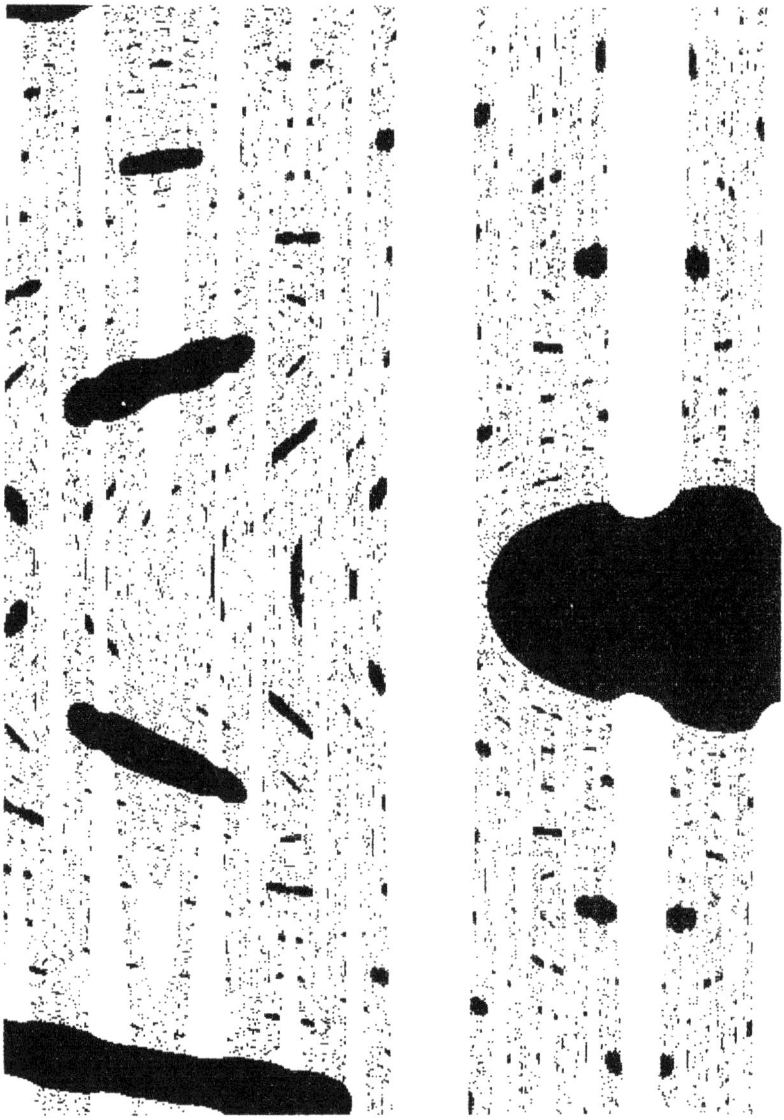

Fig. 2b

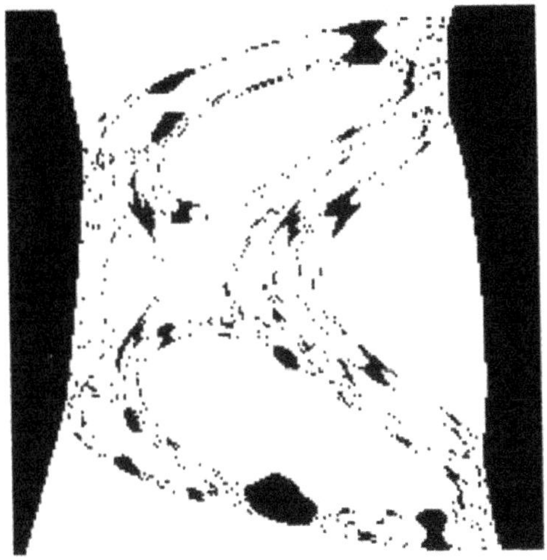

Fig. 3. In this picture, the parameters are b = 0.3, a = 4.03, and c = 1. The third parameter refers to an added term, $c(y_n - 0.5)$, on the right hand side of the third line of Eq.(1), cf. text. The possibility of a subsequence among the visible black objects possessing a non-differentiable surface structure has yet to be ruled out.

systems is to some extent known, largely due to work done in recent years in the chaos community, cf. [1]. At first sight, the new paradigm again suggests that higher-dimensional dynamical systems like neural networks can be understood by extrapolating from the known features of chaotic, that is, 3-D dynamical systems, just as one did in a previous decade with the known features of limit cycle oscillators, that is, 2-D dynamical systems.

It now turns out that the next - the fourth - dimension is again good for a surprise. The qualitative behavior of four-dimensional systems contains a new possibility once more (and so forth presumably through every added dimension). Of course, whether or not the new behavior is "important" again - as important as the paradigm of chaos was, perhaps - is an open question. Only time can tell. Recently it was proposed that hyperchaotic time series, even when generated by an attractor, can be artificially "color coded" in such a way as if they belonged to different basins of attraction, segment-wise [13]. Hereby, the fact described in [4] was made use of that color-coding trajectorial segments by the region (one of two arbitrarily defined regions on the attractor) entered next - a symbolic dynamics technique - yields the same self-similar structure under certain conditions as if two genuine attractors lurked in the two target regions.

The present example system will make it possible to exhaustively investigate the types of fractal basin boundaries of self-similar type arising in a realistic system. The results will enable one to differentiate also between different qualitative types of self-similar coloring behavior that different types of higher-dimensional chaotic time series can and cannot support. In other words, the present system may help generate a new "staining technique" in the sense of histology, for neurally generated data.

Another potential interest of the above result has to do with its transportability to parameter space. Mandelbrot [14], as well as Gumowski and Mira [15], saw that any complexity found in phase space can necessarily be recovered in parameter space. Now "most" 4-variable continuous systems (see [9] for an example involving a single quadratic term and [16] for a recent analogue in reaction kinetics) possess the complexity of the map of Eq.(1). Scanning natural systems - described by equations of similar complexity - by letting them run many different times, from the same initial condition as far as realizable, but with a neighboring point in parameter space in charge each time, is easily possible in principle in many cases. Self-similar structures, if they exist, have the desirable feature of being recognizable on different levels of resolution. With a given amount of noise present, only a certain number of levels will be resolvable pictorially. Thus it may turn out that certain experimentally favored systems, like the Belousov-Zhabotinsky reaction (cf. [17]) and certain (intrinsically much faster but so far noisier) solid-state devices [18], can be controlled to a sufficient degree to reveal some of the self-similar structures hidden in their parameter spaces. Again, Eq.(1) offers itself as a "guinea pig" for numerical pilot studies.

To conclude, a new class of generic dynamical phenomena, self-similar limiting structures in differentiable systems, was described. A unification of chaos theory on the one hand and fractal theory on the other, within the nonlinear dynamical paradigm, may be in sight.

Acknowledgments

O.E.R. thanks Scott Kelso and Gottfried Mayer-Kress for discussions. Work supported in part by the N.S.F.

REFERENCES

1. Gleick, J., Chaos, Making a New Science, Viking-Penguin, New York 1987.

2. Julia, G., J. Math. Pures et Appl. 4, 47 (1918).

3. Barnsley, M.F. and Harrington, A.N., Physica 15D, 421 (1985).

4. Rössler, O.E., Kahlert, C., Parisi, J., Peinke, J. and Röhricht, B., Z. Naturforsch. 41a, 819 (1986).

5. Pickover, C.A., Comput. & GRaphics 11, 217 (1987).

6. Harrison, J., Bull. Amer. Math. Soc. 13, 147 (1985).

7. Kaneko, K., Collapse of Tori and Genesis of Chaos in Dissipative Systems, World Scientific, Singapore 1986.

8. Rössler, O.E., Kahlert, C. and Hudson, J.L., "Bifractal Basin Boundaries in Invertible Systems", in: Chaos in Biological Systems (A.V. Holden and L.F. Olsen, eds.), Plenum Press, New York 1987.

9. Rössler, O.E., Phys. Lett. 71A, 155 (1979).

10. Stefański, K., Preprint Nicolas Copernicus University 1983.

11. Smale, S., Bull. Amer. Math. Soc. 73, 747 (1967).

12. Kelso, S., Preprint July 1987.

13. Rössler, O.E., Talk presented at the international conference on neurodynamics, Berlin, August 1987.

14. Mandelbrot, B., Ann. N.Y. Acad. Sci. 357, 249 (1980).

15. Gumowski, I. and Mira, C., Dynamique Chaotique, Transformations Ponctuelles, Transition Ordre-Désordre, Cepadues Editions, Toulouse 1980.

16. Killory, H., Rössler, O.E. and Hudson, J.L., Phys. Lett. 122A, 341 (1987).

17. Hudson, J.L. and Rössler, O.E., "Chaos and Complex Oscillations in Stirred Chemical Reactors", in: Dynamics of Nonlinear Systems (V. Hlavacek, ed.), pp. 193-219, Gordon and Breach, New York 1985.

18. Peinke, J., Röhricht, B., Mühlbach, A., Parisi, J., Nöldeke, C., Huebener, R.P. and Rössler, O.E., Z. Naturforsch. 40a, 562 (1985).

AN UNSTABLE SINGULARITY THEORY OF MOLECULAR BIOLOGICAL CODING; CALCITONIN'S STRUCTURES AND POTENCIES

Arnold J. Mandell

Biological Dynamics and Theoretical Medicine
University of California, San Diego (M-003)
La Jolla, California 92093

Motivation and Orientation

Both varied and reliable mechanisms of information transport in molecular biological systems can be obtained from the universality properties of continuous and discrete dynamical systems whose output has the general properties of a set which combines sensitivity to initial conditions and indecomposability with a positive density of periodic points. This counterintuitive approach requiring a revision of information theory for macromolecular systems is dictated by their failure to fulfill the conditions of Shannon's existence proof: there exists a one-to-one mapping that makes the channel capacity finite. Required by the theorem are additive fluctuations, noise with a density distribution of gaussian character-the power spectrum white with a cut-off, and sequential events in the series which are mutually independent. Relaxation studies of polypeptides, proteins, and nucleotides demonstrate a lack of characteristic times (with a range of $C(\tau)_i$ from picoseconds to hours) and studies of ion conductance fluctuations of membrane receptor-channel proteins in vivo reveal a fractional power law noise spectrum consistent with their characteristic intermittent bursting behavior. A more appropriate statistical context for macromolecular coding is a hyperbolic distribution with an infinite multiplicity of moments, multiplicative noise with a "colored" spectrum, belonging to the family of convolutionally stable distributions without finite mean or variance,

$$p(x,k,\alpha) = \frac{\alpha k^{\alpha}}{x^{1+\alpha}} \tag{1}$$

in which $0 < \alpha < 1$ and k_i represents one of many singularities in the distribution with x^{α} decay, the output of a single maximum, nonlinear, thermodynamic generating function acting on a one dimensional manifold,

$$\phi_{r,\varepsilon}{}^{t}\,(x) \colon M \rightarrow M \tag{2}$$

and x, the absolute distance from k_i bounded by the neighboring singularity, k_j.

The condition of convolutional stability leads to an aggregate of distributions, $p\,[p\,(x,k,\alpha)_i\,]$, being distributed like $p\,(x,k,\alpha)$. This mix of unstable modes and algebraic decays, when compared with information transport via iterative averaging in a gaussian noise context which goes like n, speeds up and makes more reliable the communication between these systems in weak resonance which go like n^2. Periodic points with positive topological entropy, $h_\mu = \Sigma\,\overline{\lambda}_i$, the growth rate of new orbits equivalent to the sum of the characteristic exponents of the system, are required since stable periodic orbits signify desensitization in biological systems specialized for information transport [1-3].

The use of generalized, infinite moment distribution function (1) fits the data since, as with all biological time series, the sequences of macromolecular information transport systems have characteristic unstable modes and fluctuations which do not converge to a central limit. One might argue that a reason for this is that the sample lengths of relative stationarity are short: polypeptide sequences range from 3 to 200 hundred amino acids and roughly three times that many in the relevent nucleotide coding chains. Pseudosingularities and nonexponential convergence can emerge from short sample length in a gaussian process. Using the theory of independent random events, even a twenty amino acid peptide with the twenty amino acids as possibility at each position would require a very long chain indeed to even see all 20^{20} possibilities.

If, however, the absence of finite moments, $\alpha < 1$, is built into the dynamics of the system such that the parameter dependent generater of h_μ is reflected in specific, universal, unstable modes and a small sample distribution function that is renormalizable across scale, a biologically unrealistic sample length becomes unnecessary. Growth hormone releasing factor is 44 amino acids in length yet we have evidence that it and all members of its family including glucagon, secretin, and VIP can be characterized as an unstable, four residue, Fourier mode in hydrophobic mass energy (kcal/mole/residue) [4]. We argue that this is an example of cross scale conjugacy in a convolutionally stable distribution with invariant measure, $\mu = p\,(x,k,\alpha)$. This, of course, is a probabilistic expression of the same topological entropy, h_μ, generated by (2) in the context of a deterministic thermodynamical system expressed as $h_\mu = \Sigma \bar{\lambda}_i$. A conjecture we make is that there is a formal path from the Fourier transformation of (1) called a Levy distribution of singularities and their decays proved convolutionally stable by Gnedenko and Kolmogorov to the mode ennumeration role of the Artin-Mazur zeta function on the output of (2) defined as the formal power series,

$$\zeta(t) = \exp \Sigma_{1,\infty} (1/m) N_m t^m \tag{3}$$

in which N_m is the number of periodic points, not necessarily of least period, in (2).

It is here that Shannon's information theoretic conditions break down in molecular biological coding and with them his definitions of information bearing redundancies: compositional redundancy (divergence from equiprobability of occurance of the elements of the alphabet), R_1; markovian redundancy (divergence from independence of single members of the alphabet in a sequence), R_2 [5]. Shannon's definition of entropy,

$$h = -k \Sigma p_i \log p_i \tag{4}$$

involves the statistical expectation (first momement) of the random variable, $-k \log p_i$, in which both of the p_i's in (2) are probabilities associated with random elementery events, x_i, $i = 1, a$, in which a represents the number of letters in the alphabet and k is some arbitrary constant. Such a system is ergodic, has invariant measure μ, in the same way as the number of heads in n trials divided by n goes to one-half as $n \rightarrow \infty$. We know that this is not true for all experiments and convergence is not uniform. The composite of singularities and algebraic decays of macromolecular sequences generated by deterministic, nonlinear thermodynamical system (2) and not inadequate sample length in a gaussian random process, does not fit within this context [6,7]. Information cannot be defined as R_1, R_2, deviations from finite expectation values that don't exist.

The statistical properties of real macromolecular sequences can be *approximated* by the composition of two ergodic sets. The first is the outcome of coin tosses that always come out heads. The second is a circle map, $\phi^t: S^1 \rightarrow S^1$, in the x, y plane which rotates an initial condition on the one dimensional manifold repeatedly by an angle incommensurable with 2π and scored as heads in locations above or on the x axis and tails below it. In the first case, $\mu = \delta$; in the second, $\mu = \Lambda$ where Λ

represents the entire manifold of S^1 and the convergence toward one-half heads as the invariant measure of the binary transformation of the dynamics is both consistent and uniform. Beyond this idealization, the characteristic saddle-sink generators of the real time series of macromolecular coding have neither measures δ nor Λ (or their composition) because they are without singular characteristic time or space scales, as seen in (1). In the language of extremal statistic α, we would say that respect to ergodic measure μ, $\alpha(\Lambda) < \alpha(\mu) < \alpha(\delta)$. What would seem nonstationary in a gaussian context achieves an invariant extremal measure in the context of hyperbolic distribution (1).

Coexistent saddle nodes and sinks on S^1 as the disappative limit of an annulus in R^2, demonstrate power spectra composed of singularities and their decays near the boundary of loss of invertability in nonlinear circle maps (Shenker) as do the classical strange attractors in which periodic points are also dense such as the Rossler, Lorenz, as well as historically the first discovered by Cartwright, Littlewood, and Levinson which served as the source of Smale's theory. Power spectra of these attractors contain parameter dependent singular modes. In addition to sensitivity to initial conditions and undecomposability, these sets have periodic points which are dense. We have hypothesized that the universal borderline between tangent-intermittency and these horseshoes serves as the parameter space of the thermodynamical system that encodes nucleotide and polypeptide sequences [6,8,9]. Its simplest expression can be seen in quadratic endomorphisms of the real line past the accumulation point of Feigenbaum in which unstable frontier sets of the stable Metropolis, Stein, and Stein U-sequence of singularities serves as reliable and varied sources of unstable hydrophobic modes for biological coding on a parabolic manifold whose variational properties are conjugate to a thermodynamic relation. This property can be measured as the excess heat capacity induced in acqueous solvent by the hydrophobic moieties, hph_i composing the amino acids and nucleotides of macromolecular polymers (Privalov, Sturdevent, Lumry) [10,11].

First Principles and an Evolutionary Equation

The specific heat capacity of water is minimal at 34–35° (Paul, Nature 323: 300; 1986), the body temperature of all nonpoikilothermic animals. Hydrophobic moieties, hph_i, lets consider here the variety of side chains, Σhph_i, of the twenty amino acids, $A_i(\Sigma hph_i)$, composing polypeptides and proteins, increase $\int_0^T \Delta C p dT$ as a function of their surface area which scales dimensionally with their mass like $m^{2/3}$. The immiscible, oily droplets induce strain in the surrounding water by compacting it into inhomogeneous regions of negative excess volume quantifiable calorimetrically as an increase in heat capacity. This property leads to a polypeptide chain minimal surface and the use of the phrase, globular protein, for their shape in acqueous solution.

The unit element of the flickering water network gels constituting the hph_i induced, inhomogeneously dense pattern of connectivities is the hydrogen bond. The larger the electronic orbital overlaps, the smaller the bending and rotation about the bonds and the larger the moment of inertia of the network -- and the inverse. When ice melts, less than 20% of the hydrogen bonds break, the fraction of intact bonds decreases about 1% with every 2.5°C of heating, gelation thresholds are in the range of 70% connectivity in $\mathbf{R}^3$, and polypeptides and proteins "freeze" the surrounding water motions. Hydrogen bond network cage strain, the result of dissipative-forcing crudely analogous to a Reynold's stress, r, generated by the intercalated variously hydrophobic amino acid side chains of polypeptides generates a long range "field" dynamic involving the interaction of (again crudely stated) Reynold's forcing, "bulk viscosity" and an entropy-enthalpy compensated collective free energy of the acqueous solvent at 35°. Estimates of the length scales of this non-van der Waal, acqueous hydrophobic mass energy force range up to 20 angstroms [6-8].

In reduced dimensional theories of hydrodynamic turbulence, the evolution of this increasing stress on the sheer and bulk viscosity of water has been represented by a one parameter group of transformations in codimension one or two in parameter regions with sensitivity to initial conditions, indecomposability, and periodic points (singularities represented physically by the vortices of turbulence). The vortices of polypeptide chains are helices with a range of pitches; the most common being the β-strand of wavelength 2.2 $A_i(\Sigma hph_i)$ and α-helix of about 3.6 $A_i(\Sigma hph_i)$.

Universality in the behavior of maps and flows in one and two dimension such as flip, Hopf, secondary Hopf, saddle, period doubling, tangent, and horseshoe bifurcations, U-sequences, and Farey modes make alternatives to this abstract hydrodynamic model for acqueous hydrophobic mass energy dynamics such as the nonlinear Laplace waves of surface tension, the density instabilities of Raleigh number dependent turbulent flows, and critical behavior in microemulsions no more than topologically equivalent mathematical representations with other arbitrarily chosen physical specifics. A variegated array of findings from the experimental physics of multicomponent chemical reactions, gas counters, lasers, Josephson-junctions, bistable optical devices, Rayleigh-Benard flow, acoustic experiments, and those involving crystallization indicate that complex systems of high physical dimension achieve topological conjugacy between the parameter-dependent orbits observed along their evolution through universal itineraries among singularities in one dimension. The ordering property of the one dimensional scenerio generates a finite set of unique, parameter dependent stable modes with unstable frontier sets manifesting nonexponential decay which can account for the observables of macromolecular coding. It is of more than casual interest that the mysterious sequence inversion occuring in the middle of long nucleotide chains, palindromes, occur naturally in the U-sequences of non-monotonic manifolds in (for examples) the antisymmetric cubic map, the trapezoid with a flat tail, some ODE's as well as Swinney's nonlinear chemical oscillator experiments (Beyer, Mauldin, Stein, Hao).

Although most attempts to develop algorithms for the prediction of relationships between amino acid sequence, protein structure and function involve reconstructions in R^3, it is clear that given the constraint of continuity and a three dimensional order parameter of helicity as determined by a parameter dependent two-dimensional vorticity set, the evolution equation of the tertiary set is a one dimensional nonlinear map. It is natural to chose a quadratic family in codimension two: r, ε; the first parameter representing forced-disappation of and by hph_i, the second, the coupling between these two aspects of the dynamics. ε controls the rate of evolution of the universal scenerio across increasing r.

$$\phi_{r,\varepsilon}{}^{t}[A_i(\Sigma hph_{n+1})] = r \cdot [A_i(\Sigma hph_i)] - r \cdot [A_i(\Sigma hph_i)]^{\varepsilon} \tag{5}$$

Stating (5) in physical terms, approximately once per second for about a minute, the next amino acid added is a function of the past one and the parameter values of the hydrophobic mass energy acqueous field as configured by the aggregate of nucleotide chains and their protein modulators in the polysome tunnel. Together these factors determine the order parameter, helicity, a vorticity cylinder generated by a single maximum, endomorphism of the real line. Recurrance times determine the helicities which range as primary, broad band Fourier modes from the 2.2 residue wavelengths of β-strands to the 13.3 $A_i(\Sigma hph_i)$ values found in the transformations of the polypeptide hormones, MSH and ACTH.

Nucleotide Redundancy and the Metric Space of hph_i

We express the peptide-water mass action free energy, ΔG, as the difference between the heat of the hydrophobic solvation reaction and the associated change in the entropy of the polypeptide-water system:

$$\Delta G(-RT \ln k) = \Delta H^\circ_T - T\Delta S^\circ_T \tag{6}$$

the second law adopted for chemical equilibria using heat of reaction, ΔH°_T, and the Clausius entropy term, $T\Delta S^\circ_T$, first proposed as the "thermodynamic foundation of the life sciences" by Planck in 1910. The heat capacity at constant pressure of water, increased by elongating polypeptide side chains, decomposes the $T\Delta S^\circ_T$ term into the bond making and breaking $T\int_0^T \frac{\Delta Cp}{T} dT$ *and* an additional term generated by induced solvent negative volume, $\int_0^T \Delta Cp dT$. The addition of each new amino acid side chain as its hydrophobic mass energy, Σhph_i, results in a characteristic tendency to curl into polypeptide vortices called helices and/or bends. High hydrophobic residues roll up on themselves into tight vortices maximizing surrounding water "freedom"; low hydrophobic or charged residues spread out in space "freezing" more water into negative volume, reducing the solvent entropy. Of course, the differences in the entropy of the two prototypic polymer configurations are reciprocal as dictated by the thermodynamic conservation law involving enthalpy-entropy compensation. The hydrophobic mass energy per amino acid side chain, Σhph_i, has entropic, hph_S, and enthalpic, hph_H, components that vary together [6-11].

Since with each additional amino acid, hph_i,

$$\Delta G' = (\Delta H + hph_H) - (T\Delta S + hph_S) = 0 \tag{7}$$

We use Tanford's transformation of hph_i as the water/ethanol partition in kcal/mol

$$hph_S = hph_H = RT \ln \frac{\text{mole fraction water}}{\text{mole fraction ethanol}} \tag{8}$$

to derive the "negative volume" inverse entropy code of polypeptide sequences:

Hydrophobic Mass Energies $A_i(\Sigma hph_i)$

Ala	0.87	Glu	0.67	Leu	2.17	Ser	0.07
Arg	0.85	Gln	0.00	Lys	1.64	Thr	0.07
Asn	0.09	Gly	0.10	Met	1.67	Trp	3.77
Asp	0.66	His	0.87	Phe	2.87	Tyr	2.67
Cys	1.52	Ile	3.15	Pro	2.77	Val	1.87

A fundemental confusion exists between what is a metric on the set of $A_i(\Sigma hph_i)$, nonnegative real numbers as a distance, $d(hph_i - hph_j)$, as represented by the Tanford transformation (as well as those of Hopp-Woods, Fraga, Bull-Breese, Fauchere-Pliska, Kyte-Doolittle, and Wolfenden), and values assigned from the results of the temporal-spatial dynamics of generating function (5) as obtained from the phase portrait in the form of X-ray crystallographic data (Chothia, Janin, Karplus-Schulz, Guo, and Parker). Metric space, algebra, and generating transformation are separate, specifiable components of a dynamical system and should not be confounded. In addition, the classification of amino acids by their tendency to be buried or on the surface of globular proteins does not hold up with respect to systems that directly measure hph_i. On one hand, there are two molecules of water on the average per amino acid on the interior of proteins; on the other, the surfaces of globular proteins consist of nearly equal numbers of polar and hydrophobic $A_i(\Sigma hph_i)$). The Tanford code (we use the one summarized by Manavalan and Ponnuswamy) differs generally from the other metric systems only for threonine, Thr, which some assign the value of 0.87, (kcal/mol/residue) [10,11].

Earlier work by Gatlin [5], Jukes, Kimura and others found Shannon redundancy, $\overline{R}$, conserved in nucleotide sequences such that $\overline{R} = R_2 + R_1$ = a constant and higher vertebrate evolutionary status was reflected in a greater proportion of markovian redundancy, R_2. No studies showed that increasing sample length converged to finite first and second moments as required by the gaussian assumptions of Shannon's first and second theorems so that an interpretation of these findings is difficult.

A physically realistic, information bearing redundancy, R_{hph_i}, can be found in a natural partition into four groups of $A_i(\Sigma hph_i)$ in the table of Tanford transformations above. It is known that the third nucleotide in the amino acid mRNA triplet code "wobbles" (Crick) such that it accounts for little of the variance with respect to amino acid coding sequences. In contrast, the middle nucleotide accounts for about 80% and the first one for the remaining 20%. Generally, if the middle nucleotide of mRNA is uridine, the amino acid has high values for Σhph_i, Phe, Leu, Ile, Val; if it is adenine, there is a mix of high and low ones, Tyr, His, Gln, Lys, Asp, and Glu; if the second nucleotide is cytosine, the amino acids are mostly low ones, Ser, Thr, Ala, and Pro (a disrupter of helicity due to its unique chemical character and not hph_i); if guanine, a mix of mostly low and middle values, Gly, Ser, Arg, and Cys. This leads to a set of binary vectors which can be equated with the four amino acid *hph* families listed below as well as serving as the nucleotide transform of the Markoff partition (0 if ≤ 0.5 *else* 1) on the itineraries of the codimension two polypeptide generating endomorphism of the unit interval (5):

Nucleotides	Amino Acid Side Chain Hydrophobic Families
1 1 *	Leu,Ile,Phe,Tyr,Trp,Pro
0 1 *	Cys,Lys,Met,Val
1 0 *	Ala,Arg,Asp,Glu,His
0 0 *	Gln,Asn,Ser,Thr,Gly

Nucleotide redundancy, R_{hph_i}, derived from $A_i(\Sigma hph_i)$, reduces the alphabet from twenty letters to four and map (5) followed by symbolic dynamic transformation defines a second information bearing redundancy in the form of the finite set of universal modes in one dimension via parameter dependent itineraries.

Support for the existence and significance of R_{hph_i} comes from evolutionary studies. Dayhoff's Atlas of Protein Sequence and Structure (see MDM78, Figure 85 in supplement 3) uses the method of ancestral sequences to estimate the rates of replacement of amino acids by other amino acids in a twenty-by-twenty matrix. The diagonal represents the identity transformation with the distance of the off-diagonal amino acids permutable such that it would be a function of the likelyhood of substitution. Using this data base, it is easy to show that the Ile, Leu, Val group is most distant from that of Glu, Asn, Asp, Gln and Ser, Thr, Ala from Phe, Tyr, and Trp. A variety of matrices of distances using substitution data by McLachlan, Grantham, French-Robson, Argyle, Swanson and others confirm the existence of the hph_i metric in this general way.

Dynamical Redundancy and the Universal Mode Structure of R_ϕ

Our use of the discrete, Fast Fourier Transform algorithm on the Tanford transform of the amino acid sequences of about 600 proteins and 150 polypeptide hormones has revealed only a small set of primary, broad band, Fourier modes, $G_\omega\,[A_i(\Sigma hph_i)] = 2.2, 3.6, 4.0, 4.4\ 5.0\ 7.2, 8.0, 10.0, 13.3$ where ω represents the mode wave length in amino acid residues. This led to the conjecture that in the same way that Shannon's redundancy of composition, R_1, is replaced by a thermodynamical property that is encoded by sequences of nucleotides, R_{hph_i}, the n-order markovian redundancy of R_2 is at best a probabilistic approximation to a nonlinear deterministic, thermodynamical process with universal mode behavior and dynamical redundancy, R_ϕ.

Map (5) can be considered a discrete time evolution equation of emergent polypeptide sequences in the mRNA translation process in the polysome tunnel, which adds an amino acid on the average once per second in chunks varying from 35 to 50 residues in length, $n = t$ in (5). It is for this reason that we use a 50 residue amino acid sequence sample length for Tanford and discrete Fourier transformation. A single large protein may be the product of several polysome tunnels.

R_{hph_i} allows the coarse-grained description of the set of points generated by (5) in a sequence of 0's and 1's as described above. This amounts to labeling each of the two monotonic branchs of the parabolic manifold by a symbol and the dynamics (and an equivalence relation in terms of itineraries--Sarkovskii, Steffan, Bowen, Thurston, Milnor, Collet, Eckmann, Guggenheimer and others) by symbol sequences. The two-by-two, $\{0,1\}$, Markoff matrix, ψ, generated by (5) serves to count the periodic points composing the modes as $Tr\,\psi$, a topological invariant of transformation (5) (Williams). This is the counting method implicit in (3) which was inspired by the Weil zeta function of an algebraic variety over a finite field. Categorization based on a bookkeeping function that keeps track of singularities is roughly analogous to the way physicists use Green's functions. The number of fixed points of ϕ^t goes as $Tr\,\psi^t$; the number of unstable periodic orbits of length t equals the number of paths through ψ^t of length t.

Very specific macromolecular coding by dynamical system, $\{\phi^t, \psi_{0,1}, A_i(\Sigma hph_i)\}$, in parameter regions with $h_\mu > 1$ fixed points is made possible by a universal property of unimodal maps: at each parameter value there exists, at most, one stable periodic orbit. A given period exists on a finite interval of the parameter line in the middle of which is a superstable period, the orbit whose itinerary contains the critical point, $\phi' = 0$, as one of its periodic points. We have recently calculated the average hph_i of the amino acids in 209 proteins (using the Georgetown University Protein Sequence Bank) and found the value expected for the middle of the range of the Tanford code if the very rare Trp was removed (1.33 kcal/mol/residue). This dictates an initial condition for map (5) as 0.5 (a point in the superstable orbit) as a reasonable one. This finds the parameter dependent periodic orbit immediately with $h_\mu > 0$, also a parameter dependent property. Said another way, we find the required polypeptide modes in the unstable frontier sets of the U-sequence of Metropolis, Stein, and Stein.

Dynamical Structures and Potencies of the Calcitonins

Fourier Transformation of Calcitonin-Related Peptide Hydrophobic Sequences

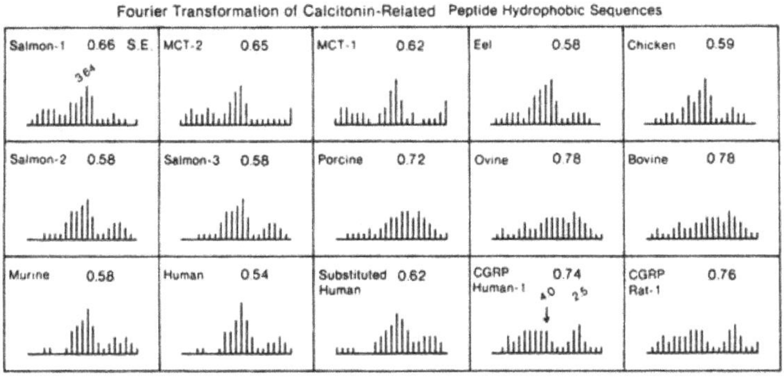

We have elsewhere summarized our findings for several polypeptide hormones and their receptors. We address here the problem of elucidating the coding invariants of the calcitonin family, a linear dotriacontapeptide. It inscribes its $A_i(\Sigma hph_i)$ broad band, unstable Fourier mode at its beginning (at the N-terminal end) where a seven membered loop of amino acids defined by a sulfide bridge between two cysteine residues is composed of two G_ω of ~ 3.5 : [Cys, Gly, Asn, Leu, Ser, Thr, Cys] = [1.52,

0.10, 0.09, 2.17, 0.07, 0.07, 1.52] in the human congener. This 3.5-3.6 amino acid wavelength continues to dominate throughout the 32 membered sequence. This is seen as a discrete Fourier broad band mode in the Tanford transformed amino acid sequence in all members of the family with the exception of the joint gene product without the actions of calcitonin called calcitonin gene related protein (CGRP) in which the 3.64 residue is replaced by $G_{\omega} \sim 2.5, 4.0$. This consistency in the behavior of the order parameter, helicity, is maintained in spite of some (non-CGRP) congeners being different from one another by as much as half, 16, of their individual amino acids.

As one of the most systematically characterized polypeptide series using osteoclastic, renal, plasma calcium and cAMP related bioassays, it was found that inspite of their common actions, the calcitonins manifested a wide range of potencies. Salmon-1 is 10^2 to 10^3 fold more potent on human tissue than the murine or human congener. MCT-2 and MCT-1 artificial analogues as well as the eel and chicken forms are more potent than the salmon-2 and salmon-3, porcine, ovine, and bovine congeners.

Our hypothesis was that the weaker congeners would have a higher average hph_i leading to more self-association, a reduced minimal surface, and less effect on the hydrogen bond network gels of the surrounding solvent. This would also lead to a generally higher center crossing frequency around the average $A_i(\Sigma hph_i)$ in the sequence. We find some support for these hypotheses in the ranking of the average hydrophobicity of the congeners (salmon-1, MCT-2, and MCT-1, the least) and the one-dimensional fractal dimension of the hydrophobic sequence as a measure related to center crossings.The greater the global "vorticity", the greater the complexity of the hph_i orbit along the line and the higher the dimension.

Some Statistical Properties of a Calcitonin Generating Function

The range of potencies among members of this family defined by the dominant broad band mode of its helicity order parameter dictates exploration of the parameter space for different itineraries with the same broad band Fourier mode. Whereas the itineraries of period 2 and 3 are singular, there are 9 stable periods seven in the U-sequence from which to chose candidates for polypeptide calcitonin activity. This hormone prevents the loss of bone mass in aging and primary and metastatic neoplastic

Calcitonin Analogue	Mean Hydrophobicity (kcal/residue/mol)	Fractal Dimension of Hydrophobic Sequence
Salmon-1	0.96 ± 1.01	1.37
MCT-2	1.03 ± 1.05	1.39
MCT-1	1.08 ± 1.14	1.39
Eel	1.03 ± 0.99	1.39
Chicken	1.08 ± 0.96	1.41
Salmon-2	1.16 ± 1.04	1.45
Salmon-3	1.16 ± 1.04	1.45
Porcine	1.21 ± 1.16	1.44
Ovine	1.24 ± 1.13	1.43
Bovine	1.25 ± 1.14	1.44
Murine	1.10 ± 1.12	1.43
Human	1.15 ± 1.14	1.44
Sub. Human	1.10 ± 1.15	1.43

conditions but must be given by daily injection because it is so quickly broken down by the proteolytic enzymes of the upper gastrointestinal track. The design of a new sequence maximizing physiological activity and minimizing vulnerability to enzymatic hydrolysis has some practical importance with respect to an oral route of administration. Of course, this kind of success might also auger an oral route for other polypeptide hormones as well as immunogenic fragments. We are currently in the process of exploring this possibility with Leonard Deftos. It has as its premise that the macromolecular message is in the emergent order parameter, helicity, of the dynamical system representable by a variety of specific amino acid and nucleotide sequences.

As an example, we chose the unstable frontier set of the stable period seven, $r = 3.8860459$, $\varepsilon = 2.00$ $hph_i = 0.50$, at which $r = 3.88602$. The binary sequence is, 1001011..., which with 14 bit precision was inscribed without error for the first 66 interations (well within the length scale of many neurohormonal peptides) in spite of the the logarithm of the exponential rate of separation of nearby orbits, the characteristic exponent, $\bar{\lambda}$ being positive, 0.2897. Seven errors in the binary sequence were made in the first 100 iterations and further decay of the seven mode with intermittent appearance of the next rank of unstable fixed points occuring as the polypeptide chain was lengthened. This behavior is characteristic of most of the proteins and large polypeptides that we have studied. We characterize the Markoff transition matrix, ψ, with

$\rho = \log Tr\ \psi / \log maximum(0,1)\ partition$. We note that ρ grows asymtotically slower across increments in powers of ten of the length of the polymer.

Residues	$\rho(\psi)$	$\bar{\lambda}$
100	0.9107	0.2897
1000	0.9383	0.4242
10000	0.9540	0.4141
100000	0.9668	0.4140

Although the ergodic theorem as applied to transformations like (5) by Hopf, Halmos, and others required as a condition for the existence of invariant measure μ that ϕ^{t} be one-to-one, Doob and Riesz showed that this condition could be dropped (allowing the many-to-one singularities of the U-sequence) if the inverse image of every measureable set is a measureable set of the same measure. This is the convolutionally stable condition of hyperbolic distribution function (1) as proved by Gnedenko and Kolmogorov with invariant extremal statistic $\alpha(\mu)$ serving as the invariant measure. With an infinite multiplicity of moments and a nonconvergent tail, the Shannon information theorems lose their applicability to the problems of macromolecular coding to be replaced by a finite collection of parameter dependent unstable singularities and their decays.

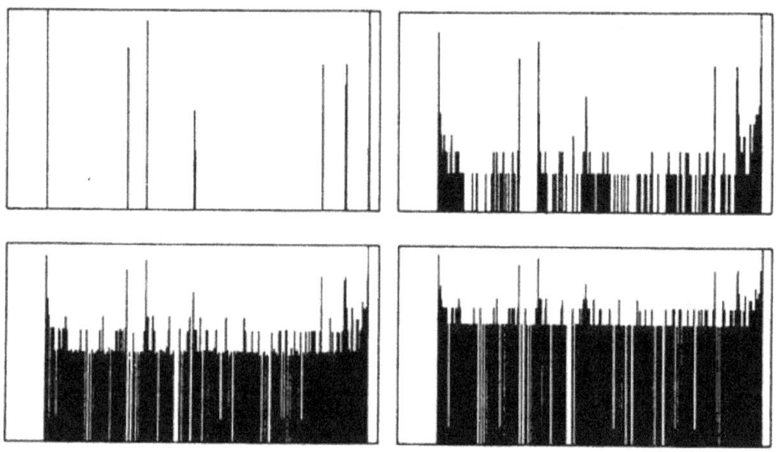

When (5) was iterated by increasing powers of ten under parameter conditions described above and the logarithmic density partitioned into 1000 bins, we see the gradual decay of the prominance of the seven mode with polymer length via the emergence of logarithmic ranks of other values in what would eventually be an absolutely continuous measure. The emergence of new modes can be seen as darker densities at the bottom of plots of the higher iterations.

Statistically equivalent representations of the unstable modes and their decays (1) generated here by saddle-sink dynamics (2,5) have been studied in the most varied of more probabilistic contexts by Shlesinger and his associates: the continuous time random walk formalism applied to slow, nonexponential relaxation near the polymer glass transition, the defect diffusion model of dielectric relaxation, variable barrier height jumping in anomalous transport in semiconductors, the representation of the behavior of the Weierstrass continuous but nowhere differentiable function used by Moser in the KAM problem, and as a reformulation of the Riemann zeta function approach to the distribution of prime numbers. In the same way that universality in one and two dimensions makes the details of the nonlinear equations of motion irrelevant for the predictive process, the limit law of convolutionally stable distributions with $0 < \alpha < 1$ captures a large class of multiplicative statistical dynamics. Shlesinger has related the renormalization group formalism, so commonly used to deal with deterministic systems like (5), to hierarchically diffusive processes through this same limit law. Intuitively, we can imagine a snap-back repeller in place of the attractive central limit in the hyperbolic distributions which are best characterized by the marginal statistics of their multiple unstable singularities; the poles, zeros, and residues of their complex transformations.

An Example of Calcitonin Sequence Design

Our goal is a highly hydrophobic congener which we know will be weak but better prepared to both resist proteolytic digestion by stomach and small bowel enzymes as well as more quickly absorbed by the intestinal mucosa. Using a phase space reconstruction technique described in [4], the differences in potencies of the important calcitonin analogues can be seen to be related to the global "tightness" of the vorticity.

We shop within the unstable frontiers of the nine available period sevens of the U-sequence for good representatives of this property. For example, using the parameter values for the stable and unstable seven modes studied above following 100 preiterations (polypeptides usually appear within the context of their much larger prohormones) leads to the characteristics of the structures below: average hydrophobicity (kcal/mol/residue); $\bar{\lambda}$ for 100 and 1000 iterations; the fixed points on the parabolic manifold of map (5); phase space reconstructions of the nucleotide binary vector transformed amino acid sequences; the discrete Fourier transformation of the calcitonin family peptide generated by the map; its "spectral entropy"--the aggregate of the variances of the modes of the power spectrum. Amino acids within each of the four groups are assigned randomly weighted by the amino acid distribution found in the cerebrospinal fluid. The prominance of the 3.6 $A_i\,(\Sigma hph_i)$ unstable mode signifying its family membership is displayed in the power spectrum. This and other congeners will be tested for biological activity and resistence to gastrointestinal inactivation by Leonard Deftos's group at UCSD.

Parameter r	$\bar{\lambda}$ (n = 100	1,000)	$\overline{\text{Hph}}$	Manifold	Phase Space Structure	Spectrum	SE
3.8860459	-0.9165	-0.9389	1.08	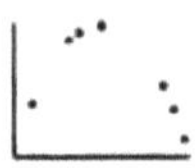		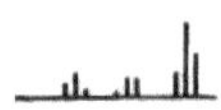	0.30

Parameter r	Manifold	Peptide Sequence	Phase Space Reconstruction	Spectrum	SE
3.88602	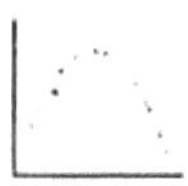	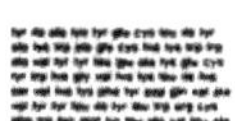		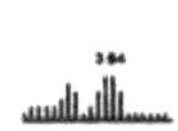	0.70

Acknowlegements

Appreciation is expressed to the Office of Naval Research (Physics and Systems Neuroscience), the U.S. Army Research Office (Neuroscience), and the National Institutes of Child Health and Human Development for support and Professor R.M. Kiehn for a helpful lemma with respect to the dimension of an evolutionary equation for helicity.

References

1. Mandell, A.J. Non-equilibrium behavior of some brain enzyme and receptor systems. Ann. Rev. Pharm. Tox. 24: 237-274, 1984.

2. Mandell, A.J. Toward a neuropsychopharmacology of habituation: a vertical integration. Math. Model. 7: 809-888, 1986.

3. Mandell, A.J. Dynamical complexity and pathological order in the cardiac monitoring problem. Physica 27D: 235-242, 1987.

4. Mandell, A.J. The source and characteristics of normal modes in molecular biology. Chaos and Order in the Natural Sciences. Ed: M. Velarde. Springer-Verlag. 1988.

5. Gatlin, L.L. Conservation of Shannon's redundancy for proteins. J. Mol. Evol. 3: 189-196, 1974.

6. Mandell, A.J., Russo, P.V., and Blomgren, B.W. Geometric universality in brain allosteric protein dynamics: complex hydrophobic transformation predicts mutual recognition by polypeptides and proteins. Ann. N.Y.A.S. 504: 88-117, 1987.

7. Mandell, A.J. From chemical homology to topological temperature: a notion relating the structure and function of brain polypeptides. Synergetics of the Brain. Ed: E. Basar, H. Flohr, Haken, H. and Mandell, A.J. Springer-Verlag, 1983.

8. Mandell, A.J. Desensitization, emergent differentiability, and loss of fractal scaling in the power spectra of brain proteins. Fractal Aspects of Materials. Ed: R.B. Laibowitz, B.B. Mandelbrot, and D.E. Passoja. Mat. Res. Soc. 1985.

9. Mandell, A.J. The hyperbolic helix hypothesis: Stapleton's fractal measure on the hydrophobic free energy mode distributions of allosteric proteins. Fractals in Physics. Ed: L. Pietronero and E. Tosatti. Elsevier. 1986.

10. Mandell, A.J. Polypeptide hormones and receptors: participants in and products of a two parameter, dissipative, measure preserving, smooth dynamical system in hydrophobic mass energy, dU/dt. Chaos in Biological Systems. Ed: H. Degn, A.V. Holden, and L.F. Olsen. Plenum Press. 1987.

11. Mandell, A.J. Prime times: the distribution of singularities in hydrophobic free energy of proteins. Perspectives in Nonlinear Science. Ed: M.F. Shlesinger, R. Cawley, A.W. Saenz and W. Zachary. World Scientific Press. 1986

Fractal Models in Physiology

Bruce West
Physical Dynamics, Inc.
10280 North Torrey Pines Road, Suite 265
La Jolla, California 92037

1. INTRODUCTION TO FRACTALS

Some of the fundamental questions in anatomy and physiology revolve around problems of scale, size and shape. For example, is there a characteristic scale factor that governs the decrease in mean bronchial dimensions from the trachea to the terminal bronchioles, such as the average length, diameter of volume? What is the "true" surface area of irregular, wrinkly structures such as the brain, the small bowel lumen, or the endocardium of the heart? Does the type of architecture seen in the pulmonary tree share any basic morphogenetic link with branching networks in the heart, vascular tree, kidney, and liver? Does the answer depend on the degree of resolution of the measuring instrument?

Questions like these, which have been of traditional interest in medicine and biology, are now being re-examined using concepts developed in an active branch of the basic services called *nonlinear dynamics*. In particular, a number of these concepts derive from the study and application of *fractals*: a mathematical idea that seems to permeate the description of physiological structure.[1] Herein we present the details of a few studies in which the fractal notion has played a key role, but more importantly we discuss why fractal models capture more of the essential features of biological systems than do the classical scaling models.

The term "fractal"[2] refers to objects (processes) having a fractional dimension, of which there are three kinds. The first kind is geometrical and has to do with the static structure of an object. Classical geometry deals with regular smooth forms uniformly filling spaces of integer dimensions, e.g., a point has zero dimension, a curve has one dimension, a surface has two dimensions, etc. However, structures in nature are perversely non-classical, as can be verified by tracing the outline of a tree, viewing a coastline on a map, watching a breaking wave, etc. The simplest example of a fractal is a fractal curve, which compared with a classical line appears to be wrinkled. If one were to magnify each wrinkle in such a line, it would be seen to consist of smaller wrinkles and these in turn of even smaller wrinkles. At each successively smaller scale, more and more levels of irregular structure are revealed. Clearly the length of such a curve is a function of the scale used to resolve it, i.e. the smaller the ruler used to measure a fractal curve, the longer the wrinkled line appears to be. In the limit of a vanishing ruler-length the length of the fractal curve becomes infinite.

If ε is the unit of measure of a curve of dimension D, then $N(\varepsilon) \propto \varepsilon^{-D}$ is the number of units required to measure the length of the curve. The length of the curve is $L(\varepsilon) = N(\varepsilon)\,\varepsilon \propto \varepsilon^{1-D}$ and since for a fractal curve $D > 1$, the length diverges as $\varepsilon \to 0$. If the $\ln L(\varepsilon)$ is plotted versus $\ln \varepsilon$ then an *inverse power law* results with a slope given by $D-1$. The fractal dimension of the wrinkly line is determined by the slope, and it turns out that, for a fractal line such as a coastline on a map, this value lies between one and two. Thus a fractal curve of this kind will have a dimension greater than that of a classical line (unity) but less than that of a smooth rectangular surface (two) which could be drawn to enclose the wrinkly curve.

The second kind of fractal is statistical and was originally studied by P. Lévy starting in the 20's. He was interested in processes comprised of a sum of identically distributed random variables for which the distribution of the sum is the same as that for any of the contributors. The class of distributors having this property are called Lévy (stable) distributors.[3] If $X(t)$ represents a Lévy process and $P(x,t)dx$ is the probability that $X(t)$ lies in the interval $(x, x+dx)$ at time t, then in terms of the characteristic function $\phi(k,t)$ we have

$$P(x,t) = \int_{-\infty}^{\infty} e^{ikx}\phi(k,t)\,dk \quad , \tag{1.1}$$

and for a free Lévy process

$$\phi(k,t) = e^{-\gamma|k|^{\alpha}t} \quad ; \quad 0 \le \alpha \le 2 \quad . \tag{1.2}$$

When $\alpha = 2$, $P(x,t)$ becomes an Einstein diffusion process, i.e. a Gaussian distribution with second moment, $\langle x^2 \rangle = \gamma t$. For general $0 < \alpha < 2$, $\langle x^p \rangle < \infty$ if $p < \alpha$ and $\langle x^p \rangle = \infty$ if $p \ge \alpha$, i.e. central moments of the distribution diverge. One consequence of this property is that the random variables $X(t)$ is intermittent i.e., the time trace of $X(t)$ manifests bursting behavior rather than a regular random fluctuation. The Lévy index α can be related to the fractal dimension of the underlying process, e.g., $D = 2 - 1/\alpha$ for the time trace of the process.[4]

We observe that, just as for the geometrical fractal, the statistical fractal displays a kind of self-similarity. The scaling relation for the dynamic variable is $X(\beta t) = \beta^{1/\alpha} X(t)$ indicating that the statistics of the time trace in a time interval expanded by the factor β, is identical to that in the original interval but with an overall amplitude change of $\beta^{1/\alpha}$. Stated differently, it is easy to demonstrate using Equations (1.1) and (1.2) that the probability density satisfies the scaling relations $P(x,t) = \beta^{1/\alpha} P(\beta^{1/\alpha} x, \beta t)$. It has also been shown[5] that the asymptotic form of the Lévy distribution is an inverse power law:

$$P(x,t) \approx \frac{A(t)}{|x|^{\alpha+1}} \quad . \tag{1.3}$$

The third and final kind of fractal is also related to dynamical processes and is correlational in nature. Consider a complex process that can not be characterized by a single characteristic time scale, but instead is regulated over many orders of magnitude in time by a self-similar mechanism. The multiplicity of time scales will be manifest in a power spectrum with a broad profile of response times. One discovers that the process in question shows structure (fluctuations) over multiple orders of temporal magnitude, e.g., minutes to milliseconds in such physiological processes as say the heart beat, in the same way that fractal forms exhibit detail over several orders of spatial magnitude in such physiological structures as the lung. The fractal scaling between variations of different time scales lead to a frequency spectrum having an inverse power law distribution, i.e., for a frequency f the spectrum is $1/f^{\mu}$ where μ is related to the fractal dimension of the process.

We have mentioned geometrical, statistical and temporal fractals each of which shares the common properties of fractal forms: *heterogeneity*, *self-similarity* and the *absence of a characteristic length scale*. The features of heterogeneity, multiple scales and self-similarity are characteristics of a variety of seemingly unrelated biological forms: the tracheo-bronchial tree, the His-Purkinje system, the chordae-tendeae, the biliary network, the vascular tree, and the urinary collecting system, to name a few. The fractal concept has proven to be a useful way to characterize the geometrical structure and dynamic processes in a number of these biological systems.

2. FRACTALS IN PHYSIOLOGY

What is the significance of the fractal structure and what are the links between fractal structure and physiological function? From a developmental viewpoint, fractals provide a new morphogenetic principle underlying the construction of many apparently unrelated, highly complex, irregular structures by means of a simple code. The basis of this code is self-similarity, so that a complex, branching structure like the bronchial tree or His-Purkinje system can be generated by adding the same structure on progressively smaller scales. Such a fractal algorithm will also serve, as we show later, to minimize constructional error, since it relies primarily on a self-similar, iterative mechanism. We now consider these latter two systems in some detail.

Consider the number of branches emanating from the trachea in a mammalian lung. If z is the generation number, then since there are two stems leaving each vertex, there is a geometric increase in the number of branches with generation number, i.e., $n(z)=2n(z-1)$ which yields $n(z)=2^z n(0)$. This simple scaling argument was used by Weibel and Gomez[6] to determine the size of the airways in the human lung. They assumed that the diameter of an airway scales as $d(z)=qd(z-1)$ between successive generations. This scaling assumption leads to an exponential decrease in the size of the diameter, i.e., $d(z)=d(0)e^{-\alpha z}$ with $\alpha=\ln q$, where $q<1$. A more detailed argument yields $q=1/2^{1/3}$. For the first ten generations the diameter of the airways in a human lung seems to satisfy the exponential relation [cf. Figure 1]. Subsequently there is a marked deviation from this exponential form. Weibel and Gomez assume that a new diffusive mechanism is initiated around $z=10$ and leads to a second exponential. We[7] wondered if in fact there is a better fit to the data that does not require the introduction of a second mechanism. We re-plotted the data on log-log graph paper and observed that there is a dominant inverse power-law behavior in the average diameter with z [cf. Figure 2]. But not just in humans, but also in dogs, rats and hamsters. In all of these data sets there appears to be a harmonic variation of the data points about the inverse power-law curve.

The first question to ask is what has been left out of the classical scaling that might possibly account for the inverse power-law behavior and the harmonic variation? What was omitted in the theoretical analysis was the multiplity of scales present at each generation. If we denote these omitted scales by α then $d(z)$ in fact becomes $d(z,\alpha)$ and the quantity being graphed has for a given z been averaged over the α-scales:

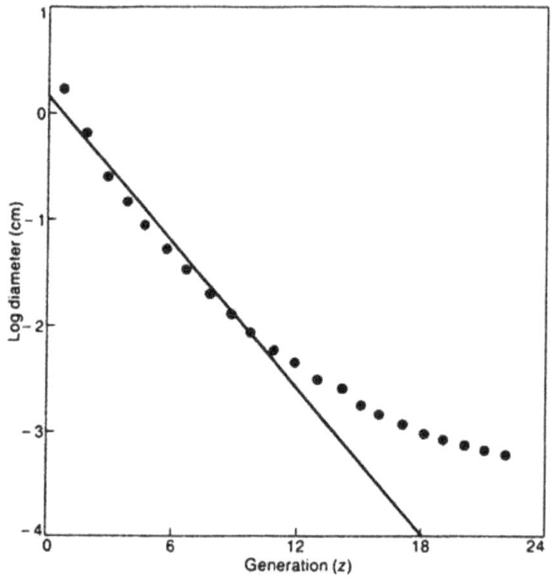

Figure 1. This semi-log graph shows measurements from Weibel and Gomez[6] for 23 generations of bronchial tubes in the human lung. The classical scaling prediction is the straight line. (Adapted from Ref. 7.)

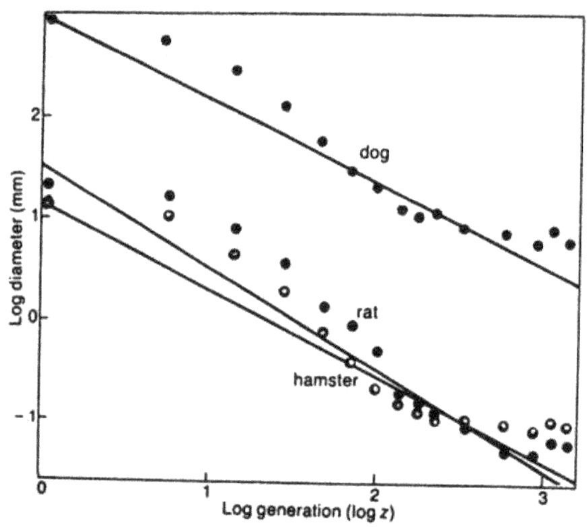

Figure 2. We plot on a log-log graph the average bronchial tube diameter for dog, rats, and hamsters taken from Raabe et al.[12] We see that the average diameter follows an inverse power law, not exponential, relation. In addition there appears to be a harmonic (periodic), variation of the data points about the power-law regression line. (Adapted from Ref. 1.)

$$d(z) = \int_0^\infty d(z,\alpha)\,p(\alpha)\,d\alpha \tag{2.1}$$

The distribution of scales $p(\alpha)$, in part, determines the functional form of $d(z)$. Weibel and Gomez, of course, use $p(\alpha)=\delta(\alpha-\alpha_0)$, i.e. a single scale α_0 to characterize the bronchial tree. A maximum entropy argument has been used to select a least biased form for $p(\alpha)$.[8] In adopting this formalism one uses experimental information to constrain the form of $p(\alpha)$ and the maximum entropy condition to completely randomize the distribution with respect to everything else. We construct the variational equation

$$I = -\int p(\alpha)\ln p(\alpha)\,d\alpha + \text{constraints} \tag{2.2}$$

where here we choose the theoretical constraint that there is no basic scale in the statistical structure. The constraint which accomplishes this is the average logarithm of α:

$$\int_0^\infty p(\alpha)\,d\alpha\,\ln\alpha = \text{constant} \; . \tag{2.3}$$

The variation $\delta I = 0$ then yields the hyperbolic distribution of scales

$$p(\alpha) \propto \frac{1}{|\alpha|^{\lambda_1}} \; , \tag{2.4}$$

where λ_1 is a Lagrange multiplier explicitly determined by the constraint.

The hyperbolic distribution (2.4) satisfies the scaling relation

$$p(\alpha/\beta) = \beta^{\lambda_1} p(\alpha) \tag{2.5}$$

so that the average diameter can be shown using (2.1) to satisfy a similar relation

$$d(z) = p\,\beta\,d(\beta z) \tag{2.6}$$

where $p = \beta^{-\lambda_1}$. The solution of the functional scaling relation is given by[7-10]

$$d(z) = \frac{A(z)}{z^{\mu}} \tag{2.7}$$

where

$$\mu = 1 - \frac{\ln(1/p)}{\ln\beta} = 1 - \lambda_1 \tag{2.8}$$

and

$$A(z) = A(\beta z) = \sum_{n=-\infty}^{\infty} A_n \exp\left[i 2\pi\mu \frac{\ln z}{\ln \beta} \right] \quad . \tag{2.9}$$

The average diameter therefore possess the two properties of having: (1) a dominant inverse power-law in z and (2) a harmonicly varying coefficient in $\ln z$. This solution gives an excellent fit to experimental data for dogs, hamsters and rats [cf. Figure 3].

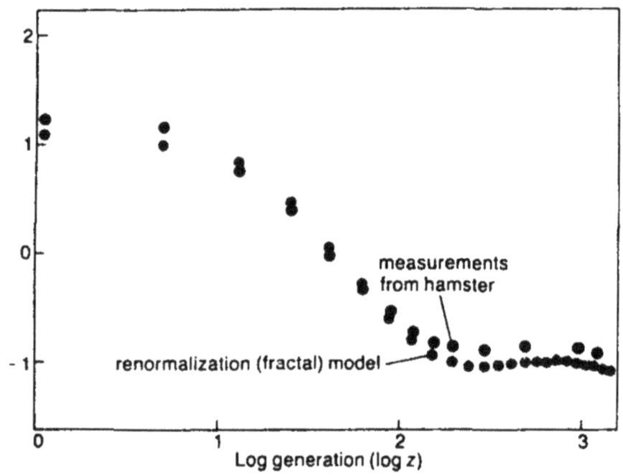

Figure 3. Here we show the close fit between the mean diameter observed in the bronchial tubes of a hamster and those predicted by the renormalization (fractal) model. The fit to the other data sets displayed in Figure 2 are equally good. (Adapted from Ref. 1.)

From the above consideration we see that the human lung is representative of a fractal structure due to the hyperbolic distribution in scales.[1,2] A fractal dynamic process is also found in physiology in the His-Purkinje conduction system of the heart.[10] Here we see a voltage pulse emanating from the pacemaker nodal region of the heart become shattered into a number of equal amplitude pulses. Each pulse travels a different path length to reach the myocardium and there superimposes to form the classical QRS pulse. The distribution in path lengths resulting from the fractal nature of the branches gives rise to a distribution of decorrelation times among the individual pulses impinging on the myocardium. Using an argument for the distribution of decorrelation times $p(\tau_c)$ that parallels that used for the spatial scales in the lung leads to a correlation function for the time series of the QRS complex having the scaling relation

$$C(t) = \frac{1}{a} C(bt), \quad a, b > 1 \ . \tag{2.10}$$

The solution to this functional equation is

$$C(t) = A(t) t^{\alpha} \tag{2.11}$$

Where $\alpha = \ln a / \ln b$ and again $A(t)$ is a harmonic function in $\ln t$. We can use a Tauberian Theorem to yield the power spectrum for the correlation function (2.11)

$$S(\omega) \sim \frac{1}{\omega^{\alpha+1}} \tag{2.12}$$

if $A(t)$ is a constant or only slowly varying in time. Thus the QRS complex should have an inverse power-law spectrum and we see in Figure 4 that it in fact does.

3. TOLERANCE TO FLUCTUATIONS (ERRORS)

Why is it that nature prefers such fractal structures to the more traditional scaling? A reason which suggests itself is the tolerance that fractal structures (processes) seem to posses over and above that of classical structures (processes); that is to say that fractal processes are more adaptive to internal changes and to changes in the environment than are classical ones.[8] To make this statement more quantitative, consider the theoretical average diameter of the lung as given by classical scaling, $d(z) = d(0) e^{-\alpha z}$. Let us assume that the scaling parameter α is made up of two pieces; a constant α_0 and a random piece ξ that can arise from fluctuations initially produced in the environment during morphogenesis or as errors in the code generating the structure of the lung. Thus no matter if the errors are produced internally or externally the average diameter of an airway is given by

$$\langle d(z) \rangle_{\xi} = d(0) e^{-\alpha_0 z} \langle e^{-\xi z} \rangle_{\xi} \ . \tag{3.1}$$

To evaluate the average on the right hand side of (3.1) we assume ξ is a zero-centered, Gaussian random variable with variance $2\sigma^2$. Thus (3.1) can be written as

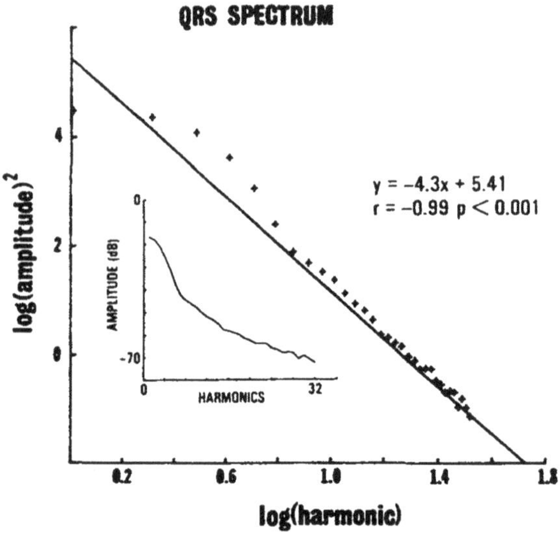

Figure 4. The power spectrum (inset) of normal ventricular depolarization (QRS) waveform (mean data of 21 healthy men) shows a broadband distribution with a long, low amplitude high frequency tail. An inverse power-law distribution is obtained by replotting the same data in log-log format. Fundamental frequency = 7.81 Hz. (Adapted from Ref. 10; copyright Biophysical Society.)

$$\langle d(z)\rangle_\xi = d(0)e^{-\alpha_0 z}e^{\sigma^2 z^2} \tag{3.2}$$

so the error grows as $\exp[\sigma^2 z^2]$. The assumed statistics for ξ have no significance except they provide a specific functional form for the error for which we can compare the classical and fractal models.

In the fractal model of the lung let us assume that the power-law index consists of a constant piece μ_0 and a random piece ξ. Here again the average diameter is averaged over ξ as follows

$$\langle d(z)\rangle_\xi = \frac{A(z)}{z^{\mu_0}}\langle e^{-\xi \ln z}\rangle_\xi \tag{3.3}$$

so that with the same statistical assumption as used to obtain (3.2) we have,

$$\langle d(z)\rangle_\xi = \frac{A(z)}{z^{\mu_0}}e^{\sigma^2(\ln z)^2} \tag{3.4}$$

The error in the fractal model therefore grows as $\exp[\sigma^2(\ln z)^2]$.

The deviation of the average diameter from that predicted by its noise-free counterpart are indicated in Figure 5. Here we see than the growth in error in the classical model

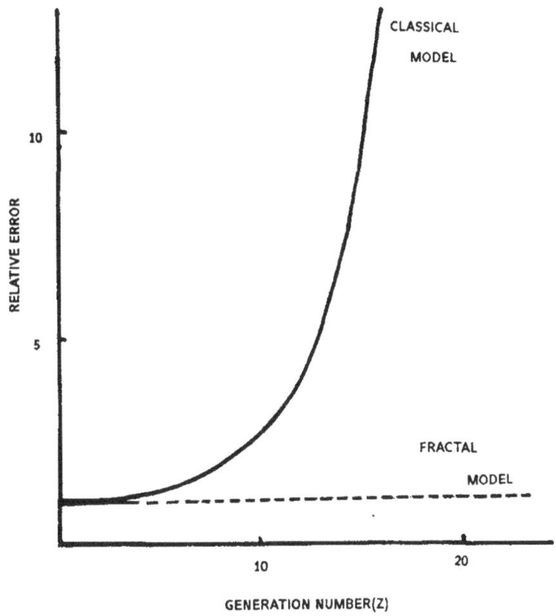

Figure 5. The relative error in the average diameter of a bronchial tube using classical scaling (—) and the fractal model (- -) is depicted as a function of the generation number z.

given by (3.2) is quite dramatic. For the parameter values chosen the average diameter of the airway at $z = 10$ is a factor of two greater than predicted by the error-free classical model, and for $z = 13$ it is already a factor of five greater. This implies that the classical model is quite sensitive to the value of the scaling parameter, and fluctuations (errors) in that parameter would be selected against by the process of evolution. On the other hand the fluctuations (errors) have negligible effect on the average dimensions of the lung in the fractal model. The reason for this latter insensitivity is that the fractal model has already incorporated such variability into its structure and is not responsive to redundancy.

4. CONCLUSIONS

From the above analysis we draw a number of conclusions:

(a) Bronchial architecture is a consequence of nature having selected for a structure with no fundamental scale of length.[7]

(b) The structure of the His-Purkinje conduction system of the heart is selected for by nature so as to not have a fundamental scale of time.[10]

(c) The modulated inverse power-law describes systems that are more error-tolerant than are those described by classical scaling.[11]

(d) The lack of favored scales promote the tolerance of variability in physiological systems and favors fractal structures and processes.[1]

ACKNOWLEDGEMENTS

This work was supported by Armstrong Aerospace Medical Research Laboratory, Human Systems Division (AFSC), United States Air Force, Wright-Patterson Air Force Base, Ohio, Contract No. F33614-87-C-538.

REFERENCES

1. West, B.J., and Goldberger, A.L., "Physiology in Fractal Dimensions," Am. Sci. *75*, 354-365 (1987).
2. Mandelbrot, B.B., *The Fractal Geometry of Nature*, W.H. Freeman (1982).
3. Lévy, P., *Theorie de L'Additions des Variable*, Aleatoires, Gautheirs, Villers, Paris (1937).
4. West, B.J., *An Essay on the Importance of Being Nonlinear*, Lect. Notes in Biomath. *62*, ed. S. Levine, Springer-Verlag (1985).
5. Montroll, E.W. and West, B.J., "An Enriched Collection of Stochastic Processes," in *Fluctuation Phenomena 2nd Ed.*, eds. E.W. Montroll and J. Lebowitz, North-Holland, Amsterdam, (1987).
6. Weibel, E.R., and Gomez, D.M., "Architecture of the Human Lung," *Science, 137*, 577-585 (1962).
7. West, B.J., Bhargava, V., and Goldberger, A.L., "Beyond the Principle of Similitude: Renormalization in the Bronchial Tree," *J. Appl. Physiol., 260*, 189-197 (1985).
8. West, B.J., "Fractals, Intermittences and Morphogenesis," in *Chaos in Biological Systems*, ed. A. Holton, NATO (1987).

9. Montroll, E.W., and Shlesinger, M.F., "On 1/f Noise and Distributions with Long Tails," PNAS *79*, 3380-3383 (1982).

10. Goldberger, A.L., Bhargava, V., West, B.J., and , A.J., "On a Mechanism of a Cardiac Electrical Stability: The Fractal Hypothesis," Biophys. J. *48*, 525-528 (1985).

11. West, B.J., and Salk, J., "Complexity, Organization and Uncertainty," Eur. J. Op. Res. *30*, 117-128 (1987).

12. Raabe, O.G., Yeh, H.D., Schum, G.M., and Phalen, R.F., *Tracheobronchial Geometry: Human, Dog, Rat, Hamster,* Albuquerque; Lovelace Foundation for Medical Education and Research (1976).

SUDDEN DEATH IS NOT CHAOS

Ary L. Goldberger

David R. Rigney

Cardiovascular Division, Department of Medicine
Beth Israel Hospital and Harvard Medical School
Boston, MA 02215

ABSTRACT

Sudden death due to cardiac arrhythmia is a major cause of mortality among Americans. A widely held view is that immediately before sudden death, the heart has chaotic dynamics. Work from this laboratory and others, however, supports a different view, namely, that the process leading to sudden death is often relatively periodic, not chaotic, and it is characterized by a narrow frequency spectrum. Patients at high risk of sudden death show periodic trajectories in the heartbeat phase space with occasional bifurcations that may be of diagnostic and prognostic value. In contrast, the electrical activity of the heart under healthy conditions appears to be fractal since normal heartrate has a 1/f-like spectrum. In fact, two-dimensional phase space plots of interbeat interval trajectories in healthy subjects suggest the presence of a strange attractor in a noisy environment.

1. INTRODUCTION: THE CHAOS DEBATE IN PHYSIOLOGY

The nonlinear revolution (1) in the physiological sciences has undergone two phases in its short lifespan. The first was heralded by a sudden upsurge of publications demonstrating that nonlinear models were essential for understanding complex physiological phenomena (2). The concepts of this first phase are only now beginning to percolate into the medical sciences, which had overlooked abrupt dynamical transitions, sustained oscillations, and chaotic behavior since they are not easily explained by traditional reductionist methods of analysis.

A controversy arose shortly after initial appreciation of the significance of nonlinear concepts in physiology and this dispute is the subject of the present paper (Table 1, refs. 3-8). In general, the debate centers on the question: "Is chaos health or disease?" In cardiology, this question can be rephrased more specifically: "Is sudden death characterized by a chaotic or a relatively periodic type of dynamics?"

Table 1.

THE ROUTE TO SUDDEN DEATH: TWO CONFLICTING MODELS

1. Old Model
   periodic, regular healthy dynamics → chaotic arrhythmias

2. New Model
   chaotic, fractal healthy dynamics → periodic arrhythmias

Table 2.

EVIDENCE THAT SUDDEN DEATH IS NOT A CHAOTIC PROCESS

1. Narrowband spectrum of body surface and epicardial electrocardiographic waveforms during ventricular fibrillation (Fig. 6) (6,19,20).

2. Organized waveforms of ventricular activation recorded from the outer surface of the heart at onset of fibrillation (21).

3. Periodic electrical activity recorded from the innermost surface of ventricles during fibrillation (22).

4. Periodic electrical activity seen during torsades de pointes (Fig. 3) (11).

5. Periodic sinus rhythm and ectopic heart beat dynamics and loss of physiologic heart rate variability in patients at high risk of sudden death (Figs. 8 and 9) (29,31-33).

2. SUDDEN DEATH: DEMOGRAPHICS AND DYNAMICS

The question is of more than theoretical interest since over 400,000 Americans die suddenly each year. Sudden cardiac death is the leading cause of mortality among American men aged 20-60 years, killing about one person every minute. Therefore, understanding the dynamics of this process is a major public health priority. The most common proximal cause of sudden death is a class of abnormal heart rhythms called ventricular arrhythmias. These electrical disturbances include: ventricular tachycardias (Figs. 1-3), ventricular flutter (Fig. 4), and ventricular fibrillation (Fig. 1). The conventional model is that sudden death, and in particular, ventricular fibrillation, has chaotic dynamics in the nonlinear sense of the word (5,9). Ruelle, in a frequently quoted article (10), cited fibrillation as the prime physiological example of a strange attractor. The most explicit recent formulation of the chaos theory of sudden death has been by Smith and Cohen (5) who modeled fibrillation as a chaotic process that followed a Feigenbaum-type period-doubling sequence. Their computer model was a descendant of an earlier analysis by Moe and coworkers (9) describing fibrillation as having turbulent dynamics that could be characterized by a Reynolds number.

Support for this notion of sudden death as a chaotic process came from several lines of evidence. The first was intuition and observation. The hand-held fibrillating heart with its apparently unpredictable gyrations certainly gives the appearance of chaos in the vernacular sense. This impression is supported by the electrocardiographic waveforms recorded from the body surface of fibrillating animals and man. These oscillations also appear, at first glance, to represent completely erratic fluctuations. Furthermore, computer simulations of reentrant-like pathways, the mechanism thought to underlie fibrillation, yielded irregular oscillations that were thought to be analogous to actual fibrillation (5,9). This conventional and historically-rooted theory of fibrillatory chaos continues to dominate present day discussions of the dynamics of cardiac death; but not without challenge.

3. THE DYING HEART: A NEW PERSPECTIVE

Over the past several years, we (6,7) have contested the notion that the dying heart is best described as a chaotic attractor. Instead, we proposed the opposite view of cardiovascular dynamics summarized by two propositions:

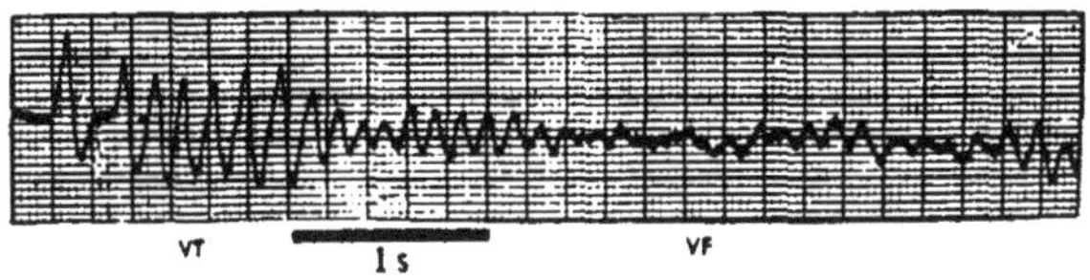

Fig 1. Ventricular tachycardia (VT) bifurcates to ventricular fibrillation (VF) in this electrocardiographic recording made during a cardiac arrest.

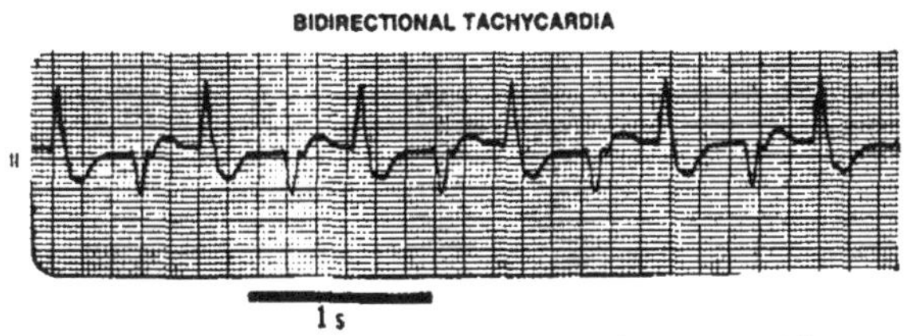

Fig 2. Bidirectional ventricular tachycardia is characterized by ventricular premature beats that alternate in their vectorial orientation. This relatively rare <u>bimorphic</u> arrhythmia contrasts with the common type of <u>monomorphic</u> ventricular tachycardia shown in Fig. 1 and with the <u>polymorphic</u> arrhythmia called torsades de pointes (Fig. 3). Bidirectional ventricular tachycardia represents a subharmonic bifurcation phenomenon. This patient's arrhythmia was due to digitalis toxicity and low serum potassium concentration.

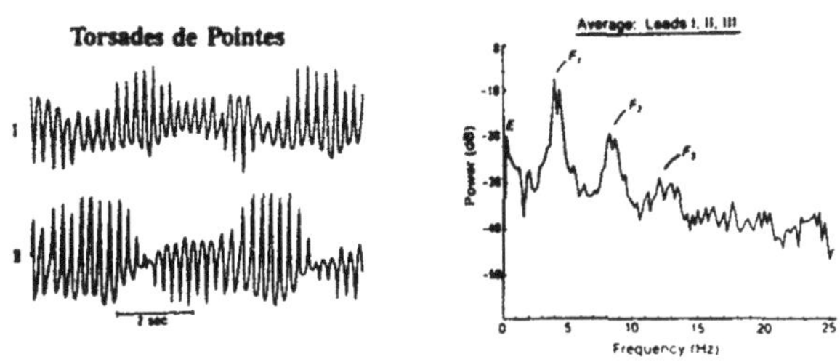

Fig 3. Torsades de pointes type of ventricular tachycardia (left panel) is characterized by relatively periodic oscillation of the ventricular depolarization vector giving the surface electrocardiogram a spindle-like appearance. Spectrum (right panel) shows spikes corresponding to the low "envelope" (E) frequency (repetition rate of the spindles) and higher frequencies corresponding to the actual heartrate (F1) and its harmonics (F2, F3, etc.). This arrhythmia is an important mechanism of sudden death (11). (Adapted from ref. 18.)

1. Ventricular fibrillation and related tachyarrhythmias causing sudden death are relatively periodic processes, not cardiac chaos (Table 2) (6,11).

2. Healthy cardiac dynamics are chaotic (Table 3) (4,6,7,12).

Sudden death, therefore, may be viewed as a bifurcation out of, not into chaos. Furthermore, period-doubling behavior, although common in perturbed cardiovascular dynamics (Figs. 2 and 5; Table 4; refs. 13-18) does not appear to follow the classical route to chaos as proposed by Smith and Cohen (5) based on their computer simulations. Instead, the term inverse bifurcation (13,18) was proposed to describe this transition from healthy chaotic dynamics to pathologic periodic dynamics associated with sudden cardiac death and to distinguish this transition from the classical bifurcation sequence that proceeds from periodicity to chaos in physical systems. (In using this term, however, we do not imply that all cardiac arrhythmias are periodic or that all pathologic states that may intervene between health and sudden death are necessarily marked by increasing regularity.)

These counterintuitive concepts---that health is a strange attractor and that sudden death is a periodic syndrome---are supported by the following lines of evidence (Tables 2 and 3). Spectral analysis (6,19,20) of fibrillatory waveforms from animals and humans has consistently revealed narrowband, not broadband spectra (Fig. 6). In a complementary finding, electrophysiological recordings from both the inner and outer surfaces of the fibrillating heart also demonstrated unexpectedly periodic behavior (21,22). In contrast, normal cardiac function is characterized by fractal structure-function relationships. For example, the normal activation sequence involves depolarization of a tree-like branching network, the His-Purkinje system. The output of the system is the normal ventricular depolarization waveform (QRS complex) on the surface ECG. Spectral analysis of individual QRS waveforms reveals a 1/f-like spectrum consistent with activation of a fractal depolarization network (23). Further, not only is each individual heartbeat dependent on a fractal mechanism of ventricular depolarization, but the neurohumoral clock that regulates beat-to-beat heartrate variability also has no characteristic time scale and appears to follow a chaotic or fractal rhythm (Fig. 7) (24,25).

Table 3.

MARKERS OF CARDIAC CHAOS IN HEALTHY INDIVIDUALS

1. Strange-like attractors seen in phase space plots
   Example: normal sinus rhythm in healthy subjects (Figs. 10,11)

2. 1/f-like spectra
   Examples: spectra of normal QRS complexes (23)
   spectra of healthy heartrate variability (23,25,26)

3. Fractal structures (12,24,25)
   Examples: coronary arteries
   His-Purkinje system
   chordae tendineae
   tracheobronchial-pulmonary arteriovenous interface

Table 4.

PERIOD-DOUBLING SEEN IN PATHOLOGIC CARDIAC DYNAMICS

Electrical alternans in pericardial tamponade (16)
Sick Sinus Syndrome (17)
Bidirectional ventricular tachycardia (Fig. 2)
ST segment-T wave alternans preceding ventricular fibrillation (5,6)
T-U wave alternans preceding torsades de pointes (12)
QRS and R-R alternans with atrio-ventricular reentrant tachycardia in Wolff-Parkinson-White syndrome (Fig. 4)
Pulsus alternans in heart failure and cardiac tamponade (16)
Catecholamine-induced arrhythmias in animals (15)
Electrically perturbed chick heart embryo cells (14)

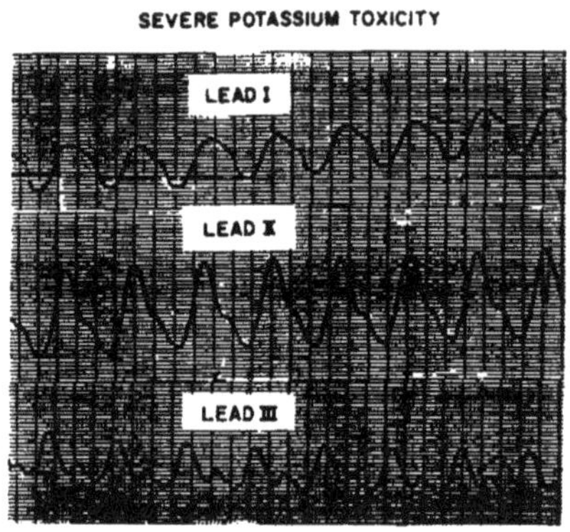

Fig 4. Slow ventricular flutter-like rhythm due to extreme elevation of extracellular potassium (hyperkalemia). Note that the oscillations are regular but not purely sinusoidal, which is consistent with a periodic nonlinear wave mechanism.

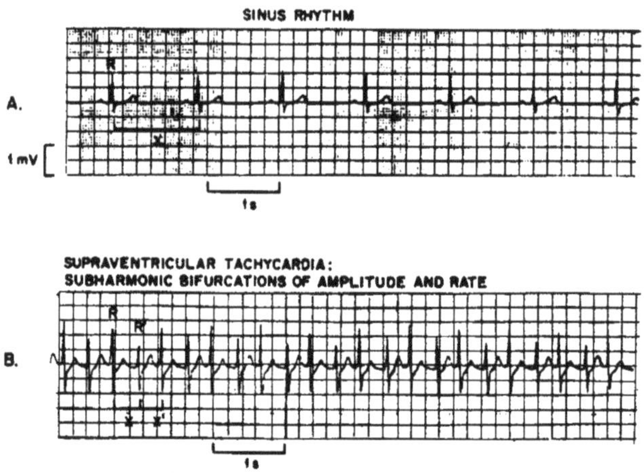

Fig 5. Subharmonic bifurcations of the QRS complex and interbeat interval in a patient with probable atrio-ventricular reentrant tachycardia involving a concealed bypass tract (Wolff-Parkinson-White variant). Note periodic alternation of the interbeat interval between longer (x=380 ms) and shorter (x'=320 ms) values, as well as beat-to-beat oscillation of the amplitude of the QRS. This type of *electrical alternans* phenomenon is not normal (top panel). Subharmonic bifurcations are common in other examples of perturbed cardiac dynamics (Fig. 2; Table 4).

4. HEALTHY CHAOS, CHAOTIC HEALTH

A common misconception among physicists and physiologists is that the normal heartbeat is metronomically regular. This mistake has been perpetuated by clinicians who use the term "regular sinus rhythm" to describe the normal pulse of their healthy patients. However, most cardiologists and not a few dynamicists are surprised to find that the normal heartbeat is not perfectly regular. In actuality, it fluctuates in a highly erratic fashion (Fig. 7) and its spectrum is 1/f-like (Fig. 8) with superimposed spikes corresponding to physiological oscillations associated with respiration, baroreceptor control, thermoregulation and the renin-angiotensin system (24-27). This type of broadband spectrum with superimposed peaks is reminiscent of the "noisy periodicity" described by Lorenz in chaotic attractors (28).

What is the mechanism of this 1/f-like spectrum of heartrate variability? At present the details of the nonlinear control system that is responsible for it are uncertain. We (7,23-25,29) have proposed that physiological heart-rate variability is governed by a regulatory process leading to self-similar fluctuations in interbeat intervals across multiple scales of time (Fig. 7). These fractal dynamics involve the complex interaction of parasympathetic and sympathetic branches of the autonomic nervous system impinging on the sinus node (25). The push-pull interplay of neurohumoral inputs leads to a stable broadband state far from equilibrium during normal sinus rhythm.

To avoid potential confusion, we would like to emphasize the distinction between the generation of a single ventricular depolarization waveform (QRS complex) by the fractal His-Purkinje system on the one hand (23), and the organization of sinus rhythm interbeat interval variability by a neurohumoral fractal feedback system, on the other. In the first case, that of the single QRS complex, a direct link can be made between the fractal structure (His-Purkinje conduction tree) and fractal function (depolarization pulses with a 1/f-like spectrum) (23). In the second case, that of heartrate variability, there is no evident physical fractal structure that underlies the healthy variability of sinus rhythm (25). According to this formulation, there is chaos both <u>within</u> the heart (fractal conduction system) and, on a larger scale, chaos <u>outside</u> the heart that

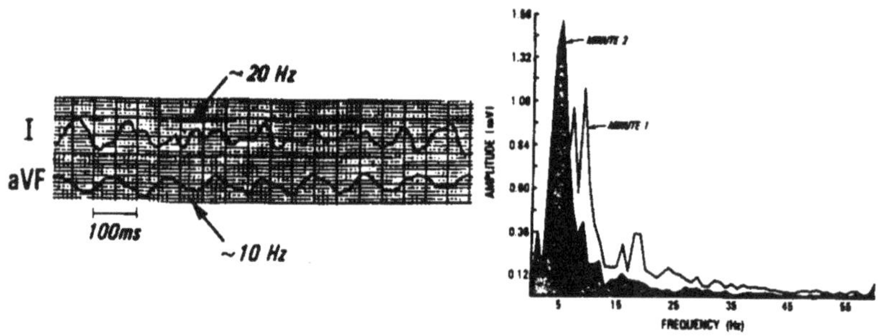

Fig 6. Ventricular fibrillation (canine experiment) looks "chaotic" at first glance (Fig.1). However, closer inspection reveals periodic oscillations (left panel), and the spectrum (right panel) of the epicardial or body surface electrocardiogram shows a narrowband pattern during the first minute of fibrillation with even further narrowing during the second minute. (Adapted from ref. 6.)

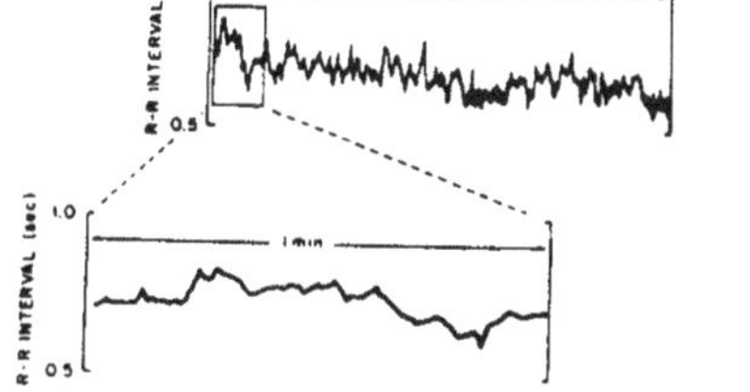

Fig 7. The fractal nature of normal heartrate variability is revealed by examining interbeat interval time series with varying temporal resolution. Fluctuations are seen over multiple orders of temporal magnitude, a feature characteristic of chaotic processes since they lack a characteristic time scale. (Adapted from ref. 23.)

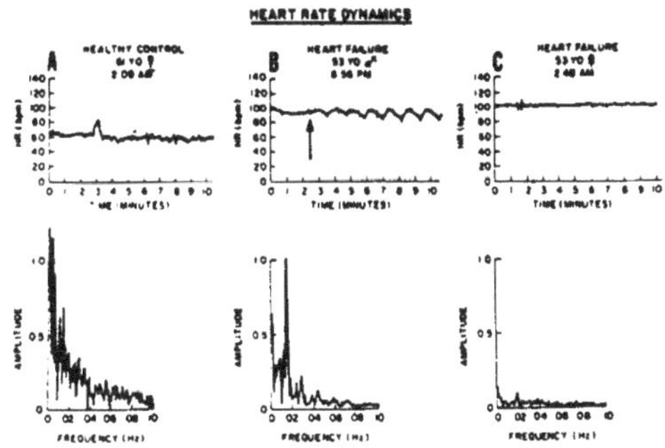

Fig 8. Healthy subjects (A) show an erratic beat-to-beat heart rate pattern, not a metronomically regular pulse rate. The spectrum is 1/f-like. Patients with heart failure, a group at particularly high risk of sudden death, characteristically show either low frequency (.01-.04 Hz) oscillations (B) or a marked overall reduction in beat-to-beat heart rate variability with a flat spectrum up to .1 Hz (C). Note the abrupt onset of oscillations (arrow) in panel B. (Adapted from ref. 12.)

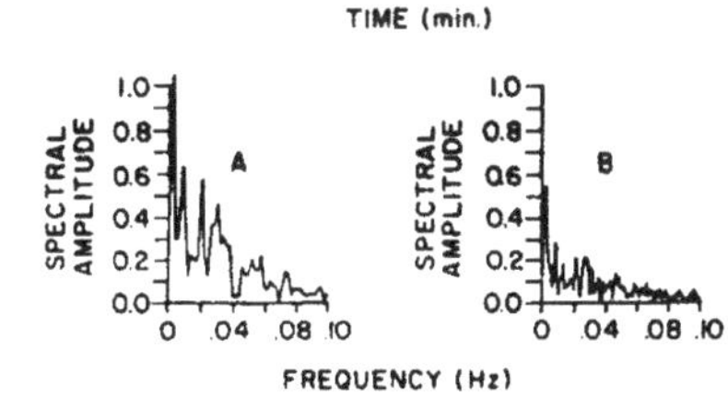

Fig 9. Heartrate time series and corresponding amplitude spectra from the ambulatory electrocardiographic record of a patient who died suddenly. Low frequency interbeat interval oscillations (A) were followed by an abrupt change to a different frequency pattern (B) followed a few minutes later by ventricular tachyarrhythmias and cardiac arrest.

tethers the heartbeat to the fractal dynamics of the nervous system. But there is no direct link between the two types of fluctuations.

5. LOSING CHAOS: PATHOLOGIC PERIODICITIES

The chaos of normal heartrate variability contrasts with the loss of variability seen in several pathologic conditions, including diabetic neuropathy (30), multiple sclerosis (30), fetal distress (31) and patients susceptible to sudden death (29,32,33). We recently identified two patterns of abnormal sinus rhythm heart rate dynamics in patients at increased risk of sudden cardiac death (Fig. 8) (29). One we have labeled the flat pattern because it is characterized by a virtual absence of the normal beat-to-beat variability. The second is a regular oscillatory pattern characterized by relatively low frequency (.01-.04 Hz) heartrate oscillations that may accompany the oscillations in breathing rate and amplitude referred to as Cheyne-Stokes respiration (4). These cardiopulmonary oscillations are observed to start and stop abruptly (Figs. 8 and 9) which is diagnostic of perturbed nonlinear systems. In addition to sinus rhythm heart rate oscillations, low frequency oscillation of ectopic beats is observed in some of these patients (29,34). Our preliminary data suggest that these pathologic heartrate dynamics---loss of variability and low frequency oscillations--- are sensitive markers of increased susceptibility to sudden death (29).

6. PHASE SPACE PLOTS OF HEARTRATE:

Topological features of normal and pathologic heartrate time series may be demonstrated using the type of phase space plots that have been used to characterize other nonlinear systems (35,36). Such plots represent a time series as a vector, elements of which correspond to the time-series separated by equally spaced time delays. Figures 10-13 depict representative two dimensional phase space trajectories of the heartbeat in two healthy subjects and two patients who had cardiac arrests shortly after the recordings. The original time series and Fourier spectra of the heart rates for both subjects are also shown. Note the erratic beat-to-beat heartrate variability in the healthy individuals, characterized by a 1/f-like spectrum. The phase space trajectories of the normal interbeat interval follow a complex pattern that could be a strange attractor in the

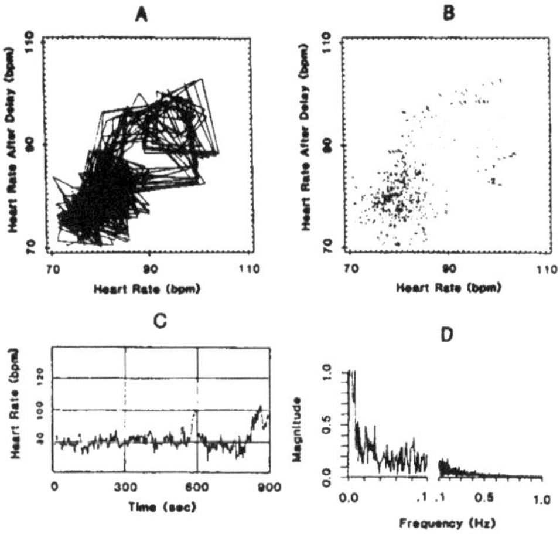

Fig 10. Two-dimensional phase space representation of 900 seconds of a normal individual's heartrate (62 year-old female). Panel A shows the trajectory of the heartrate vector where the first variable is the current heartrate, and the second variable is the heartrate after a fixed delay of 12 seconds. Panel B shows the same data as Panel A, except that for clarity, the trajectory at each point is shortened to be proportional to the distance that the trajectory travels after an evolution time of 12 seconds. Panel C shows the original heartrate data from which the phase space plots were calculated, and Panel D shows the amplitude spectrum of the time series.

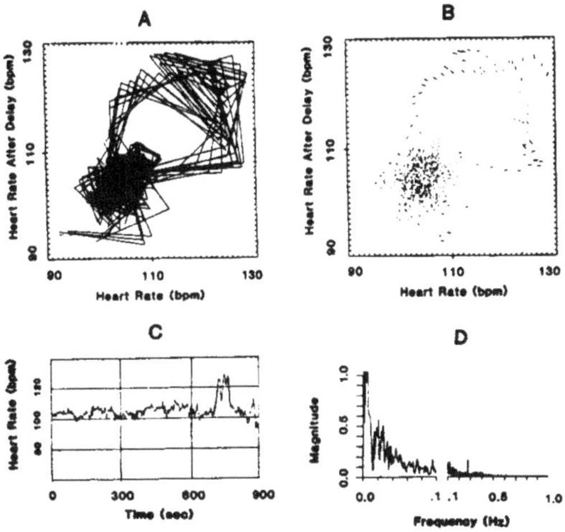

Fig 11. Two-dimensional phase space representation of 900 seconds of another normal individual's sinus rhythm heartrate (35 year-old male). See Fig. 10 legend for details. Same parameters are used in Figs. 10-13. Note the similarity of the strange-like "arachnoid" attractors in the two normal subjects as compared with the torus-like orbits of the patients with sudden death syndrome (Figs. 12 and 13).

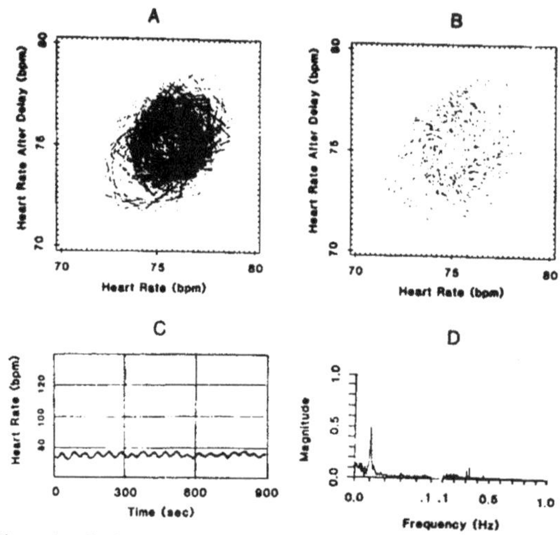

Fig 12. Two-dimensional phase space representation of 900 seconds of sinus rhythm heartrate data from a patient who sustained a cardiac arrest due to ventricular tachyarrhythmias about 2 hours later. Compare with Figs. 10,11 and 13.

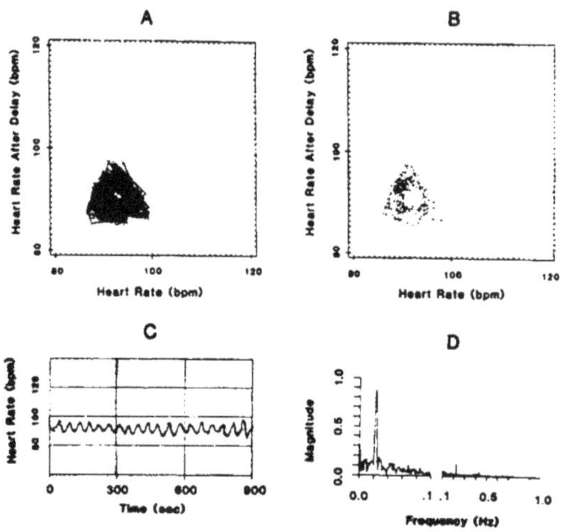

Fig 13. Two-dimensional phase space representation of 900 seconds of sinus rhythm heartrate data from a patient who died suddenly 8 days after this recording. Compare with Figs. 10,11 and 12.

presence of noise. The trajectories appear to be driven toward a dense central region, leave the center along irregular paths, and then return to the center. The chaotic character of the trajectories is seen in the central region around the "average" heartrate where behavior becomes unpredictable and trajectories diverge.

In contrast, in the patients with heart disease and impending cardiac arrest, the sinus rhythm beats appear highly ordered. The time series shows a relatively regular oscillatory pattern having a narrowband spectrum with a sharp peak at about .02 Hz and the phase space trajectories of these heartrate vectors follow torus-like orbits. Unlike the healthy sinus rhythm phase plots in which trajectories converge toward a central node and then diverge outward, the pathologic sinus rhythm trajectories are center-avoiding and tend to follow relatively periodic elliptical or circular orbits.

7. WHY ARE HEALTHY DYNAMICS CHAOTIC?

Chaotic dynamics appear to underlie the variability and adaptability necessary for responding to a fluctuating environment. Severely perturbed dynamics, in contrast, may lead to fatal outcomes if the organism progresses from healthy chaos (Figs. 10 and 11) to pathologic regularity (Figs. 12 and 13) (4,6,7,12,18,24,25).

The practice of medicine would be impossible without such periodic syndromes. It is, to a large extent, the periodicities and patterns, the loss of chaos, in pathology that allow physicians to identify and classify many features of the abnormal appearance and behavior of their patients, including their heartbeat records (bigeminy, trigeminy, Wenckebach, torsades, electrical alternans, etc.). Health with its broadband spectrum and strange attractor dynamics is, necessarily, much harder to classify. Patients with severely pathologic dynamics, in contrast, are remarkably similar. Periodic behavior in disease is likely to be more the rule (12) than the exception (37). Indeed, the more extreme the pathology (e.g. impending sudden death) the more marked the periodicities often become (Figs. 1-6,8,12,13).

This formulation is not intended to imply that all aspects of disease are periodic or that the inverse bifurcation from health to sudden death is always characterized by a progressive increase in periodicity. The apparently random heartrate fluctuations in atrial fibrillation serve as an obvious counterexample to the notion of periodic dynamics in disease. However, the type of pathologic randomness seen in the ventricular response to atrial fibrillation is reminiscent of white noise (38) and is probably not chaotic in the technical sense. Such examples of aperiodic pathologies do not contradict the dynamical theory of fractal health and periodic behavior in the sudden death syndromes discussed here.

8. SUMMARY

We presented evidence supporting the counterintuitive theory that cardiac arrest follows a progression, often involving bifurcations, from the fractal, chaotic dynamics of the normal heartbeat to the pathologic periodicities and loss of variability of the dying heart. This schema is at odds with the conventional notion of sudden death as chaos and health as a periodic, regular state. Loss of physiologic sinus rhythm heartrate variability and sometimes the emergence of highly periodic oscillations may be sensitive markers of increased risk of sudden death and may precede the appearance of potentially fatal cardiac arrhythmias (Figs. 8 and 9). An analogous loss of physiologic chaos appears to be a common feature of pathologic dynamics in other organ systems as well (38-40).

Acknowledgements: This work was supported by a grant from the Whitaker Health Sciences Fund. The authors acknowledge the technical assistance of Joseph Mietus and the secretarial skills of Robyn Hickox.

REFERENCES

1. Gleick, J., Chaos: Making a New Science. New York, Viking Press, 1987.
2. Koslow, S.H., Mandell, A.J. and Shlesinger, M.F., eds. Perspectives in Biological Dynamics and Theoretical Medicine. Ann. N.Y. Acad. Sci. 504, 1987.
3. Mackey, M.C. and Glass, L., Oscillation and chaos in physiological control systems. Science 197:286, 1977.
4. Goldberger, A.L., Findley, L.J., Blackburn, M.R. and Mandell, A.J., Nonlinear dynamics in heart failure: Implications of long-wavelength cardiopulmonary oscillations. Am. Heart J. 107:612, 1984.

5. Smith, J.M. and Cohen, R.J., Simple finite-element model accounts for wide range of cardiac dysrhythmias. Proc. Natl. Acad. Sci. U.S.A. 81:233, 1984.
6. Goldberger, A.L., Bhargava, V., West, B.J. and Mandell, A.J., Some observations on the question: Is ventricular fibrillation "chaos"? Physica 19D:282, 1986.
7. Goldberger, A.L and West, B.J., Chaos in physiology: health or disease? In Chaos in Biological Systems, eds. Degn, H., Holden, A.V., and Olsen, L.F. New York, Plenum Press, 1987, p.1.
8. Glass, L., Goldberger, A.L., Courtemanche, M. and Shrier, A., Nonlinear dynamics, chaos and complex cardiac arrhythmias. Phil. Trans. Roy. Soc. A, 1987. In press.
9. Moe, G.R., Reinboldt, W.C. and Abildskov, J.A., A computer model of atrial fibrillation. Am. Heart J. 67:200, 1964.
10. Ruelle, D., Strange attractors. Math. Intelligencer 2:126, 1980.
11. Bhargava, V., Goldberger, A.L., Ward, D. and Ahnve, S., Torsades de pointes: a characteristic spectral pattern in sudden death. IEEE Trans. Biomed. Eng. 33:894, 1986.
12. Goldberger, A.L., Nonlinear dynamics, fractals, cardiac physiology and sudden death. In Temporal Disorder in Human Oscillatory Systems. Rensing, L., An der Heiden, U. and Mackey, M.C., New York, Springer, 1987, p. 118.
13. Schwartz, P.J. and Malliani, A., Electrical alternation of the T-wave: Clinical and experimental evidence of its relationship with the sympathetic nervous system and with the long Q-T syndrome. Am. Heart J. 89:45, 1975.
14. Guevara, M.R., Glass, L. and Shrier, A., Phase locking, period-doubling bifurcations, and irregular dynamics in periodically stimulated cardiac cells. Science 214:1350, 1981.
15. Ritzenberg, A.L., Adam, D.R. and Cohen, R.J., Period multupling-evidence for nonlinear behaviour of the canine heart. Nature 307:159, 1984.
16. Goldberger, A.L., Shabetai, R., Bhargava, V., West, B.J. and Mandell, A.J., Nonlinear dynamics, electrical alternans and pericardial tamponade. Am. Heart J. 107:1297, 1984.
17. Goldberger, A.L., Bhargava, V., West, B.J. and Mandell, A.J., Nonlinear dyanmics of the heartbeat II. Subharmonic bifurcations of the cardiac interbeat interval in sinus node disease. Physica 17D:207, 1985.
18. Goldberger, A.L. and West, B.J., Applications of nonlinear dynamics to clinical cardiology. Ann. N.Y. Acad. Sci. 504:195, 1987.
19. Battersby, E.J., Pacemaker periodicity in atrial fibrillation. Circ. Res. 17:296, 1965.
20. Herbschleb, J.N., Heethaar, R.M., van der Tweel, I., et al., Signal analysis of ventricular fibrillation. In Computers in Cardiology. Long Beach, CA, IEEE Computer Society, 1978, p. 245.
21. Ideker, R.E., Klein, G.J., Harrison, L., et al., The transition to ventricular fibrillation induced by reperfusion after acute ischemia in the dog: a period of organized epicardial activation. Circulation 63:1371, 1981.
22. Worley, S.J., Swain, J.L., Colavita, P.G., et al., Development of an endocardial-epicardial gradient of activation rate during electrically induced, sustained ventricular fibrillation. Am. J. Cardiol. 55:813, 1985.
23. Goldberger, A.L., Bhargava, V., West, B.J. and Mandell, A.J., On a mechanism of cardiac stability: the fractal hypothesis. Biophys. J. 48:525, 1985.

24. West, B.J. and Goldberger, A.L., Physiology in fractal dimensions. Am. Scientist 75:354, 1987.
25. Goldberger, A.L. and West, B.J., Fractals in physiology and medicine. Yale J. Biol. Med. 60:421, 1987.
26. Kobayashi, M. and Musha, T., 1/f fluctuation of heartbeat period. IEEE Trans. Biomed. Eng. 29:456, 1982.
27. Axelrod, S., Gordon, D., Ubel, F.A., et al., Power spectrum analysis of heartrate fluctuation: a quantitative probe of beat-to-beat cardiovascular control. Science 213:220, 1981.
28. Lorenz, E.N., Noisy periodicity and reverse bifurcation. Ann. N.Y. Acad. Sci. 357:282, 1980.
29. Goldberger, A.L. and Rigney, D.R., Defending against sudden death: Fractal mechanisms of cardiac stability. Proceedings of the Ninth Annual Conference of the IEEE Engineering in Medicine and Biology Society. 1:313, 1987.
30. Neubauer, B. and Gunderson, H.J.G., Analysis of heartrate variations in patients with multiple sclerosis. A simple measure of autonomic disturbances using an ordinary ECG. J. Neurosurg. Psychiatry 41:417, 1978.
31. Modanlou, H.D. and Freeman, R.K., Sinusoidal fetal heart rate pattern: its definition and significance. Am. J. Obstet. Gynecol. 142:1033, 1982.
32. Myers, G.A., Martin, G.J. and Magid, N.M. Power spectral analysis of heart rate variability in sudden cardiac death: Comparison with other methods. IEEE Trans. Biomed. Eng. 33:1149, 1986.
33. Klieger, R.E., Miller, J.P., Bigger, J.T., Jr., et al., Decreased heart rate variability and its association with increased mortality after acute myocardial infarction. Am. J. Cardiol. 59:256, 1987.
34. Findley, L.T., Blackburn, M.R., Goldberger, A.L. and Mandell, A.J., Apneas and oscillation of cardiac ectopy in Cheyne-Stokes breathing during sleep. Am. Rev. Respir. Dis. 130:937, 1984.
35. Grassberger, P. and Procaccia, I., Characterization of strange attractors. Phys. Rev. Lett. 50:346, 1983.
36. Wolf, A., Swift, J.B., Swinney, H.L. and Vastano, J.A., Determining Lyapunov exponents from a time series. Physica 16D:285, 1987.
37. Reimann, H.A., Periodic Diseases. Philadelphia, F.A. Davis, 1963.
38. Goldberger, A.L., West, B.J. and Bhargava, V. Nonlinear dynamics in physiology and pathophysiology. Toward a dynamical theory of health and disease. Proceedings of the 11th International Association for Mathematics and Computers in Simulation World Congress, Oslo, Norway. Wahlstrom, B., Henrickson, R. and Sundby, N.P., eds. Amsterdam, North Holland Publishing Co., 1985, vol. 2, p. 239.
39. Goldberger, A.L., Kobalter, K. and Bhargava, V. 1/f-like scaling in normal neutrophil dynamics. Implications for hematologic monitoring. IEEE Trans. Biomed. Eng. 33:874, 1986.
40. Babloyantz, A. and Destexhe, A., Low-dimensional chaos in an instance of epilepsy. Proc. Natl. Acad. Sci. U.S.A. 83:3513, 1986.

Predicting Chaotic Dynamics

J. Doyne Farmer and John J. Sidorowich[1]
Theoretical Division and Center for Nonlinear Studies
Los Alamos National Laboratory
Los Alamos, NM 87545.

Abstract

We discuss new approaches to forecasting and noise reduction. When random behavior is caused by low dimensional chaos our methods make predictions that can be several orders of magnitude better than those of standard linear forecasting techniques. We give error estimates for the accuracy of our methods, in terms of the attractor dimension and entropy, the number of data points, and the extrapolation time. These estimates demonstrate that it is better to iterate a short-term estimate than to recalculate long-term estimate. We also discuss a technique that under favorable conditions can reduce noise by several orders of magnitude.

Our methods provide an example of an alternative approach to adaptive networks, and can be applied to some of the problems currently addressed by neural networks.

1 Introduction

The great promise of chaos lies in the hope that randomness might become predictable. Although chaotic dynamics puts limits on long term prediction, it implies predictability over the short term. Applications of modern nonlinear data analysis techniques indicate that chaotic dynamics is quite common, and that in many cases random behavior is due to low dimensional chaos rather than complicated dynamics involving many irreducible degrees of freedom. Until recently, however, there has been no way to exploit the presence of low dimensional chaos to actually make predictions.

Most forecasting is currently done with linear methods. Linear dynamics cannot produce chaos, and linear models cannot produce good forecasts for chaotic time series. While nonlinear forecasting is an active field of investigation [1,2,3,4,5], as far as we know, the word "chaos" is not mentioned in this literature. In this paper we re-examine forecasting problems in terms of the new paradigm that chaos offers, amplifying the results of a previous letter [6]. In this letter we demonstrated that local approximation can be used to build forecasting models for dynamical systems such as the Mackey-Glass differential delay equation, and a time series produced by

---

[1]Permanent address: Physics Department, UC Santa Crus 95064

a convection experiment [7]. For short times our forecasts were roughly 50 times as good as those made using a standard linear autoregressive forecasting model.

In this paper we present other results that we did not have the space to report in [6]. We apologize if some of our remarks are speculative; given the current interest in these problems, we have decided to report work in progress along with completed work. A more thorough description, including omitted derivations and more extensive numerical results, will be published elsewhere [8].

1.1 Chaos and randomness

Chaos [9,10,11,12] has caused a fundamental change in the way we think about randomness. This influence is felt strongly in the nature of models for random phenomena. For example, a model for "excess noise" in Josephson junctions [13] has been formulated in terms of a simple system of differential equations, very different than the kind of model that arises by thinking in statistical terms. Similarly, in this paper we show how chaos leads to a new approach to the forecasting problem.

Chaos changes how one thinks about a randomness. Until recently it was usually assumed that randomness was caused by extreme complication, *i.e.* the presence of many irreducible degrees of freedom. This naturally led to notions of probability, which Kolmogorov [14] merged with dynamics in his theory of random processes. He defined a random process in terms of its probability distribution $\mathcal{P}$. For a series $\{x_t\}$, the dth order distribution is

$$\mathcal{P}(\xi_1,\ldots,\xi_d) = Probability\{x_1 < \xi_1,\ldots,x_d < \xi_d\}. \tag{1}$$

$\{x_t\}$ can represent events at discrete times, or samples of a continuous function. This can also be discussed in terms of the probability density function,

$$\int_0^{\xi_1}\cdots\int_0^{\xi_d} P(x_1,\ldots,x_d)dx_1\ldots dx_d = \mathcal{P}(\xi_1,\ldots,\xi_d). \tag{2}$$

This is *deterministic* if there exists some d such that the probability density approaches a delta function in the limit where we have perfect measurements of $\{x_i\}$.

It is common to speak of random processes as though they were a fundamental *source* of randomness. This is misleading. It is important to keep in mind that the theory of random processes is nothing but an empirical technique for coping with inadequate information. It makes no statements about the *causes* of randomness. As far as we know, the only truly fundamental source of randomness is the uncertainty principle of quantum mechanics; everything else is deterministic, at least in principle. Most interesting random phenomena, such as fluid turbulence or economics, derive their apparent randomness from causes other than quantum mechanics.

It is probably true that many phenomena seem random simply because of complication. Nonetheless, as originally pointed out by Poincaré [15], many of the classic examples of randomness are not complicated. The dynamics of a flipping coin or a roulette ball, for example, involve only a few degrees of freedom. Their randomness comes from ***sensitive dependence on initial conditions*** – a small perturbation causes a much large effect at a later time. When sensitive dependence on initial conditions occurs in a sustained way it is called *chaos*[2]. Since chaos is defined in the context of deterministic dynamics, in some strict sense it might be incorrect to say that chaos is random – ultimately uncertainty originates from something external to the dynamics, such as measurement error or external "noise". But sensitive dependence exaggerates uncertainty, so that small uncertainties become large ones. Since chaos amplifies noise exponentially any uncertainty at all is amplified to macroscopic proportions in finite time. Short-term determinism becomes long-term randomness.

Up until now, we have avoided saying what we mean by "random". This word means different things to different people. Some would say that something is random if it is not deterministic. This leads to the empty conclusion that, strictly speaking, quantum mechanical measurements are the only random events. Usually, however, this is a matter of practice rather than principle. A more practical viewpoint is that randomness occurs to the extent that something cannot be predicted, which usually depends on our information. Randomness is in the eye of the beholder[3].

With this point of view, chaos *creates* randomness by strongly amplifying what we don't know. Even with only a few degrees of freedom and a very small source of uncertainty, points in a chaotic time series that are far apart in time cannot be deterministically related, unless the source of uncertainty is drastically reduced. Chaotic systems pass many tests of randomness; for example, there are simple chaotic maps that produce uncorrelated time series, with $\langle x_t x_{t+j} \rangle = 0$ unless $j = 0$.

In contrast, if the dynamics is not chaotic errors grow slowly and the main requirement for determinism is that d be large enough. As long as this is true forecasts can be made for long times into the future.

Dissipative dynamical systems often have the property that undisturbed trajectories approach a subset of the state space, called an *attractor*. This can sometimes causes a drastic reduction in the number of degrees of freedom. Fluid flows, for example, have an effectively infinite dimensional state space, but can have low dimensional chaotic attractors [19,20,7].

Thus, we should not distinguish chaos and randomness, but rather we should distinguish systems with low dimensional attractors from those with high dimensional attractors. If a time series is produced by motion on a very high dimensional attractor, then from a practical point of view it is impossible to gather enough infor-

---

[2] A trajectory is *chaotic* if it has positive Lyapunov exponents, *i.e.*, if on average it is locally unstable. The motion of a roulette ball is sensitive to initial conditions, but strictly speaking not chaotic; since it comes to rest, it does not have any positive Lyapunov exponents. Chaos is a special case of sensitive dependence to initial conditions, where there is also sustained motion.

[3] Another notion of randomness is due to Kolmogorov and Chaitin [16]. Whether or not chaotic systems are random in this sense is controversial [17,18].

mation to exploit the underlying determinism. If we model the dynamics in a state space whose dimension is lower than that of the attractor, we only see a projection of the dynamics, and determinism is invisible – the dynamics look random. Even if the dimension of the model is large enough, the amount of data needed to make a good model for a high dimensional attractor can be prohibitive. This problem gets exponentially worse as the dimension increases [21,22]. This is also apparent from our forecasting error estimates, discussed in Section 3.

With many degrees of freedom, the statistical approach is probably as good as we can do. If the dimension is really high, linear models may even be optimal. But if random behavior comes from low dimensional chaos, we can make forecasts that are much better than those of linear models. Furthermore, the resulting models can give useful diagnostic information about the nature of the underlying dynamics.

2 Model Building

2.1 State space reconstruction

Suppose we have some data in the form of a time series, $\{v_t\}, t = 0, 1, \ldots, N$. Unless otherwise stated we assume that $\{v_t\}$ is stationary, which is automatic if it comes from an attractor.

Very often v_t is a projection of dynamics in a higher dimensional state space. In order to make use of any determinism in $\{v_t\}$ we must represent it in a way that makes it evident. To do this we must *reconstruct* a state space.

A typical example occurs in fluid flow experiments. We know that fluid flows can be modeled in terms of the Navier-Stokes equations and other classical partial differential equations. In practice, however, we may not be able to use these models because the available data is inappropriate. For example, suppose we use a probe to measure a given component of the velocity at a fixed point in space. This data is simply inadequate to provide initial conditions for the Navier-Stokes equations. To work with the data at hand, we are forced to reconstruct a state space from a single time series.

A method for doing this was introduced by Packard *et al.* [23] and put on a firm mathematical foundation by Takens [24]. Suppose we create a state vector $x(t)$ by assigning coordinates

$$\begin{aligned} x_1(t) &= v(t), \\ x_2(t) &= v(t-\tau), \\ &\vdots \\ x_d(t) &= v(t-(d-1)\tau), \end{aligned} \tag{3}$$

where τ is a delay time. If the dynamics takes place on an attractor of dimension D, then a necessary condition for determinism is $d \geq D$. If q is the dimension of a manifold containing the attractor, Takens showed that $d = 2q + 1$ is sufficient, at least in principle.

In principle, τ is arbitrary as long as it is not rationally related to $x(t)$. In practice, if τ is too small the coordinates become singular, so that $x_j \approx x_{j+1}$. If τ is too big, chaos makes x_1 and x_d causally disconnected. Taken together, these two considerations imply an effective upper bound on the embedding dimension d. d is usually chosen by trial and error, starting with a low value and increasing, searching for optimal results.

The use of delay coordinates to reconstruct a state space is not original to dynamical systems theory. It goes at least as far back as Yule [25], who in 1927 made a model for sunspot activity based on a linear combination of past values. The important contribution from dynamical systems theory is the demonstration that reconstruction preserves geometrical *invariants* of the dynamics, such as attractor dimension, metric entropy, and the positive Lyapunov exponents.

Delays are not the only way to embed data in a high dimensional space. For example, taking derivatives of the time series also provides an embedding. For clean data derivatives produce nicer embeddings than delays, but since differentiation amplifies noise, high dimensional embeddings are impractical [23,21].

Broomhead and King [26] have suggested an alternative approach. They apply the Karhunen and Loeve principal value decomposition to the delay coordinate representation, and produce embeddings that have the nice properties of derivatives, but without the numerical problems. The simplest way to implement their procedure is to compute the covariance matrix $\langle x_i x_j \rangle$ and compute its eigenvalues α_i and eigenvectors. The eigenvalues α_i are the average root-mean-square projection of the d-dimensional time series onto the eigenvectors. Ordering them according to size, the first eigenvector has the maximum possible projection, the second has the largest possible projection for any fixed vector orthogonal to the first, and so on. Thus, the eigenvectors form a basis which is an optimal linear embedding. The numerical calculations of Broomhead and King demonstrate that α_i falls off exponentially with i, until it reaches a floor determined by the noise level. This fall off steepens as the sampling time Δt decreases.

A nice feature of the Broomhead and King procedure is that the global embedding dimension d is computed automatically, simply by counting the number of eigenvalues above the noise floor. This eliminates the usual trial and error procedure to determine the dimension of the state space. The only undetermined parameter is the *lag window* $\tau_L = \tau(d-1)$. If τ_L is too small, the embedding makes incomplete use of the available information, but if τ_L is too large, the determinism is overwhelmed by the amplification of noise. Our suggestion to overcome this problem is to first construct weighted delay coordinates, according to

$$\begin{aligned} x_1(t) &= v(t), & (4)\\ x_2(t) &= e^{-h\tau} v(t-\tau), \end{aligned}$$

$$\vdots$$
$$x_d(t) = e^{-h(d-1)\tau}v(t-(d-1)\tau),$$

where h is the metric entropy, and then perform a principal value decomposition. h is still an undetermined parameter, but it can be calculated by a variety of different algorithms [27]. The resulting coordinates are linearly optimal from an information theoretic point of view. This technique can also be extended for assessing the relevance of spatial samples or multivariate time series, as will be discussed in more detail in future work [28].

Up until now we have assumed that the state $x(t)$ is constructed directly in terms of values of the time series. This is is usually referred to as autoregressive (AR) modeling. For a deterministic dynamical system it is the most natural approach. An alternative is to make predictions in terms of the residuals, $\delta(t) = x(t) - \hat{x}(t)$, where $\hat{x}(t)$ is a prediction of $x(t)$. This approach, originally due to Slutsky [29], is called the moving average (MA) model. An alternative, called the ARMA model [5], extends the state space to include both. AR, MA, and ARMA models are formally equivalent, but not necessarily in practice. The ARMA approach may offer some advantages, even for modeling fully deterministic dynamics. We intend to investigate this further.

2.2 Learning nonlinear transformations

Once we have found a state space representation, the next task is to fit a model to the data. There are several approaches. The simplest is to make time discrete, and write the dynamics as a map in the form

$$x(t+T) = f_T(x(t)) + n(t), \tag{5}$$

where the current state is $x(t)$, and $x(t+T)$ is a future state. $n(t)$ can be thought of as "noise"; it is a fudge factor to take into account high dimensional phenomena that we cannot hope to model. f and x are d-dimensional vectors. The problem is to make an estimate $\hat{x}(t,T)$ for $x(t+T)$, based on $x(t)$, of the form

$$\hat{x}(t,T) = \hat{f}_T(x(t)). \tag{6}$$

An alternative approach is to write the dynamics as a differential equation

$$\frac{dx}{dt} = f(x(t)) \tag{7}$$

and find an approximation in these terms. In this paper we mainly discuss the discrete-time approach, since it is faster to implement numerically, and since we have studied it more. Differential models may offer improvements in accuracy, and we intend to compare the two approaches in a future paper.

2.2.1 Representations

Chaotic dynamics occurs only when f is nonlinear. There are an infinite number of ways to represent nonlinear functions, which makes nonlinear approximation difficult. As soon we pick a representation, we make an *ad hoc* assumption that may or may not be appropriate. With prior information we may be guided to a good representation, but we are usually forced to make arbitrary choices. There is a theory for the approximation of functions [30], but unfortunately, it seems to be of little practical use for this problem.

Finding a good representation is largely a matter of trial and error. We make a guess and then test to see whether it produces a good model. Convenience plays an important role, since fitting parameters is usually time consuming unless it can be reduced to a linear problem. It seems that certain representations are usually better than others, but this is more a matter of numerical lore than theory.

We now discuss a few interesting representations.

Polynomials are a good representation because they are convenient and arise naturally in Taylor expansions. The use of polynomials for forecasting has been suggested by Weiner [2], who proposed them for moving average models, and Gabor [1,31], who proposed them for autoregressive models. Polynomials have also independently recently been used for fitting deterministic equations of motion by several authors [32,33,6]. When polynomials are used in more than one dimension the most general form is

$$A_m(x_1, \ldots, \ldots x_d) = \sum_{i_1=0,\ldots,i_d=0} a_{i_1 \ldots i_d} {x_1}^{i_1} \cdots {x_d}^{i_d}. \tag{8}$$

$a_{i_1 \ldots i_d}$ are parameters, with $\sum_{j=1}^{d} i_j \leq m$, where m is the degree and d the dimension. The number of parameters is $\frac{(d+m)!}{d!m!} \approx d^m$. Fitting this many parameters rapidly becomes impractical when m and d are large. Polynomials have the disadvantage that they do not extrapolate well beyond their domain of validity since $|A| \to \infty$ as $|x| \to \infty$.

Rational approximation by the ratio of two polynomials provides an appealing alternative [34]. Rational approximations extrapolate better than polynomials, particularly when the numerator and denominator are of the same degree, since they remain bounded as $x \to \infty$. Furthermore, like polynomials, fitting parameters by least squares is a linear problem. The work of Bayly *et al.* [35] suggests that rational approximation does a good job of fitting chaotic attractors, and preliminary investigations of Lee and Lee [36] suggest that they perform well for forecasting problems.

Radial basis functions [37] only depend on the Euclidean distance between points.[4] They are particularly effective as local representations, and often produce approximations that are comparable to those of polynomials, with less free parameters. A special case of radial basis functions are thin plate splines, but there are many other possibilities. Preliminary numerical investigations suggest that this is a promising approach.

Neural networks can be viewed as just another class of functional representations. For example, one popular form for a neural network uses neurons whose outputs are *tanh* functions of their summed inputs. Lapedes and Farber [38] have recently applied a neural net with two hidden layers to some of the same forecasting problems that we studied in [6], Their neural net can be written as

$$\hat{x}(t,T) = \sum_k w_k z_k - a_o \tag{9}$$

$$z_k = \tanh\left(\sum_j w_j y_j - a_k\right), \tag{10}$$

$$y_j = \tanh\left(\sum_i w_i x_i - a_j\right), \tag{11}$$

$$x_i = x(t - i\tau), \tag{12}$$

where y_j and z_k are the values of the "neurons" in the "hidden layers". The parameters a and w are determined using back-propagation, which is essentially a nonlinear least squares algorithm. The tanh representation has several nice properties, but it has the disadvantage that fitting the parameters cannot be reduced to a linear problem, and the back propagation algorithm is extremely time consuming. As a result, forecasts take several orders of magnitude more computer time.

For a representation to do a good job of approximating a function f, it must conform naturally to its variations. If f is well-behaved, any complete representation will provide a good approximation, as long as it has enough free parameters, and enough data to stably solve for them. However, if f is complicated, there is no guarantee that any given representation will give an efficient approximation.

Note: Even when the least squares problem is linear, the solution is not well behaved if the data are singular. This can be improved through singular value decomposition [34], which is the method we use throughout this paper.

2.2.2 Local approximation

The dependence on representation can be reduced by *local approximation*. The basic idea is to break up the domain into local neighborhoods and fit parameters in each neighborhood separately. When f is smooth the neighborhoods can be small

[4]We would like to thank Martin Casdagli for making us aware of the literature on radial basis functions.

enough so that f does not vary sharply in any of them, making the choice of a representation less important.

Local approximation has the advantage that it automatically improves as the neighborhoods get smaller. In contrast, global approximations are limited by the number of free parameters. This comparison is not really fair, since the accuracy of global approximation improves as more parameters are added. Just as global approximations require enough data points for a stable fit of their parameters, local approximations require enough data for a stable fit in each neighborhood. However, it seems that adding more local neighborhoods is usually more efficient than adding more parameters and going to higher order. The rule of thumb from numerical lore is that it takes roughly N data points to fit $\sqrt{N}$ free parameters.[5] Local approximation can be much better. For example, for m^{th} order interpolation on a one dimensional domain with N equally spaced points, the approximation error decreases as $N^{-(m+1)}$. Thus splines are usually the method of choice for the approximation of complicated functions.

Although local approximation can be more trouble to implement, it is often well worth the effort, particularly for large data sets. There are three basic steps:

1. Pick a local representation.
2. Assign neighborhoods.
3. Fit parameters in each neighborhood. We will call the resulting local approximation in a given neighborhood a *chart*.

The basic idea is illustrated in Figure (1).

A simple way to assign neighborhoods is to partition the domain into disjoint sets, for example, by dividing each coordinate into equal parts. This has the disadvantage that, since there is no overlap between the neighborhoods, there is no continuity between charts. A point is that is not centered in its neighborhood may be poorly approximated compared to one in the middle.

An alternative is to overlap neighborhoods, so that each chart is constructed from a "good" set of neighbors. To make this precise we impose a metric, denoted $\|\ \|$. We say that a neighborhood is *centered* on x if its members $\{y_i\}$ minimize $\|x - y_i\|$.

In our first letter [6] we created charts as they were needed, centering them on the current state. This approach can be summarized as follows: To predict $x(t+T)$ based on $x(t)$ and its past history, we find the k nearest neighbors of $x(t)$, *i.e.*, the k points with $t' < t$ and minimum values of $\|x(t) - x(t')\|$. We then construct a chart $\hat{f}_{x(t)}$, making a map from the past to the future by regarding $\{x(t')\}$ as points in the domain and $\{x(t'+T)\}$ as the corresponding points in the range.

[5] We thank Jim Crutchfield for pointing this out.

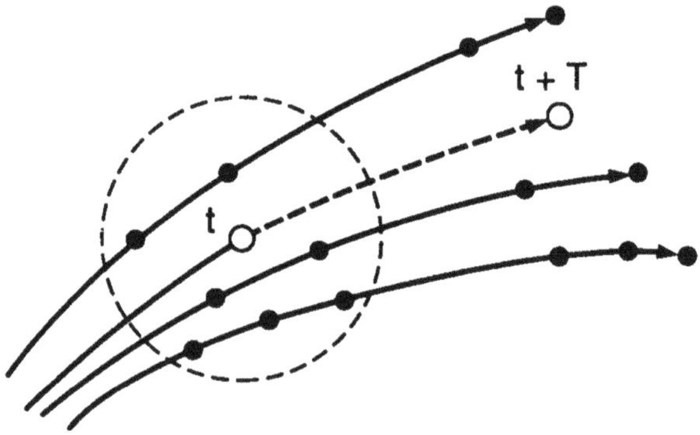

Figure 1: *Local approximation.* We find the nearest neighbors of the current state, and fit a chart to them. The points in the dashed circle define the domain of the chart, and the same points a time T later define its range. To make a prediction we evaluate the chart at the current state.

Note: Using delay coordinates it is sometimes convenient to let $x(t+T)$ be one dimensional, even though $x(t)$ is d-dimensional, as we did in [6]. For more general state space reconstructions, such as the Broomhead and King procedure [26], to make predictions we have to invert the reconstruction, to convert back to the the original time series. Thus $x(t+T)$ must be d-dimensional.

A trivial approach to constructing charts is *zeroth order approximation*, which amounts to simply looking through the data set for the nearest neighbor, and predicting that the current state will do what it did a time T later, *i.e.*, letting $k=1$ and $\hat{x}(t,T) = x(t'+T)$. For example, to predict tomorrow's weather we could search the historical record and find the weather pattern most similar to that of today, and predict that tomorrow's weather pattern will be the same as that one a day later.[6] This can be improved by finding more neighbors and averaging their predictions, for example by weighting according to distance from the current state.

A bigger improvement can be made using higher order approximation. For example, let $k \geq d+1$ and fit a linear polynomial to each pair $(x(t'), x(t'+T))$. When $k = d+1$ this is equivalent to linear interpolation[7] If the data is noisy, the *local linear approximation* procedure described above may be more stable when $k > d+1$. Again, this procedure can be improved by weighting the contributions of the neighboring points according to their distance from the current state.

Higher order approximation can give much more accurate predictions, as discussed in Section 3. Any nonlinear representation provides a possible candidate. The criteria for a good local representation are different, however, from those for a good global representation. On one hand, performing a fit within a local neighborhood is easier, because the variations of the data are less extreme. Wild variations or discontinuities can be accommodated by assigning neighborhoods properly. On the other hand, a local neighborhood necessarily has less data in it, so the representation must produce good fits with only a few points.

Another interesting alternative is to compute local charts for each data point, and then treat each chart *as though it were a data point*, to make *metacharts*. This process could obviously be continued *ad infinitum*. The metacharts may be smoother than the individual charts, resulting in a more compact or more accurate description. We intend to investigate this further.

---

[6] This was attempted by E.N. Lorens, who examined roughly 4000 weather maps. The results were not very successful because it was difficult to find good nearest neighbors, apparently because of the high dimensionality of weather. Still, it would be interesting to reexamine this question using some of the other techniques discussed here.

[7] The use of linear interpolation of nearest neighbors as a means of forecasting chaotic data was independently suggested by Peter Grassberger [39], who implemented it for the Henón map. Crutchfield and MacNamara [33] have also independently suggested local approximation with disjoint neighborhoods, which they refer to as an "atlas dynamical system".

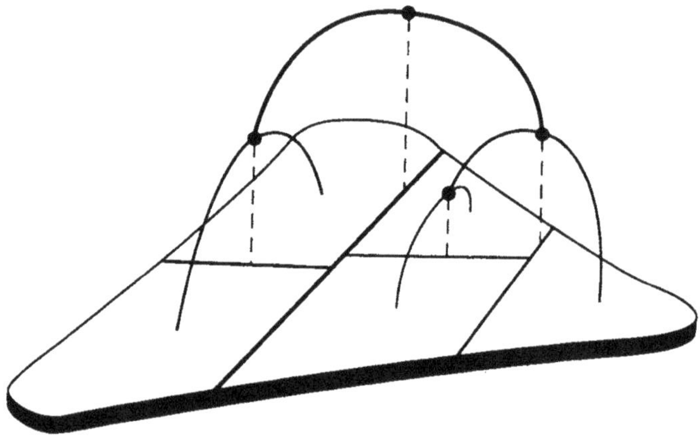

Figure 2: An illustration of the k-d tree. The data is partitioned one coordinate at a time. The nodes of the tree correspond to the partitions of the domain, as indicated by the dashed lines. The partitions can be made on any coordinate, and the tree can have different depth in different regions. The result is a flexible partition of the data space, with finer partitions in the areas where there is more data. Use of the k-d tree speeds up the search for nearest neighbors, decreasing the number of steps from N to $\log N$.

2.3 Tricks

In this section we explain how to make local approximation run fast, and how to cope with difficult situations such as nonstationary behavior or approximation across discontinuities.

2.3.1 Making local approximation fast

Finding neighbors in a multidimensional data set is time consuming if done by brute force. Finding a nearest neighbor by systematically computing the distance to each of N points takes roughly N steps. This can be reduced to roughly $\log N$ steps by organizing the data with a decision tree, such as the *k-d tree* [40,41]. The basic principle is illustrated in Figure (2). The data set is partitioned one coordinate at a time. Any criterion can be used for the partitions; for example, for the results reported here we find the coordinate with the largest range, and partition it at its median value. The values corresponding to each partition are stored in the tree as *keys*. The neighborhood of a given query point can quickly be located by comparing its coordinates to the keys. This makes it possible to find nearest neighbors quickly, since many points can be eliminated from consideration without actually computing their distance. The k-d tree has the nice property that it partitions only the parts of the space that actually contain data, and does this flexibly by adding partitions only where they are needed.

Providing the k-d tree is used to find nearest neighbors, the principal speed limitation for local approximation is the time required to fit the parameters of the charts. For real-time applications or where many predictions need to be made from a fixed data base, the charts can be computed in advance and stored in the tree, sacrificing computer memory to gain speed. Unfortunately, we cannot pre-compute a chart that is guaranteed to be centered on the current state, unless we compute a large number of different combinatorial possibilities, We cannot anticipate the proper set of neighbors. A good compromise is to compute charts for all of the points in the data base, find the nearest neighbor of the current state, and borrow its chart to make a forecast. This procedure is very fast, but consumes computer memory. For a given situation there are many possible compromises, depending on whether the application is constrained by speed, computer memory, or data.

If an application is limited by data, it may be important to update the tree after each prediction, to make use of new data in future predictions. At the opposite extreme, if an unlimited supply of data is available, and computer memory is at a premium, it may be desirable to find an optimal data set by selectively inserting new points where the approximation is bad and removing old points where it is already good. Regions where f changes rapidly should be covered densely, while regions where f is relatively smooth should only be covered sparsely. A simple criterion is to inserting states that make bad predictions, and compensate by removing states that make good predictions. Aspects of this are discussed by Omohundro [42].

For most of our numerical results, for convenience we simply use the first part of the data set to build a data base, and use the second part to make predictions. Of course, if the data is nonstationary, it could be very important to continually update the data base.

Although local approximation is certainly more trouble than global approximation, using the techniques discussed here it is quite fast. The k-d tree is easy to implement, as long as one avoids caveman computer languages such as FORTRAN. Local approximation does not require massive amounts of computer power. For example, the computations for this paper were made with a SUN microcomputer.

2.3.2 Trajectory segmenting

Most of the discussion above assumes discrete time. When time is continuous it is important to take this into account. A *trajectory segment.* is a group of data points within the same neighborhood that have adjacent sampling times. For example, the local neighborhood in Figure (1) has four trajectory segments. If the sampling rate Δt is small, finding a fixed number of neighboring data points may be very different than finding a given number of trajectory segments. If it is really small, all the neighbors of a given point will be on the same trajectory segment. This results in a highly singular chart, due to the near co-linearity of the data. We can attempt to compensate by increasing the number of nearest neighbors, but due to nonuniformities some regions may behave quite differently than others. A more reliable approach is to choose neighborhoods so that they include a given number of trajectory segments, rather than a given number of points. This gives much better control over the quality of the fits.

2.3.3 Nonstationarity

Most of the results in this paper assume that we are dealing with stationary data. In many applications this is not the case The most straightforward way to deal with nonstationarity is to extend the state space to include time. Time can be included in the metric, so that the search for nearest neighbors favors recent data. In addition, with time as a coordinate, the maps can be constructed so that they take into account trends and other time dependent effects. Similarly, a periodic function of time can be included to cope with seasonality or other periodicities.

2.3.4 Discontinuities

Discontinuities can create problems, which get worse if the neighborhoods straddle them. If a point is near a discontinuity, its neighbors in the domain may lie on both sides of it. A chart for such a neighborhood will inevitably produce a poor forecast.

A superior approach is to check neighborhood relations in both the domain *and the range*. Of course, by definition the future value of the current state is unknown, but this is not true for the rest of the database. By extending the metric to include the range values, we can make sure that neighborhoods are chosen so that they do not cross the discontinuity. We will get good predictions as long as we find the correct neighborhood. If we find a neighborhood on the wrong side, the predictions will be terrible, but at least we win some of the time, rather than losing all of the time. This procedure also makes it possible to deal with multiple valued behavior, since points can be classified according to the branch in the range that they belong to.

3 Error estimates

In this section we give a brief summary of our work on estimates for forecasting errors. We only give the final results here; the derivations will be published in detail elsewhere [8]. These estimates depend on properties of the dynamics, such as the attractor dimension D, metric entropy h and the Lyapunov spectrum $\{\lambda_i\}$. They also depend on properties of the data set, such as the number of data points, N, and the signal-to-noise ratio S, as well as the time T that we want to extrapolate into the future. Given these parameters, how good are the resulting forecasts?

As a measure of the quality of our forecasts, for our numerical results we normally use the normalized root-mean square error,

$$\langle E^2 \rangle = \frac{\langle (\hat{x}(t,T) - x(t+T))^2 \rangle}{\langle (x(t) - \langle x(t) \rangle)^2 \rangle} \tag{13}$$

Thus $E = 0$ for perfect forecasts, and $\langle E^2 \rangle = 1$ for forecasts made using $x_{pred}(t,T) = \langle x(t) \rangle$.

If the noise level is high the forecasting accuracy will be limited by the resolution of the current state. In this case prediction errors grow according to the metric entropy. Normally, however, the forecasting accuracy is limited by the number of data points. In this case, the error depends on the method used to make forecasts.

Local approximation schemes can be classified according to their order, which can be defined in terms of the scaling of the approximation error as the number of points is increased. Alternatively, an mth order scheme can be defined as one whose errors depend on $(m+1)^{th}$ order derivatives. These two definitions should coincide; if ϵ is the spacing between data points, in one dimension the approximation error is ϵ^{m+1} with either definition. Since the average spacing between N points uniformly distributed over a D dimensional space is $\epsilon \approx N^{-\frac{1}{D}}$, the average approximation error for m^{th} order local approximation should scale as $E \approx N^{-\frac{(m+1)}{D}}$.

We have tested this by making forecasts at fixed T for several different dynamical systems, using local polynomial approximation in zeroth, first, and second order. For example, in Figure (3), we show the results for the sine map, $x_{t+1} = sin(\pi x_t)$, where $-1 < x_t < 1$. Similarly, in Figure (4) we show the scaling behavior of the Mackey-Glass delay differential equation [43]. The resulting slopes are roughly $\frac{(m+1)}{D}$, in agreement with previous calculations via independent methods [44].

Understanding how approximation errors grow in time is more complicated. For convenience, let us assume that time is discrete, so that $T = 1, 2, \ldots$. If we directly construct an approximation for the Tth iterate of a map using m^{th} order local approximation, we have shown [8] that the approximation errors grow as

$$\langle |E| \rangle \approx C N^{\frac{m+1}{D}} e^{(m+1)kT}, \tag{14}$$

where k is the largest Lyapunov exponent when $m = 0$, and is the metric entropy otherwise, and C is a constant that depends on the details of the map.

For $T >> 1$ this can be improved considerably by fitting a model for $T = 1$ and then iterating it. In this case, as shown in [8], the error grows according to the metric entropy, without the factor of $m+1$ in the exponential. We feel that the

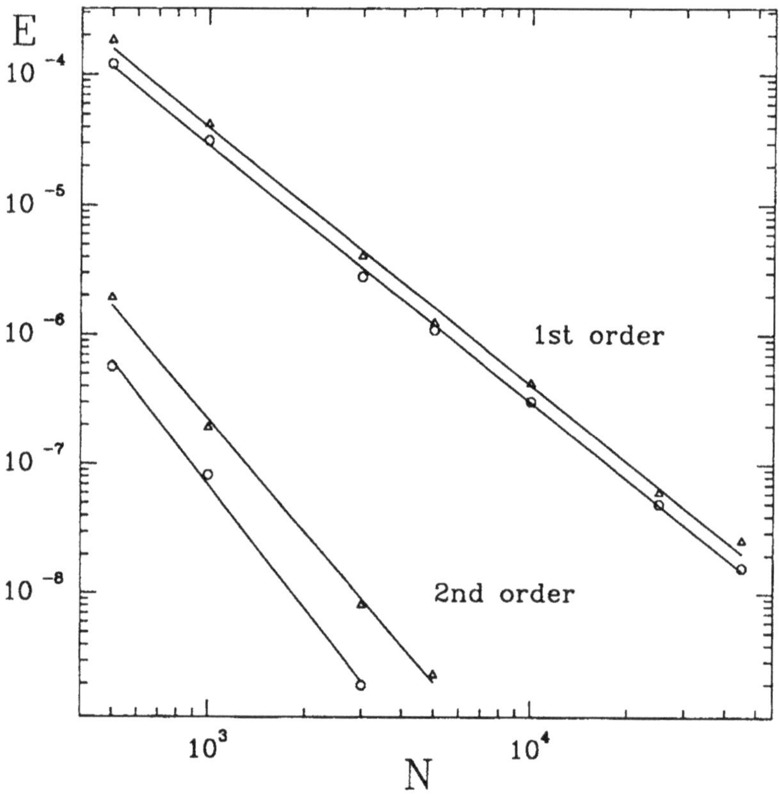

Figure 3: Local approximation by polynomials for the sine map. As predicted the slopes are roughly $\frac{1}{m+1}$. For the data points shown with triangles we used centered neighborhoods for the predictions, and for those indicated by circles we used fixed disjoint partitions.

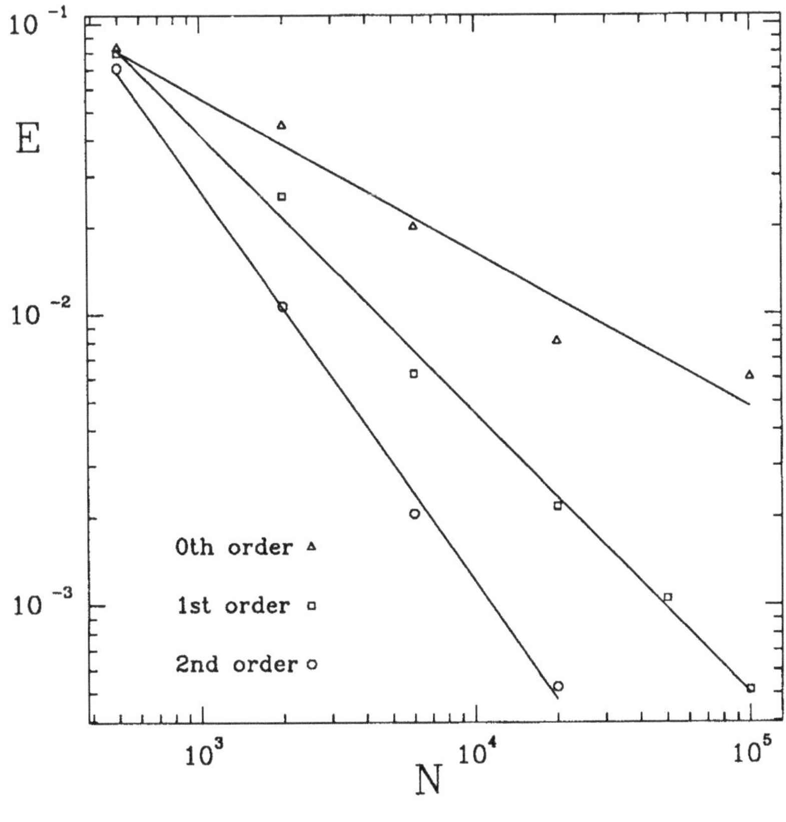

Figure 4: Local approximation by polynomials for the Mackey-Glass delay differential equation, for a fixed extrapolation time T.

underlying reason for the superiority of iteration is that it takes advantage of the recursive nature of higher iterates of chaotic maps.

The validity of these estimates is demonstrated in Figure (5), where we show the approximation error as a function of the extrapolation time T. As predicted, the normalized error grows roughly according to equation (14) when approximations are recalculated for each time T, and according to the metric entropy when an approximation is computed for $T = 1$ and then iterated to make forecasts for $T > 1$.

So far our error estimates have been for discrete-time maps, where each iteration of the map causes a large change in the state. Two problems occur in continuous-time systems.

The first problem has to do with the number of points N. In a continuous time system, the number of points in the data set can be increased by varying the sampling rate Δt. As Δt is decreased, we eventually approach a point of diminishing returns in which making Δt smaller gives little additional information. The quantity of useful data depends more on the number of characteristic time scales than on the number of data points. Unfortunately, the notion of a characteristic time scale is ambiguous. It can be estimated in many ways, for example as the correlation time, as the inverse of the average frequency of the power spectrum, or as the average time between level crossings. Yet another way to define a characteristic time comes about from trajectory segmenting; the ratio of the number of trajectory segments to the number of data points inside a local neighborhood estimates the extent to which the data is oversampled. Thus, for a continuous time system the estimates should be stated in terms of the number of characteristic times, rather than the number of data points. Note that although this introduces some absolute uncertainty into the error estimates, there is no *relative* uncertainty – scaling properties are uneffected.

The second problem occurs in iterative forecasting. Although we proved that iterative forecasts are superior, we have not said anything about how to pick the best time to construct the approximant. Clearly the effects of noise and finite precision dictate that this time should not be zero. Thus, there is some optimal time $T_{opt} > 0$ for re-constructing the dynamics, which should minimize the errors in the forecasts. This also occurs when the dynamics is approximated in terms of the derivative, writing the equations of motion in the form $\dot{x} = f(x)$, except that in this case the question becomes how to fit $\dot{x}$. The best method is to fit it in terms of its integral, but the question is how far ahead in time. These questions will be treated in more detail in a future paper [28].

4 Noise Reduction

In this section we summarize a method we have recently proposed for reducing noise in a dynamical system. The basic idea is nonlinear smoothing; once we can make forecasts, we can transport points at different times to the same point in time. We

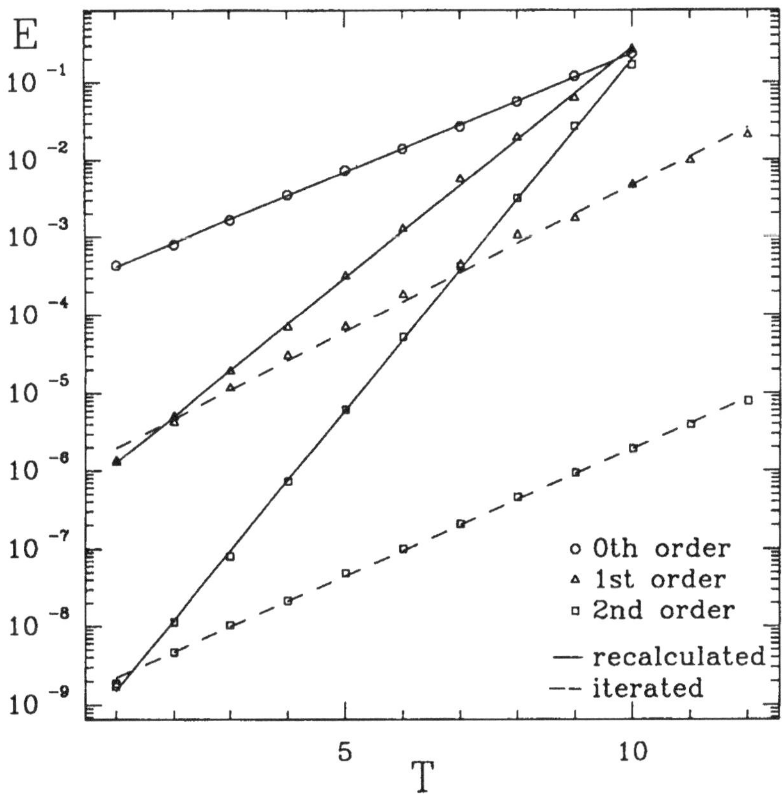

Figure 5: The normalized error for forecasts made from a time series generated by the sine map, as a function of the extrapolation time T, holding the number of data points fixed. The error for recalculated forecasts grows according to $(m+1)h$, where m is the order of approximation and h is the metric entropy, whereas the error for iterative forecasts grows according to h, independent of the order of approximation.

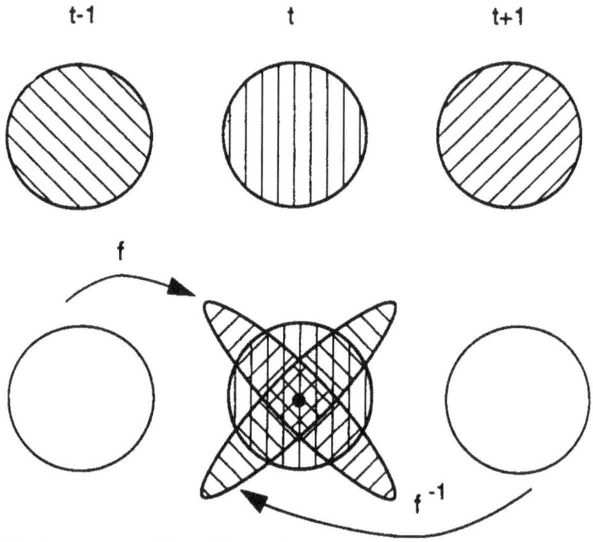

Figure 6: *Nonlinear smoothing.* The circles represent noisy measurements of a deterministic trajectory at three different times. As successive measurements are transported to the same point in time, the associated noise probability distributions distort. By weighting the values of the transported points correctly, they may be averaged together to produce a better estimate of the true value y_t. The estimates can be improved by iterating this process.

can then average them together. According to the central limit theorem, the result gives a better estimate of the true value. By applying this procedure recursively we can reduce noise substantially. The basic idea is schematically illustrated in Figure (6).

By making a local linear approximation for the joint probability distribution, and making a maximum likelihood estimate, it is possible to derive a smoothing formula. The calculations involved are presented elsewhere [8], but the final result is

$$Y_t = \left(\sum_{j=M}^{j=-N} \Theta_{tj} \right)^{-1} \sum_{j=M}^{j=-N} \Theta_{tj} f^j(x_{t-j}) \tag{15}$$

where

$$\Theta_{tj} = \left([df^j(x_{t-j})]^T df^j(x_{t-j}) \right)^{-1}.$$

Y_t is the smoothed value. Θ_{tj} is a $d \times d$ symmetric matrix, which depends on x_{t-j} as shown. It contains information about local expansion and contraction rates, which produce weighting factors that take into account the distortion of the noise as it is pulled-back or pushed-forward. The directions in which the noise distribution is compressed contain more useful information, and consequently receive higher weights.

When we know the map exactly, with even a small amount of data it is possible to reduce the noise almost down to machine precision. To demonstrate this, we have applied this to the Henón map, as shown in Figure (7). Note that we achieve a noise reduction of roughly 10^{10}, more than 100 decibels. When the true map is not known, this is limited by forecasting accuracy. Still, even when we approximate the map we have been able to reduce the noise by a factor of 1000 or more.

Please see Reference [8] for a more detailed discussion.

5 Other Uses of Adaptive Dynamics

There has been a great deal of interest recently in solving artificial intelligence problems with *adaptive networks* such as neural nets [45] and the classifier system [46]. Our methods are an example of an alternative approach.

Forecasting is a special case of a problem that is often called *learning with a teacher*. The task is to predict "outputs", based only on inputs. For forecasting, for example, the input is the present state, and the output is a future state. The record of past states provides a "teacher", a set of known input-output pairs. This hopefully allows us to generalize from the teaching set to estimate unknown outputs.

We can restate the problem more formally as follows: Given an input x_i and an output y_i, we want to find maps F and G of the form

$$\hat{y}_i = F(x_i, \alpha_i) \tag{16}$$

$$\alpha_{i+1} = G(x_i, \hat{y}_i, y_i, \alpha_i) \tag{17}$$

that minimize $\|y_i - \hat{y}_i\|$, where $\hat{y}$ is an estimate of y, and the metric $\|\ \|$ provides a criterion for the quality of estimation. The $\{\alpha_i\}$ are parameters for F, and G is a map that changes $\{\alpha_i\}$, *i.e.*, a *learning algorithm*. x, y and i can be either continuous or discrete. For a forecasting problem, for example, i corresponds to time, x is the current state x_t, and y is a future state, $y_t = x_{t+T}$.

Neural nets are a class of adaptive systems, corresponding to a particular choice for F and G. Although the form of neural nets was originally motivated by biology, there is no reason to be constrained by this in artificial intelligence problems, as reflected by many recent developments in this field. Neural nets have had success in certain problems that can be solved by learning with a teacher, for example, text to speech conversion [47] or the prediction of protein secondary structure based on amino amino acid sequence [48]. Lapedes and Farber have also shown that neural nets can be effective for forecasting [38].

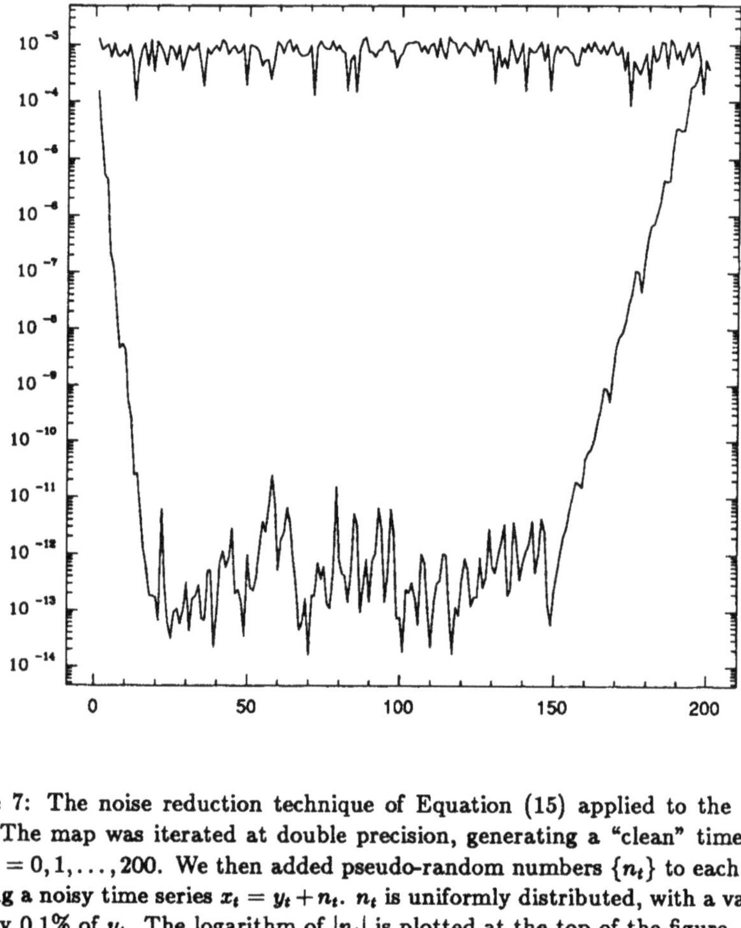

Figure 7: The noise reduction technique of Equation (15) applied to the Henón map. The map was iterated at double precision, generating a "clean" time series $\{y_t\}$, $t = 0, 1, \ldots, 200$. We then added pseudo-random numbers $\{n_t\}$ to each point, forming a noisy time series $x_t = y_t + n_t$. n_t is uniformly distributed, with a variance roughly 0.1% of y_t. The logarithm of $|n_t|$ is plotted at the top of the figure. Below it we plot $\log|Y_t - y_t|$, where Y_t is the smoothed value after applying Equation (15) 10,000 times to the entire time series, using $M = N = 3$, and using the known map to compute f^j and Θ.

However, as recently pointed out by Omohundro [42], alternative approaches that depart significantly from the usual form of neural nets may be computationally much more efficient. Our approach to forecasting (which we proposed independently), provides a good example of this; our methods give more accurate forecasts than the neural net of Lapedes and Farber [38], and are several orders of magnitude more efficient in terms of computer time. Furthermore, since the computations used in our approach can be performed in parallel, we expect that this speed discrepancy will persist even with future parallel hardware. Omohundro has pointed out that similar methods may be employed for other problems, such as associative memory, classification, category formation, and the construction of nonlinear mappings. Many aspects of the methods that we have proposed here are applicable to this broader class of problems.

Although we have assumed that x and y are continuous, with the addition of thresholds our methods are easily converted to the discrete domain. A comparison of our work with that of Lapedes and Farber makes it clear that the neural network solves problems by surface estimation. As pointed out by Lapedes, the same is true in the discrete domain, except that answers are obtained by "rounding" the surface, truncating to a discrete value. Generalization occurs through the extrapolation of the surface to regions in which there is no data. There is no *a priori* reason to constrain the functional representation to those currently popular in the study of neural networks, or to use global rather than local approximation.

Clearly these possibilities deserve more investigation.

6 Conclusions

The paradigm of chaos provides new insight into the problem of predicting random processes. It leads to a geometrical view of the problem, which can be broken into two parts: Embedding in a state space and approximating the dynamics. Both involve the choice of a representation.

The central problem of reconstructing a state space is to find a representation for the data in which any determinism is apparent. The coordinates of the state space must contain enough information to make the evolution of the system deterministic, so that in the absence of external noise, knowing the state uniquely determines the future evolution. This means that the dimension of the state space must be at least as large as that of the attractor of the underlying dynamical system. In principle any method of embedding is sufficient, but in practice some are much better than others. The weighted principal value decomposition we propose here, which is an extension of a method proposed by Broomhead and King [26], provides an embedding that is linearly optimal in information theoretic terms.

The next problem is to approximate the dynamics. The primary approach we investigate here is approximation as a discrete time map, even when the original time series is continuous. Discrete time has the advantage of being very fast. Approximation in differential terms is also promising, and we intend to compare these two approaches in a future paper. In either case, the problem boils down to approximating the graph of a nonlinear function. Success depends on picking a good representation. There are two basic approaches, global and local. Global approximation is convenient, but unless the representation is well matched to the map it may not produce good results. Many of the standard nonlinear representations undergo an explosion of parameters as the dimension increases. Local approximation has the advantage that it is less dependent on representation, and is guaranteed to get better as the number of data points increases. When used in conjunction with a data structure such as the k-d tree, it can be quite fast. The best approach depends on the details of the problem, such as the the nonlinear function being approximated, the number of data points, *etc.*

An advantage of formulating the forecasting problem in the language of deterministic chaos is that it makes it possible to derive error estimates. These estimates are couched in terms of properties of the dynamics, such as the metric entropy and the dimension, and the length of the data set and its signal to noise ratio, and the extrapolation time. We arrive at the conclusion that iterative forecasts are better than non-iterative forecasts, *i.e.*, it is better to make long-term forecasts by approximating the dynamics for a short time and iterating rather than explicitly recalculating an approximation for a longer time. The iterative approach takes advantage of the recursive form of the higher iterates of dynamical systems, and is probably the reason for its superiority. With the iterative approach approximation errors grow at the same rate as errors due to uncertainty in the initial state, *i.e.*, they grow exponentially according to the metric entropy.

The ability to approximate nonlinear dynamics naturally leads to a method for reducing external noise through nonlinear averaging. This technique is mainly limited by the ability to make good forecasts. For low dimensional systems with a modest number of data points it gives noise reductions of several orders of magnitude. An interesting aspect is that the noise reduction possible for a chaotic dynamical system is much greater than that for a nonchaotic system.

All of the methods discussed above work well for low dimensional deterministic chaos. When the dimension is low they give results that are orders of magnitudes better than those of standard linear methods. However, they lose their effectiveness when the dimension is too large. The limits can be estimated through the error estimate of Equation (14). As seen from this equation, they also depend on the method used: Higher order approximation schemes give better results with a given number of points. Of course, if it happens that we know a problem is described by a simple functional form, it may be possible to overcome the constraint of large dimensions. In general, however, this is difficult if not impossible.

The methods we have described here are new and not fully explored. We anticipate that there will be considerable progress in this area in the near future, and intend to publish another paper soon that will contain considerably more numerical results.

Acknowledgements We would like to thank Martin Casdagli, Alan Lapedes, Y.C. Lee, Steve Omohundro, Norman Packard, and David Rand for valuable conversations. We are grateful for support from the Department of Energy and the Air Force Office of Scientific Research under grant AFOSR-ISSA-87-0095.

We urge the reader to use these results only for peaceful applications.

References

[1] D. Gabor. Communcation Theory and Cybernetics. *Transactions of the Institue of Radio Engineers*, CT-1(4):9, 1954.

[2] N. Weiner. *Nonlinear Problems in Random Theory.* Wiley, 1958.

[3] M.B. Priestley. State Dependent Models: A General Approach to Nonlinear Time Series Analysis. *Journal of Time Series Analysis*, 1:47–71, 1980.

[4] H. Tong and K.S. Lim. Threshold Autoregression, Limit Cycles and Cyclical Data. *Journal of the Royal Statistical Society B*, 42(3):245–292, 1980.

[5] M.B. Priestley. New Developments in Time-Series Analysis. In M.L. Puri, J.P. Vilapiana, and W. Wertz, editors, *New Perspectives in Theoretical and Applied Statistics*, John Wiley and Sons, 1987.

[6] J. D. Farmer and J. J. Sidorowich. Predicting Chaotic Time Series. *Physical Review Letters*, 59(8):845–848, 1987.

[7] H. Haucke and R. Ecke. Mode Locking and Chaos in Rayleigh-Benard Convection. *Physica*, 25D:307, 1987.

[8] J.D. Farmer and J.J. Sidorowich. Exploiting Chaos to Predict the Future. 1987. Los Alamos Preprint.

[9] E.N. Lorenz. Deterministic Nonperiodic Flow. *Journal of Atmospheric Science*, 20:130–141, 1963.

[10] R.S. Shaw. Strange Attractors, Chaotic Behavior, and Information Flow. *Z. Naturforschung*, 36a:80–112, 1981.

[11] J. P. Crutchfield, J. D. Farmer, N. H. Packard, and R. S. Shaw. Chaos. *Scientific American*, 254(12):46–57, 1986.

[12] P. Berge, Y. Pomeau, and C. Vidal. *Order in Chaos.* 1986.

[13] B.A. Huberman, J.P. Crutchfield, and N.H. Packard. Noise Phenomena in Josephson Junctions. *Applied Physics Letters*, 37(8):750–752, 1980.

[14] A. Kolmogorov. Grundbergriffe der Wahrscheinlichkeitsrechnung. *Eng. Mat.*, 2(3), 1933.

[15] H. Poincaré. *Science et Methode*. Biblioteque Scientifique, 1908. English translation by F. Maitland (Dover, 1952).

[16] G.J. Chaitin. Randomness and Mathematical Proof. *Scientific American*, 232(5):47–52, 1975.

[17] S. Wolfram. Origins of Randomness in Physical Systems. *Physical Review Letters*, 55(5), 1985.

[18] J. Ford. *Physics Today*, 36(4):40, 1983.

[19] B. Malraison, P. Atten, P. Bergé, and M. Dubois. Dimension of Strange Attractors: An Experimental Determination for the Chaotic Regime of Two Chaotic Systems. *J. Phys. Lett. (Paris)*, 44:L897, 1983.

[20] A. Brandstäter, J. Swift, H.L. Swinney, A. Wolf, J.D. Farmer, E. Jen, and J.P. Crutchfield. Low-Dimensional Chaos in a Hydrodynamic System. *Physical Review Letters*, 51(16):1442, 1983.

[21] H. Froehling, J.P. Crutchfield, J.D. Farmer, N.H. Packard, and R.S. Shaw. On Determining the Dimension of Chaotic Flows. *Physica*, 3D:605, 1981.

[22] J. Guckenheimer. Noise in Chaotic Systems. *Nature*, 298(5872):358–361, 1982.

[23] N.H. Packard, J.P. Crutchfield, J.D. Farmer, and R.S. Shaw. Geometry from a Time Series. *Physical Review Letters*, 45:712–716, 1980.

[24] F. Takens. Detecting Strange Attractors in Fluid Turbulence. In D. Rand and L.-S. Young, editors, *Dynamical Systems and Turbulence*, Springer-Verlag, Berlin, 1981.

[25] G. U. Yule. *Philos. Trans. Roy. Soc. London A*, 226:267, 1927.

[26] D.S. Broomhead and G.P. King. Extracting Qualitative Dynamics from Experimental Data. *Physica*, 20D:217, 1987.

[27] G. Mayer-Kress, editor. *Dimensions and Entropies in Chaotic Systems*, Springer-Verlag, Berlin, 1986.

[28] J.D. Farmer. Discovering Relevance. in preparation.

[29] E. Slutsky. The Summatation of Random Causes as the Source of Cyclic Processes. *Problems of Economic Conditions*, 3:1, 1927. English translation in Econmetric, 5, 1937, p. 105.

[30] G.G. Lorentz. *Approximation of Functions*. Holt, Rinehart, and Winston, New York, 1966.

[31] D. Gabor, W.P. Wilby, and R. Woodcock. A Universal Nonlinear Filter, Predictor and Simulator which Optimizes Itself by a Learning Process. *Proceeding of the IEEE*, 108B:422, 1960.

[32] J. Cremers and A. Hübler. Construction of Differential Equations from Experimental Data. *Z. Naturforsch.*, 42a:797–802, 1987.

[33] J.P. Crutchfield and B.S. McNamara. Equations of Motion from a Data Series. *Complex Systems*, 1:417–452, 1987.

[34] W.H. Press, B.P. Flannery, S.A. Teukolsky, and W.T. Vettering. *Numerical Recipes*. Cambridge University Press, 1986.

[35] B.J. Bayly, I. Goldhirsch, and S.A. Orszag. *Independent Degrees of Freedom of Dynamical Systems*. Technical Report, 1987. Applied and Computational Mathematics, Princeton U. preprint.

[36] K. Lee and Y.C. Lee. System modeling with rational functions. in preparation.

[37] M.J.D. Powell. *Radial Basis Function for Multivariable Interpolation: A Review*. Technical Report, University of Cambridge, 1985.

[38] A.S. Lapedes and R. Farber. *Nonlinear Signal Processing Using Neural Networks: Prediction and System Modeling*. Technical Report LA-UR-87-, Los Alamos National Laboratory, 1987. submitted to Proc. IEEE.

[39] P. Grassberger. *Information Content and Predictability of Lumped and Distributed Dynamical Systems*. Technical Report WU-B-87-8, University of Wuppertal, 1987.

[40] J.H. Bentley. Multidimensional Binary Search Trees Used for Associative Searching. *Communications of the ACM*, 18(9):509–517, 1975.

[41] J.H. Bentley. Multidimensional Binary Search Trees in Database Applications. *IEEE Transactions on Software Engineering*, SE-5(4):333–340, 1979.

[42] S. Omohundro. Efficient Algorithms with Neural Network Behavior. *Complex Systems*, 1:273–347, 1987.

[43] M.C. Mackey and L. Glass. *Science*, 197:287, 1977.

[44] J.D. Farmer. Chaotic Attractors of an Infinite-Dimensional Dynamical System. *Physica*, 4D:366–393, 1982.

[45] D. Rummelhart and J. McClelland. *Parallel Distributed Processing.* Volume 1, MIT Press, Cambridge, MA, 1986.

[46] J. Holland. Escaping Brittleness: the Possibilities of General Purpose Machine Learning Algorithms Applied to Parallel Rule-Based Systems. In Michalski, Carbonell, and Mitchell, editors, *Machine Learning II*, Kaufmann, 1986.

[47] T.J. Sejnowski and C.R. Rosenberg. NETtalk. *Complex Systems*, 1, 1987.

[48] A. S. Lapedes and R. M. Farber. Neural Net Learning Algorithms and Genetic Data Analysis. in preparation.

Adaptation Toward the Edge of Chaos

Norman H. Packard
Center for Complex Systems Resarch
and the Physics Department
University of Illinois, 508 South Sixth Street, Champaign, IL 61820, USA.
n%complex@uxc.cso.uiuc.edu, ...[ihnp4,ucbvax]!uiucdcs!complex!n

Abstract.
A simple model for adaptation is explored through simulation. The model consists of a population of cellular automaton rules that evolves under the action of the genetic algorithm combined with survival of the fittest, where fitness is measured by the cellular automaton's ability to accomplish a computational goal. The population of rules is seen to move toward a region in the space of all rules that marks the boundary between chaotic rules and non-chaotic rules.

1. Adaptation in the space of cellular automata.

Biological evolution is the quintessential example of adaptation,, so it is not surprising to find a tradition of borrowed concepts and vocabulary in the construction of models for adaptation, even when the models are abstract, with no biological context as their basis. We will follow this tradition, beginning with the idea that adaptation occurs in a population by constantly changing the makeup of the population, and allowing only the fittest individuals to survive. All that remains for the description of the model, then, is to specify what individuals make up the population, how they can change, and how fitness is defined.

The individuals are cellular automaton rules. Each rule is a map from the space of patterns of ones and zeros on a one dimensional lattice to itself. Repeated action of the rule creates a sequence of patterns. Cellular automaton rules are local in the sense that the evolution of a given site depends only on the site values in a fixed neighborhood around it. The local function is repeated for each site of the lattice. Thus, each cellular automaton rule is specified by a rule table with an entry for each local configuration giving the evolution of that configuration. For a fixed neighborhood size, the number of entries in the rule table is finite (2^{2r+1} entries for neighborhoods of radius r) and the space of rules is finite ($2^{2^{2r+1}}$ rules). The population will be a population of rules, and typically the population will represent only a miniscule sample in the space of all possible rules.

Cellular automata may be thought of as algorithms, programs that act on data specified by an initial condition. In fact, some cellular automata have been proven to be Turing universal, in the sense that initial conditions may be constructed so that the action of the rule can perform an arbitrary computation. The output of the program is the pattern obtained after some number of time steps. We will think of cellular automata as programs in this way, and the fitness of a cellular automaton will be defined in terms of its ability to perform a given computational task.

It is useful to keep track of the biological analogy at this point: cellular automata provide a natural genotype-phenotype distinction. The rule table is analogous to the genotype, and the action of the rule through time may be thought of as the phenotype (this may be visualized as the pattern on a space-time lattice formed by the rule). The initial condition may be considered either as the environment or as part of the genotype. The point here is that selection takes place not directly on the rule itself, but on what is produced by the action of the rule, just as selection in biology takes place not on the genome itself, but on the organism produced by the genome.

The biological analogy is also useful for understanding how the population is to be changed. Cellular automaton rules may be changed in much the same way that the genome can be changed during reproduction, with point mutations and crossover. Point mutations are straightforward; one entry in the rule table is simply altered. In the present case, since the rule table entries are all ones and zeros, this simply means that one of the bits is flipped. Crossover takes two rule tables and produces two more, by cutting each into two pieces and switching the pieces to form two new rules, each with pieces of the old rules. The biological analogy breaks down a bit here. The bases on a string of DNA have a natural linear ordering to them, but a cellular automaton rule table has no such *a priori* ordering. The general case would be to take two arbitrary subsets of rule table entries; in the present experiments, we assign an ordering to the rule table by using the binary numerical values of the neighborhood configurations. Point mutation and crossover applied in this way to abstract rules are examples of *genetic operators*, and using these operators in the context of adaptation through selection of the fittest is known as the *genetic algorithm* [1].

To complete the description of this model for adaptation, the only remaining point is the specification of the computational goal. The goal is suggested by a cellular automaton rule constructed by Gacs [2] to do a specific task: decide whether the density of ones in an initial condition is greater than .5 or less than .5. The rule "decides" this by mapping all initial patterns with density greater than .5 eventually to a pattern of all ones, and all patterns with density less than .5 to a pattern of all zeros.[1] The Gacs rule has a neighborhood of radius three, and its action is illustrated in figure 1.

This, then, is the computational goal. The test for "fitness" of a rule

[1]To be precise, "all" should be replaced by "almost all", and the lattice considered should be infinite. I know of no proof that the Gacs rule satisfies this property, but there is very strong empirical evidence.

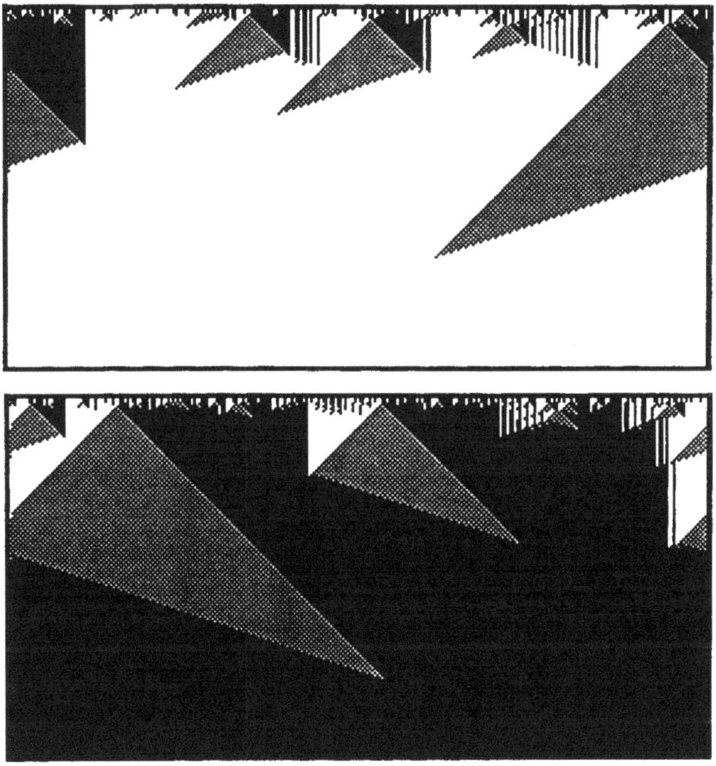

Figure 1: An illustration of the Gacs rule, a one dimensional cellular automaton with radius three. It has been constructed to have the property that initial states with densities less than .5 are mapped to all zeros asymptotically, and initial states with densities greater than .5 are mapped to all ones asymptotically. The action of the rule is illustrated in two cases, the first with an initial condition having density .45, the second with initial density .55. The one dimensional lattice stretches horizontally, and time moves down.

simply consists of trying the rule out on many initial conditions of different densities, and observing whether the rule tends to increase or decrease the density. The actual fitness measure used was normalized so that if the rule ended up with all ones (zeros) given initial densities greater than .5 (less than .5) it received one unit of fitness.

The algorithm for adaptation may be summarized as follows:

- Start with an initial set of cellular automaton rules.
- Test the fitness of each one. Rank all the rules in order of their fitness.
- Remove some fraction of the rules, choosing those that rank lowest.
- Replace those removed with new rules, obtained from the best remaing rules by the action of the genetic operators of point mutation and crossover.
- Retest this new set of rules, and proceed.

There are many questions that remain to be answered about this particular form of adaptation, *e.g.* how hard should the system be driven for maximal evolution rate? What is the optimum combination of genetic operators? How does the evolution rate depend on the size of the space of cellular automaton rules (*i.e.* on the size of the neighborhood)? These, and other questions will be answered in future work. The point to be made here is that the process of adaptation takes the population of rules to a particular region in the space of cellular automaton rules. I must now digress to describe a little of the structure of the space of rules.

2. Structure in the space of cellular automata.

Wolfram examined a large set of one dimensional cellular automata, and found that the qualitative long-time behavior of the rules acting on random initial conditions seems to fall into three classes: *(i)* spatially homogeneous asymptotic states, *(ii)* spatially inhomogeneous but temporally constant asymptotic states, *(ii)* chaotic asyptotic states, and *(iv)* other. The class *(iv)* rules were characterized by a rich variety of propagating structures and erratic behavior of difference patterns [3]. This classification turns out to be incomplete, but will suffice for the present discussion. Langton later discovered that the space of cellular automaton rules for a given neighborhood appears to be foliated into level sets of an "activity parameter," a real valued function on rules that is has low values when the rule table has a large proportion of a single symbol, and high values when it is filled evenly. Inactive rules tend to have simple temporal behavior (*i.e.* like Wolfram's class *(i)* and *(ii)*), and very active rules typically have chaotic behavior (Wolfram's class *(iii)*). For a small range of intermediate activity, the rules show complex temporal behavior (Wolfram's class *(iv)*) [4]. Langton also voices the hypothesis that cellular automaton rules capable of computation lie in the range of intermediate activity.

Paths may easily be created from inactive rules (*e.g.* the rule table filled with zeros) to active rules by successively turning on random bits [5]. Using the Hamming distance metric in the space of rules, we see that the smallest increment along such a path is due to the change of a single entry in a rule table. Paths formed in this way are discrete analogs of generic arcs in spaces of smooth dynamical systems, along which bifurcations may be observed. The change of a rule table by one unit of Hamming distance is exactly the change obtained by the mutation genetic operator.

Much work is now being done to elucidate the transition from inactive rules to active rules, and to characterize the complex behavior of the rules at the transition [5,6,7]. The transition is clearly seen in difference patterns, informational measures, and various other statistics. Rather than try to tell the whole story here, we will comment only on features of the transition that appear to be relevant to to the process of adaptation.

3. Adaptation toward the edge of chaos.

Figure two shows the evolution of a population of rules, viewed in a histogram of density of ones in the rule table. Recall that the activity of a rule table is zero when the density of ones is zero, maximal when the density is one half, and zero again when the density is one. There are two transitions to chaos, one as the density is raised from zero, and the other when the density is lowered from one. The location of the transition may be seen in figure 3(a), where we graph the difference pattern spreading rate (a measure of chaos) [3,8] as a function of density of ones in the rule table.

We see that the population evolves so that the density of ones in the rule tables present in the population are clustered around the same density at which the transition to chaos occurs. Thus, the population appears to evolve toward that part of the space of rules that marks the transition to chaos.

It may seem initially that the region of rule space that the population evolves toward should be highly dependant on the computational goal. I suggest, however, that the particular rules that are selected are dependent on the computational goal, but that most computationally effective rules lie near the transition to chaos. There are trivial counterexamples for trivial computations (e.g. if the computational goal is to compute the zero configuration for any input, the zero rule does fine, and is far from the transition to chaos), but nontrivial computation requires communication between different parts of the lattice. Inactive rules cannot have this property because their asymptotic behavior creates frozen domains that contain local perturbations forever. Very active rules tend to communicate too much information to do successful computation. For these chaotic rules, perturbations propagate over the entire lattice, or put another way, a given site is influenced equally by a large number of sites in the past. At the transition to chaos, asymptotic behavior is dominated by complex propagating structures that can selectively transmit information from one part of the lattice to another. Thus, it is reasonable to expect that any computation that requires such communication

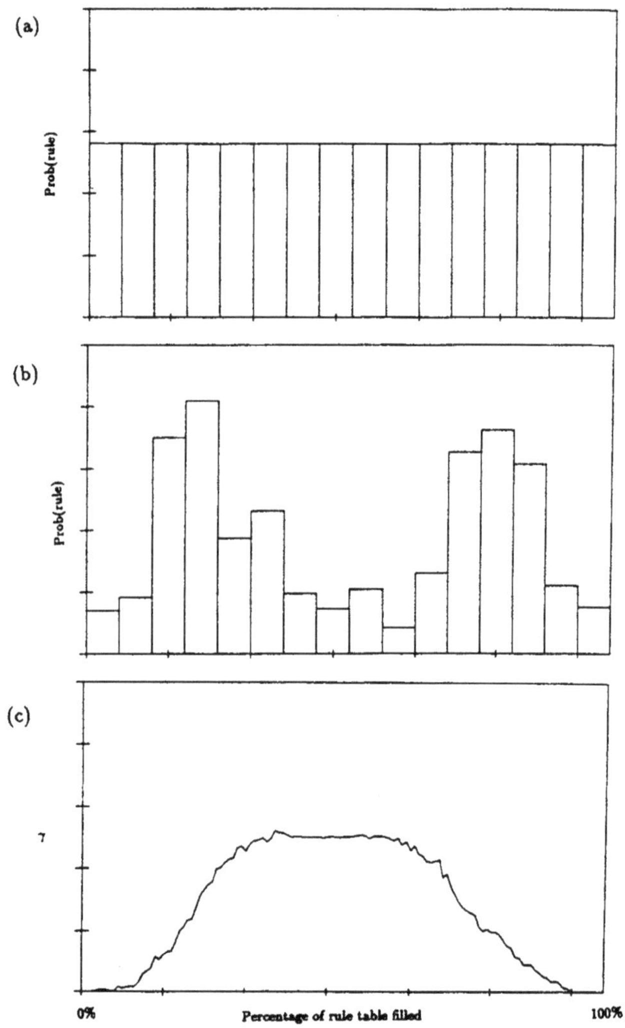

Figure 2: (a) A histogram of rule table densities (*i.e.* density of ones in the rule table)in an initial population. (b) The same histogram after one hundred generations. (c) A graph of the difference pattern spreading rate as a function of rule table density, showing the transition to chaos.

will only be accomplished by rules near the transition to chaos. Work is currently being done to explore the generality of this hypothesis. The necessity of communicating information to accompish computation has led to the exploration of mutual information in space-time as a method to quantify the computational capability of cellular automata [4,5].

It turns out that at least one rule that has been proven to be computation universal, Conway's game of life, does indeed lie in this transition region. Wolfram conjectured that all class *(iv)* rules might be computation universal; this is much stronger than what we propose here, and is probably false, though impossible to say without a good definition of class *(iv)* rules.

4. Concluding comments.

It may be usefule to consider the present model in the context of past models for adaptation. One of the earliest models was Kauffman's Boolean networks as a model for genetic regulatory control [9]. Kauffman and Smith performed adaptive selection experiments on spaces of automata very similar to those reported here [10]. In this work, the connectivity of the network of automata was fixed for an ensemble, and mutations were introduced by varying connections (these automata were not on a lattice) and the Boolean functions performed at each automata. In contrast, here the connectivity (*i.e.* the activity) was allowed to vary during adaptation. The Kauffman Smith experiments seem to be commensurate with the results reported here, though transitions in global behavior are not understood as completely for their class of automata networks as for cellular automata.

There are other models where there is no clear genotype/phenotype distinction, *e.g.* Anderson and Stein's primordial soup mode, where fitness is given a *priori* by a random (spin glass) function of the genome. This identification is also evident in many optimization problems that implement adaptive dynamics, *e.g.* the simulated annealing approach to the traveling salesman problem. The proposed picture of adaptation moving a population toward the edge of chaos does not make any sense in these cases where there is no genotype-phenotype distinction. The picture makes sence only when there is a phenotype that may be regarded as a dynamical outcome encoded by the genotype.

There are other models that have the same sort of clear genotype-phenotype distinction as the present model. One is Hollands classifier system [11]. This example also uses the genetic algorithm to move around in a space of "programs", and the fitness of the program is determined by its performance. In this case, however, the action of the program is typically not terribly dynamic, but rather the classifiers that get chosen as being fit function more like a set of production rules. The only analog to the present situation might be in the dynamics that leads to the choice of classifiers. These dynamics may use very strong, highly coupled interactions between classifiers (analogous to the dynamics of active cellular automata), or sparsely coupled dynamics (analogous to inactive dynamics). I know of no results indicating what coupling properties of classifiers tend to get selected for typical computations. Transitions between different types of dynamics as a function of the sparseness of a network is also evident two other adaptive network models, one for the immune system and one for the primordial soup [12,13,14]. Work is currently underway to discover whether the dynamics that are selected in such networks turn out to be near transitions to chaos.

Another adaptive system that has a novel genotype-phenotype distinction is Axelrod's use ofthe genetic algorithm to devise winning strategies for the prisoner's dilemma [15]. In this case, the game strategy is a look-up table, and the "phenotype" is the outcome of using the strategy in many rounds of playing the game. Fitness is determined by average performance at playing the game. Naively, it would appear that there are simple, trivial strategies that don't work well (*e.g.* the constant strategy), and very involved strategies that don't work well (*e.g.* a randomly filled strategy lookup table), and that the "good"strategies are somewhere in between. The strength of a relatively simple strategy, TIT-FOR-TAT may argue against this picture, though Axelrod has found more successful strategies using adaptive dynamics. A complete elucidation of this point will require an understanding of the structure of the space of strategies analogous to the structure described above for the space of cellular automata rules.

We have discovered a certain phenomenon in the study of adaptative dynamics in the space of cellular automaton rules: the rules that tend to get chosen by the adaptive process tend to lie near the transition to chaos. We have argued that this phenomenon makes sense because rules near the transition to chaos have the capability to selectively communicate information with complex structures in space-time, thus enabling computation. It is possible that this feature that characterizes the computational efficacy of cellular automata may be generalized to systems made up of many coupled components that are not necessarily coupled through the structure of a lattice. In these more general cases, there will still be systems with weak, sparse coupling with simple dynamics, strong coupling with chaotic dynamics, and an intermediate range of coupling where there is efficient information flow that allows computation to take place. These are the systems that will be selected by adaptive dynamics.

I am very grateful to the following people for helpful conversations and collaborations leading to this work: Doyne Farmer, Chris Langton, Wentian Li, Otto Rössler, and Rob Shaw. Part of this work was completed at the Center for Nonlinear Studies, Los Alamos National Laboratory. It has also been supported by the National Center for Supercomputing Applications at the University of Illinois, and the National Science Foundation, grant number PHY86-58062 PYI.

References

[1] J. H. Holland "Genetic Algorithms and Adaptation", Technical Report #34, Univ. Michigan, Cognitive Science Department (1981).

[2] P. Gacs, Technical Report #132, University of Rochester, (1983).

[3] S. Wolfram, "Universality and Complexity in Cellular Automata", Physica **10D** (1984) 1.

[4] C. Langton, "Studying Artificial Life with Cellular Automata", Physica **22D** (1986) 120.

[5] C. Langton, N. Packard, W. Li, "Bifurcation-like Behavior in Cellular Automata" in progress.

[6] C. Langton, Thesis, University of Michigan Department of Computer Science (1988).

[7] W. Li, Thesis, Columbia University Department of Physics (1988).

[8] N. H. Packard, "Complexity of Growing Patterns in Cellular Automata," in *Dynamical Behavior of Automata: Theory and Applications*, J. Demongeot, E. Goles, and M. Tchuente, eds., Academic Press (1984)

[9] S. A. Kauffman, J. Theo. Bio. **22** (1969) 437.

[10] S. A. Kauffman and R. G. Smith "Adaptive Automata Based on Darwinian Selection", Physica **22D** (1986) 68.

[11] J. H. Holland, "Escaping Brittleness: The possibilities of general purpose learning algorithms applied to parallel rule-based systems", in: *Machine Learning 2*, R. S. Michalski, J. G. Carbonell and T. M. Mitchell, eds. (Morgan Kauffman, Los Altos, in press), chap. 20. 1986.

[12] J. D. Farmer, N. H. Packard, and A. S. Perelson, "The Immune System, Adaptation, and Machine Learning", Physica **22D** (1986) 187.

[13] J. D. Farmer, S. A. Kauffman, N. H. Packard, "Autocatalytic Replication of Polymers", *Physica* **22D** (1986) 50.

[14] J. D. Farmer, S. A. Kauffman, N. H. Packard, A. S. Perelson, "Adaptive Dynamic Networks as Models for the Immune System and Autocatalytic Sets", Ann. N.Y.A.S. **204** (1987) 120.

[15] R. Axelrod, in **Genetic Algorithms and Simulated Annealing** p. 32-41, Lawrence Davis, ed. Morgan Kaufmann (1987).

Part IV

Action, Perception and Developing Systems

THE VITE MODEL: A NEURAL COMMAND CIRCUIT FOR GENERATING ARM AND ARTICULATOR TRAJECTORIES

Daniel Bullock† and Stephen Grossberg‡
Center for Adaptive Systems
Boston University
111 Cummington Street
Boston, MA 02215

1. MOVEMENT PLANNING IN MULTI-JOINT SYSTEMS

A major issue in research on the neural basis of motor control is the nature of movement planning in systems with many degrees of freedom; for example, an arm with many controlling muscles acting at several joints, or a speech system with many articulators. All solutions to the planning problem depend upon assumptions about both the mechanics of the effectors and the sensory and computational resources. For example, if an arm has few mechanical degrees of freedom, then the serial preplanning required to work around the arm's inherent constraints becomes a salient issue. Alternatively, if the arm has many degrees of freedom, the computational load imposed by the need for simultaneous coordination becomes a salient issue. If the arm is part of a body that grows, or if a robotic arm must remain in service without external maintenance despite unpredictable changes in its mechanical parameters, then yet another issue comes into view: autonomous recalibration.

This paper extends earlier results (Bullock and Grossberg, 1986, 1988a) that are part of a larger program of research on adaptive sensory-motor control in biological systems (e.g., Grossberg, 1978a, 1982; Grossberg and Kuperstein, 1986). In this program, emphasis is on emergent properties of a neural architecture and on neural network dynamics within systems whose effectors may have many degrees of freedom and whose mechanical parameters may change due to growth, injury, etc. Our goal is to explain how these systems exhibit such exquisite coordination in many distinct task environments, and how they maintain such coordination during all but the most rapid periods of growth.

---

† Supported in part by the National Science Foundation (NSF IRI-84-17756).
‡ Supported in part by the Air Force Office of Scientific Research (AFOSR F49620-86-C-0037) and the National Science Foundation (NSF IRI-84-17756).

Simulations of neural networks for self-organization of spatial maps and continuous autonomous calibration were presented in the book by Grossberg and Kuperstein (1986) on adaptive neural control of ballistic eye movements. The research reported in Bullock and Grossberg (1988a) built on the basis developed in that book, while focusing on automatic trajectory formation and coordination during variable-speed arm movements. In particular, a real-time neural network model, called the Vector Integration to Endpoint, or VITE, Model, was developed and used to quantitatively simulate behavioral and neural data about planned and passive arm movements. In this paper, we extend our application of the VITE model to explain additional neurophysiological and kinematic data on arm movements as well as data about speech articulator movements. In particular, after reviewing a number of basic properties of the model, we discuss data concerning the role of the globus pallidus in movement speed control, the equifinality of movement synergies, rate-dependence of velocity profile asymmetries, and observed variations in the ratio of maximum to average movement velocities. We also note the inability of various other models, notably optimization models, to explain these data, and emphasize the close relationship between a model's computational form and the possibility of correctly identifying and parsing the functional problems that are solved by a complex biological system.

2. EMERGENT INVARIANTS OF THE VITE CIRCUIT

In the VITE model for arm movements, kinematic invariants emerge through network interactions rather than through an explicitly precomputed trajectory. Parallel motor planning occurs in the form of a Target Position Command, or TPC, an array that specifies the lengths to which all arm-controlling muscles intend to move, and an independently controlled GO command, which specifies the movement's overall speed. Automatic processes convert this information into an arm trajectory with invariant properties, notably properties of synchrony among muscle synergists. These automatic processes include computation of a Present Position Command, or PPC, and a Difference Vector, or DV. The DV is the difference between the TPC and the PPC at any time. The PPC is gradually updated by integrating the DV through time. The time-varying GO signal multiplies the DV before it is integrated by the PPC. The PPC generates an outflow movement command to its target muscle groups (Figure 1). Opponent interactions regulate the PPC's to agonist and antagonist muscle groups at each joint.

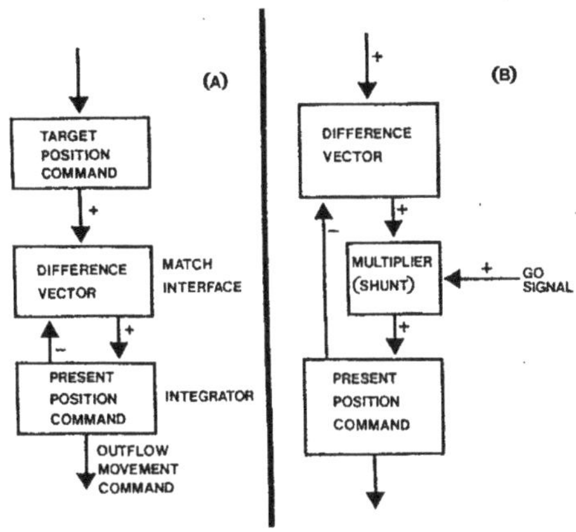

Figure 1. (a) A match interface within the VITE model continuously computes a difference vector (DV) between a target position command (TPC) and a present position command (PPC), and adds the difference vector to the present position command. (b) A GO signal gates execution of a primed movement vector and regulates the rate at which the movement vector updates the present position command.

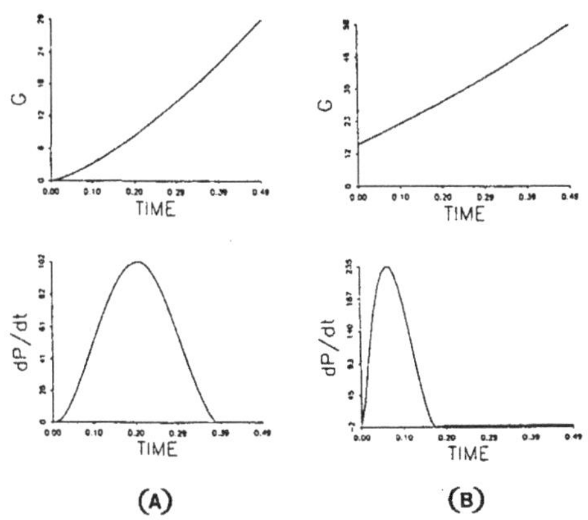

Figure 2. A much higher peak velocity is predicted by the model whenever a target, T, is activated after the GO signal has already had time to grow. (A): The control condition, in which T and the GO signal growth process are activated synchronously. (B): Same T as in (A), but here T was activated after the GO signal $G(t)$ had been growing for 300 msec.

In its simplest form, excluding terms expressing opponent interactions, the VITE circuit obeys the equations:

Difference Vector

$$\frac{d}{dt}V_i = \alpha(-V_i + T_i - P_i) \tag{1}$$

and

Present Position Command

$$\frac{d}{dt}P_i = G[V_i]^+, \tag{2}$$

where $[V_i]^+ = \max(V_i, 0)$. Equations (1) and (2) describe interactions of a generic component of a target position command $(T_1, T_2, \ldots, T_n)$, a difference vector $(V_1, V_2, \ldots, V_n)$, a present position command $(P_1, P_2, \ldots, P_n)$, and a time-varying velocity command, or GO signal $G(t)$. The difference vector computes a mismatch between target position and present position, and is used to update present position at a variable rate determined by $G(t)$ until the present position matches the target position. Such a scheme permits multiple muscles, or other motor effectors, to contract synchronously even though the total amount of contraction, scaled by $T_i(0) - P_i(0)$, may be different for each effector (Figure 3).

The VITE circuit is not sufficient in itself to accomplish all the tasks required of a variable-speed variable-load arm movement system. In concert with several parallel circuits, however, it can generate flexible and adaptive trajectories without suffering from the combinatorial explosions and rigid performance of control systems that preplan an entire trajectory. In particular, a key difference between the VITE model and many other model proposals is that its PPC is computed by outflow or feedforward, signals, rather than by inflow, or feedback, signals from the muscles. The process of guaranteeing that the PPC outflow movement command actually moves the arm to a corresponding position are accomplished, in part, by a separate adaptive linearization network that uses outflow-inflow mismatches to generate error signals that adaptively alter the gain of the total outflow movement command (Grossberg and Kuperstein, 1986). The cerebellum has been implicated as the locus of adaptive gain change by the manner in which the functional architecture of this model compares with neural data.

TABLE 1
FOR FIXED DURATION (MT), ERROR GROWS IN PROPORTION TO DISTANCE

MT	DISTANCE	ERROR
.56	10	.084
.56	20	.170
.56	40	.349
.56	80	.700

TABLE 2
FOR FIXED ERROR LEVEL, DURATION (MT) GROWS LINEARLY WITH DISTANCE DOUBLING

ERROR	DISTANCE	MT
.059	2	.39
.057	4	.49
.058	8	.59
.059	16	.70
.057	32	.80
.059	64	.91

The VITE model generates synchronous movements across synergetic muscles by automatically compensating for the different total contractions that each muscle group must undergo. The model accomplishes this in a manner that generates, in quantitative computer simulations, a wide range of properties documented by experimentalists in psychology and neurophysiology (see Georgopoulos, 1986). Among these properties are: Woodworth's law (Table 1 and Woodworth,1899); the speed-accuracy trade-off function known as Fitts' law (Table 2 and Fitts, 1954); peak acceleration as a function of movement amplitude and duration (Table 3) and isotonic arm movement properties before and after arm-deafferentation in animals deprived of visual feedback (Bizzi, Accornero, Chapple and Hogan, 1984); synchronous and compensatory "central error correction" properties of isometric contractions (Gordon and Ghez, 1987); velocity amplification during target switching (Figure 2 and Georgopoulos, Kalaska, and Massey, 1981); velocity profile invariance across different movement distances (Freund and Büdingen, 1978) and change in profile asymmetry across different movement durations (Figure 3 and Beggs and Howarth, 1972); the automatic compensation for staggered onset times of synergetic muscles (Figure 4 and Hollerbach, Moore, and Atkeson, 1986); vector cell properties in precentral motor cortex (Georgopoulos, Kalaska, Caminiti, and Massey, 1984; Evarts, 1984); and the inverse relationship between movement duration and peak velocity (Lestienne, 1979).

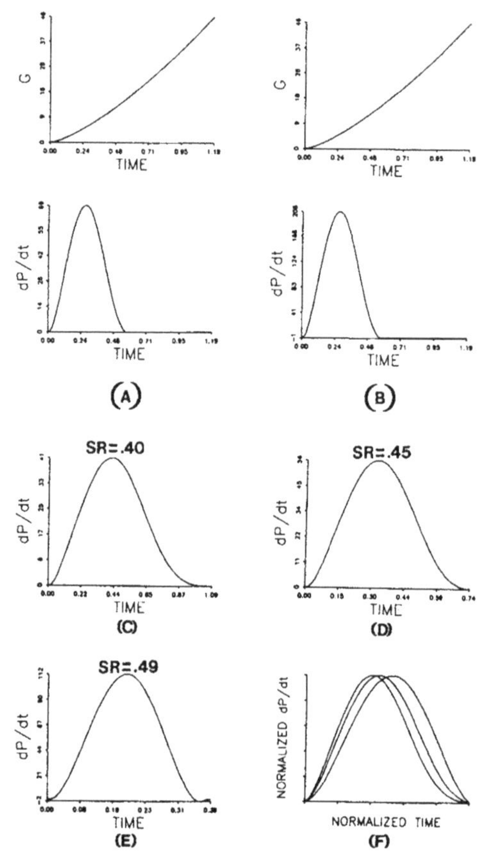

Figure 3. (A,B): With equal GO signals, movements of different size have equal durations and perfectly superimposable velocity profiles after velocity axis rescaling. Shown are GO signals and velocity profiles for 20 and 60 unit movements lasting 500 msec. (C,D,E): Velocity profiles associated with small, medium, and large GO magnitudes result in slow, medium, and fast performance of a 20 unit movement. Each SR value gives the trajectory's *symmetry ratio*; that is, the time taken to move half the distance divided by the total movement duration. These ratios indicate progressive symmetrization at higher speeds, within the range of speeds shown. (F): The velocity profiles shown in (C), (D), and (E) are not perfectly superimposable.

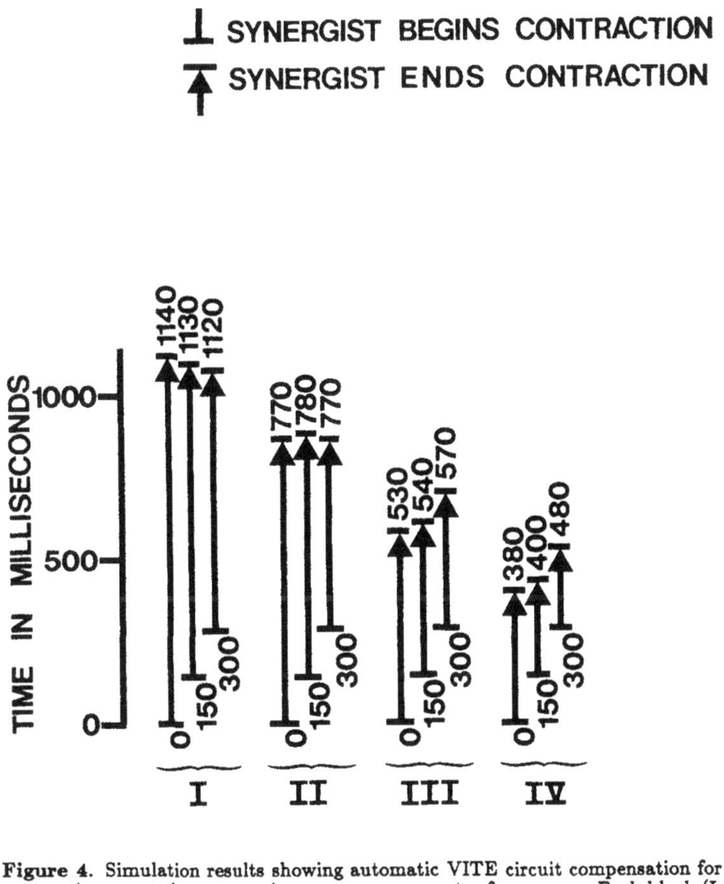

Figure 4. Simulation results showing automatic VITE circuit compensation for contraction-onset-time staggering across components of a synergy. Each block (I, II, III, IV) shows results for a different value (10, 20, 40, and 80, respectively) of the GO signal magnitude.

TABLE 3
A COMPARISON OF FOUR MODELS' ABILITIES TO PREDICT DATA ON PEAK ACCELERATION ($\ddot{P}$)

DISTANCE	MT	PEAK $\ddot{P}$	PEAK $\ddot{P}$ SOURCE
20°	.554	397°/sec^2	Bizzi *et al.* (1984)
60°	.692	1130°/sec^2	(experimental data)
20°	.554	376°/sec^2	Minimum-jerk model
60°	.692	722°/sec^2	(simulation)
20°	.554	399°/sec^2	Minimum-effort model
60°	.692	767°/sec^2	(simulation)
20°	.554	394°/sec^2	VITE model
60°	.692	854°/sec^2	(simulation)
20°	.554	396°/sec^2	VITE$^+$ model
60°	.692	1127°/sec^2	(simulation)

3. ACTIVELY GATED LEARNING OF TARGET POSITION AND PRESENT POSITION

Beyond their service in trajectory formation, TPC, PPC, and DV computations are needed to actively modulate, or gate, the learning of associative maps between TPC's of different modalities, such as between the eye-head system and the hand-arm system (Figure 5). The gating process prevents learning from occuring except when the PPC is close to the TPC; that is, except when the DV is small. Such gating helps to prevent spurious correlations from being learned, say between a fixed target position of the hand and all present positions which the eye assumes while moving to look at the hand. By using such an intermodality associative map, looking at an object can activate a TPC of the hand-arm system, as Piaget (1963) noted. Then a VITE circuit can translate this latter TPC into a synchronous movement trajectory.

Active gating is also needed to regulate learning of present position commands, notably to prevent such learning from occurring during active arm movements. In particular, an auxiliary circuit, called the Passive Update of Position, or PUP, Model, uses inflow signals to update the PPC during passive arm movements due to external forces (Figure 6), but not during active arm movements. Because the scales of outflow position command signals and inflow position sensing signals cannot be assumed to be the same, the PUP circuit incorporates a synaptic

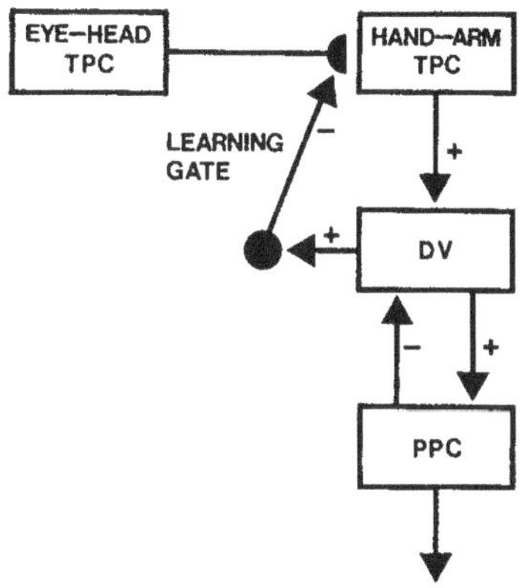

Figure 5. Learning of an intermodal associative transformation between target position maps is gated by a DV process which matches TPC with PPC to prevent incorrect associations from forming between eye-head TPC's and hand-arm TPC's. Learning only occurs when the DV is small.

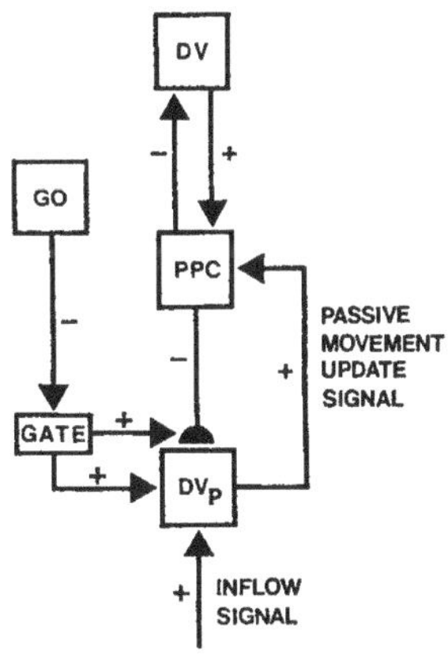

Figure 6. A passive update of position (PUP) circuit. An adaptive pathway $PPC \rightarrow DV_P$ calibrates PPC-outflow signals in the same scale as inflow signals during intervals of posture. During passive movements, output from GO equals zero. Hence the passive difference vector DV_P updates the PPC until it equals the new position caused by any passive movements that may occur due to the application of external forces.

modification mechanism for adaptively recalibrating corollary discharges of PPC outflow signals until they are computed in the same numerical scale as the inflow signals to which they are compared. Interactions of outflow and inflow signals are also needed for adaptive linearization of a nonlinear muscle plant (Grossberg and Kuperstein, 1986) and for automatically or predictively adapting to the inertial properties generated by variable loads and velocities (Bullock and Grossberg, 1988b). The equations of a typical PUP circuit are:

Present Position Command

$$\frac{d}{dt}P_i = G[V_i]^+ + G_p[M_i]^+ \tag{3}$$

Outflow-Inflow Match

$$\frac{d}{dt}M_i = -\beta M_i + \gamma I_i - z_i P_i \tag{4}$$

Adaptive Gain Control

$$\frac{d}{dt}z_i = \delta G_p(-\epsilon z_i + [M_i]^+). \tag{5}$$

Equation (3) supplements equation (2) with an update signal $G_p[M_i]^+$ that is turned on only when the passive gating function, or "pauser" signal, G_p becomes positive in the passive, or postural, state. Function z_i in (5) is an long term memory trace, or associative weight, which adaptively recalibrates the gain of outflow signals P_i until they are in the same scale as inflow signals γI_i in (4).

In summary, offset of the GO signal within the VITE circuit enables a pauser signal within the PUP circuit to drive its learning and reset functions. Such pauser-modulated learning during mismatches has been suggested to occur in several adaptive sensory-motor control circuits (Grossberg and Kuperstein, 1986).

4. PHYSIOLOGICAL EVIDENCE FOR GO-SIGNAL PATHWAY IN GLOBUS PALLIDUS

Because the VITE model proposes that trajectories are generated as the arm tracks the evolving state of a neural circuit, the model can be tested in two ways: by comparing trajectories of the neural circuit's output stage with actual arm trajectories, and by checking for the existence of the neural components postulated in the model.

In Bullock and Grossberg (1988a) we reviewed evidence of Georgopoulos and his colleagues that cell populations in precentral motor cortex could be analysed as an *in vivo* analogue of model DV stage neurons. Additional physiological support for the VITE model comes from recent experiments involving lesions and electrical stimulation of the basal ganglia. Data from a set of experiments by Horak and Anderson (1984a, 1984b) are consistent with the interpretation that the internal segment of the globus pallidus is an *in vivo* analogue of the VITE model's GO-signal pathway.

An *in vivo* candidate for a GO-signal pathway must pass three tests. First, stimulation at some site in the proposed pathway must have an effect on the *rate* of muscle contractions. Second, it must have this effect without affecting the amplitude of the contractions. Thus stimulation should have no effect on movement accuracy. Third, this rate-modulating effect should be non-specific: it should affect all muscles that are typically synergists for the movement in question.

The studies conducted by Horak and Anderson (1984a, 1984b) addressed these issues. Horak and Anderson (1984a) showed that "when neurons in the globus pallidus were destroyed by injections of kainic acid (KA) during task execution, contralateral arm movement times (MT) were increased significantly, with little or no change in reaction times (p.290)." This satisfies the rate criterion. Moreover, the rate of motor recruitment was depressed "in all the contralateral muscles studied at the wrist, elbow, shoulder, and back, but there were no changes in the sequential activation of the muscles (p.20)." This satisfies the non-specificity criterion. Finally, the authors also noted that "animals displayed no obvious difficulty in aiming accurately ... they did not miss the 1.5-cm target more often following KA injections, and there was no noticeable dysmetria around the target (p.300)." This satisfies the accuracy criterion.

Horak and Anderson (1984b) used an electrical stimulation paradigm instead of a lesion paradigm. They found that "stimulation in the ventrolateral internal segment of the globus pallidus (GP_i) or in the ansa lenticularis reduced movement time, whereas stimulation at many sites in the external pallidal segment (GP_e), dorsal (GP_i), and putamen increased movement times for the contralateral arm (p.305)." Once again, these effects were non-specific: "no somatotopic effects of stimulation were evident. If stimulation at a site produced slowing, it produced a depression of activity in all the muscles studied. Even stimulus currents as low as 25 μA affected proximal as well as distal muscles, flexor as well as extensor muscles, and early- as well as late-occurring activity (p.309)."

In the VITE model, activation of the GO-signal pathway produces movement only if instatement of a TPC different from the current PPC leads to the computation of a non-zero DV. In agreement with this property, Horak and Anderson (1984b) observed that "stimulation at sites that speeded movements did not induce involuntary muscle activation in resting animals nor did it change background EMG activity prior to self-generated activity during task performance (p.313)." In Bullock and Grossberg (1988a) we noted that "very rapid freezing can be achieved by completely inhibiting the GO signal at any point in the trajectory". This property of the model has also been shown to be a property of the GP system. In particular, Horak and Anderson reported that "stimulation with 50 or 100 μA at ... sites ventral and medial to typical GP_i neuronal activity completely and immediately halted the monkey's performance in the task (p.315)." Taken together, their experiments led Horak and Anderson (1984b) to conclude that "the basal ganglia ... determine the speed of the movement" (p.321).

5. ENSURING EQUIFINALITY IN SPACE AND TIME

The striking correspondence between the experimental results of Georgopoulos *et al.* and of Horak and Anderson and the theoretical predictions of the VITE model regarding separate DV and GO-signal processes is important because it supports the hypothesis that motor systems, like sensory systems, implement a factorization of pattern and energy (Grossberg, 1970, 1978a, 1982). In the motor system, this factorization means that a movement's speed ("energy") can be scaled up or down over a wide range without disrupting the movement's direction or spatial endpoint ("pattern"). Moreover, by using a GO-signal that grows gradually during the movement time (as exemplified in Figure 2), all synergists will complete their contractions at approximately the same time even if movement onset times of different synergists are staggered by a large amount (Figure 4). These properties of the model, together with the strong evidence for separate DV and GO-signal pathways *in vivo*, provide a basis for understanding how primates can achieve space-time equifinality—all synergists reaching their length targets at equal times—yet retain separate control of rate and position. Note that rate-control models relying on *static* stiffness adjustments (e.g., Cooke, 1980) lack the critical temporal-equifinality property.

Several other important automatic compensatory properties emerge from interactions between the DV and GO processes. In Bullock and Grossberg (1988a), we noted that in addition to compensating for muscles that begin to contract at staggered onset times, the VITE circuit automatically compensates for changes of target position during the movement time. In particular, Figure 2 shows that the model generates the amplification of peak velocity that occurs during target-switching experiments (Georgopoulos, *et al.*, 1981).

Such an amplification of velocity facilitates reaching the target after an incorrect initial TPC is replaced by an updated TPC. This speed-up occurs "on-the-fly." It is not preprogrammed, but is rather an automatic emergent property of VITE circuit interactions.

The VITE circuit can accomplish such compensations because changes in the TPC during target switching, or differences in onset times of contraction across muscles in a synergy, or updating of a PPC via inflow signals during an externally caused movement—all these circuit perturbations *flow through* the system via dynamic real-time computations. For example, during a target-switching experiment, first there is quick reset of the TPC, and then more gradual compensatory change of the DV and the DV-GO signal product that controls movement rate.

Similar catch-up phenomena are known to exist in the domain of speech motor control. Abbs, Gracco, and Cole (1984) have argued that such speech compensations cannot be explained by more common compensatory mechanisms such as stretch reflexes. Though a full treatment of how the VITE circuit contributes to an explanation of these data will not be given here, other data suggestive of a VITE circuit role in speech are treated in Sections 6 and 7.

6. RATE-DEPENDENT ASYMMETRY IN VELOCITY PROFILES

A distinctive prediction of the VITE model is that velocity profiles will generally be asymmetrical about the movement-time midpoint, and that their degree and direction of asymmetry will vary as a function of movement time (MT). Such a dependence of asymmetry on MT has been known empirically since the work of Beggs and Howarth (1972), whose results were modelled via the series of simulations shown in Figure 3C, D, and E. Recently, Zelaznik, Schmidt, and Gielen (1986) have shown that within a faster range of speeds (MT $<$ 250 msec.), the direction of asymmetry can reverse from that observed in the medium-to-slow speed

TABLE 4
KINEMATIC VARIABLES AS A FUNCTION OF THE GO-SIGNAL SCALAR (G_o) AND THE VELOCITY THRESHOLD (ϵ_i)

		Distance		Duration		V_{max}/V_{avg}		T-SR		D-SR	
G_o	V_{max}	ϵ_1	ϵ_2	ϵ_1	ϵ_2	ϵ_1	ϵ_2	ϵ_1	ϵ_2	ϵ_1	ϵ_2
10	38.2	19.9	19.6	.96	.84	1.84	1.63	.44	.45	.50	.50
20	52.4	20.0	19.8	.71	.63	1.86	1.67	.45	.46	.49	.49
40	72.8	20.2	20.1	.50	.47	1.80	1.70	.46	.47	.51	.51
80	102.5	20.6	20.6	.36	.34	1.79	1.69	.50	.50	.53	.53
160	145.7	21.5	21.5	.27	.26	1.79	1.73	.51	.51	.54	.54
320	207.7	22.9	22.9	.20	.20	1.81	1.77	.55	.54	.57	.57
640	295.8	24.9	24.8	.16	.15	1.84	1.73	.55	.55	.58	.58
1280	417.9	27.2	27.2	.12	.12	1.85	1.77	.56	.57	.58	.58

Note: Mean V_{max}/V_{avg} values are 1.82 and 1.71 in columns ϵ_1 and ϵ_2, respectively. $\epsilon_1 = 1$ unit/sec. $\epsilon_2 = 4$ units/sec. TPC = 20 for all simulations.

range studied by Beggs and Howarth. Table 4 shows results from a series of VITE model simulations that replicate the results reported in Zelaznik *et al.* (1986). The T-SR values (time symmetry ratios) were calculated as in Figure 3: the time required to move half the total distance was divided by the time required to move the entire distance. The D-SR values (distance symmetry ratios) were calculated as in Zelaznik *et al.* (1986); namely, the distance traversed at the time of the arm's peak velocity was divided by the total movement distance.

Model parameter settings were the same as those used in Bullock and Grossberg (1988a). To match the Zelaznik *et al.* measurement conditions as closely as possible, measurements of duration and distance in the simulation runs began when velocity first exceeded a 4 cm/sec threshold and ended when velocity fell below a 4 cm/sec threshold (see the ϵ_2 columns). As can be seen in Table 4, model-generated D-SR values for movement times between 250 and 125 msec. exhibit the direction of asymmetry (all D-SRs > .5) observed by Zelaznik *et al.* Moreover, their range (.54 to .58) closely matches the observed range (.54 to .60). Finally, the ordering of the model-generated values is also consistent with the overall pattern established by the observed values: "the distance traversed at the time of maximum velocity was inversely related to MT" (Zelaznik *et al.*, 1986, p.361).

Several additional remarks should be made about the T-SR and D-SR values shown in Table 4. First, these values are invariant under changes of TPC and initial DV. That is, they vary with movement rate, but not movement amplitude. Second, both these statistics are stable despite small shifts in the velocity threshold used to begin and end measurements of a given movement (compare the ϵ_1 and ϵ_2 columns). Third, it can be seen from the table that the T-SR statistic is more sensitive to change in VITE velocity profiles than the D-SR statistic, which even shows a slight violation of monotonicity. Because Zelaznik *et al.* also appeared to have difficulty proving replicability of the small D-SR differences they initially observed (compare their Experiment 1 and Experiment 2 results), we suggest that future experiments concerned with rate-dependent asymmetry use the T-SR statistic.

Two final points concern the generality and origin of rate-dependent asymmetries. Although the VITE model was developed for arm movements, there are many reasons to believe that it is applicable to speech motor control. Like reaching movements, speech movements are subject to continuous rate variations, and higher speaking rates lead to poorer accuracy. Recently it has been shown that speech movements also exhibit unimodal velocity profiles that are comparable to those shown in Figure 3 for arm movements and Ostry, Cooke, and Munhall (1987) have reported strong "evidence for ... a rate dependent asymmetry in the velocity of orofacial movement, under a variety of conditions" (p.44). Moreover, because the asymmetry was true of both speech and non-speech movements (i.e. was not task-specific) the authors concluded that "the most promising alternatives to account for the asymmetry of form seem to be articulator biomechanics and/or neural control" (p.45) rather than speech-specific task demands. Because the rate-dependent asymmetry is common to several biomechanically dissimilar systems, we are inclined to place more weight on an explanation in terms of a neural controller, viz., the VITE circuit. The asymmetric velocity profiles of the VITE model coexist with the model's asymmetric DV profiles, as equation (2) shows. The asymmetric DV profiles are compared in Bullock and Grossberg (1988a) with neurophysiological evidence from the Georgopoulos laboratory demonstrating asymmetric vector cell profiles in prefrontal motor cortex.

7. A COMMON KINEMATIC INVARIANT IN SPEECH AND ARM MOVEMENTS: V_{max}/V_{avg}

One kinematic measure that is now frequently used in the experimental literature on arm and speech articulator movements (Hasan, 1986; Munhall, Ostry and Parush, 1985; Ostry and Cooke, 1987; Ostry *et al.*, 1987; Soechting, 1984) is the ratio of a movement's maximum velocity (V_{max}) to its average velocity (V_{avg}). This ratio has been found to be virtually constant across a wide range of movement speeds and distances in both the speech and reaching domains. Moreover, the absolute values of the ratios found for speech and reaching movements—about 1.7 and 1.9, respectively—are close enough to invite speculation that this kinematic invariant might be a "signature" of a generative mechanism common to the two domains.

The VITE model predicts a mean value of 1.95 (range 2.18 to 1.86, with a .001 unit/sec velocity threshold; predicted ranges and means for two other thresholds are shown in Table 4), which stands exactly at the midpoint of the empirical range reported by Hasan (1986) for arm movements. Other models of arm trajectory formation predict similar values: Hogan's (1984) minimum jerk model predicts 1.88, whereas Hasan's minimum effort model predicts 1.97. Why are both empirical and predicted values for discrete arm movements (1.9) larger than commonly observed values for speech movements (1.7)?

A simple hypothesis is that the discrepancy is, at least partly, due to a truncation effect. Note that if you truncate the tails of a bell-shaped velocity profile, and if you measure distance and duration relative to the truncated profile, then V_{max} remains the same while V_{avg} increases, relative to values calculated from a full (un-truncated) profile. Thus the ratio V_{max}/V_{avg} varies inversely with the degree to which the velocity profile's edges get clipped. This is illustrated in Table 4 where the ϵ_2 column represents a more severe truncation than the ϵ_1 column.

We consider two kinds of evidence that truncation does contribute to the arm vs. speech discrepancy. First, Soechting (1984) reported a speech-like value of V_{max}/V_{avg} (1.75) in an arm experiment. However his experimental procedures can explain this anomalous value as an example of a truncation artifact. As he noted, "Movement time T was defined as the interval between the time at which wrist velocity began to exceed 10% of its maximal value and the time of contact with the target." This meant that movements had already achieved velocities of from

6.5 to 16.5 cm/sec before measurement began. By omitting these small velocities from the average, the average velocity was spuriously inflated. (In addition to causing an underestimation of V_{max}/V_{avg}, this severe truncation led Soechting to mistakenly conclude that velocity profiles from fast and slow movements were superimposable after time and velocity normalization—contrary to the findings of Beggs and Howarth [1972], Moore and Marteniuk [1986], and Zelaznik *et al.* [1986] regarding rate-dependent asymmetry.)

Second, Munhall *et al.* (1985) have shown that the ratio varies across speech conditions. By examining their tables of mean values as a function of speech condition, it is possible to test the truncation hypothesis. However, in this case we are speaking not of a truncation artifact introduced by experimental procedures, but of a "real" truncation caused by switching to a new TPC before the prior DV has been completely zeroed. We believe that such non-artifactual truncations are the rule in medium- and fast-paced continuous speech.

Were such normative truncation the source of the lower V_{max}/V_{avg} values, then speech V_{max}/V_{avg} values should be higher (and closer to discrete arm movement values) when the subject speaks more slowly, when the subject places greater stress on a syllable, and when the consonantal target position must be briefly held (as in a fricative). Consistent with this expectation, inspection of the Munhall *et al.* (1985, Table 2, p.464) data reveals the highest mean V_{max}/V_{avg} value, 1.83, for the slow, stressed, fricative condition, and the lowest mean value, 1.66, for the fast, unstressed, stop consonant condition (compare Table 4 simulation means). A more recent experiment (Ostry *et al.*, 1987), though conducted for other reasons, provided a more direct test of the truncation hypothesis by examining the V_{max}/V_{avg} ratio for both continuous and discrete elbow, tongue, and jaw movements. The data show that for both tongue and jaw lowering, and for elbow movements, "the ratio was greater in the discrete than in the continuous condition ($p < .001$)" (p.41).

In summary, the small but reliable discrepancy between V_{max}/V_{avg} values observed in studies of discrete reaching vs. continuous speech can be plausibly explained as a truncation effect. Thus this discrepancy should not block further pursuit of the hypothesis that a common mechanism explains trajectory formation in both systems (see also Cohen, Grossberg, and Stork, 1988).

8. KINEMATIC PROPERTIES AND COMPETING MODELS OF TRAJECTORY FORMATION

The abundance of information on kinematic properties of planned point-to-point arm movements has prompted the formulation of a number of alternative models of trajectory formation. Many of these models—such as the minimum jerk model (Hogan, 1984) or the minimum effort model (Hasan, 1986)—have attempted to rationalize kinematic properties as consequences of optimizations. Thus Hogan proposed that the bell-like shape of observed velocity profiles indicates an optimization for smooth motions, whereas Hasan proposed that the same shapes reflect an optimization that minimizes the "psychological effort" required to produce movement.

Two types of problems—empirical and conceptual—attend such optimization models. The empirical problem is exemplified by the relatively poor predictions made by these models in Table 3, just one illustration that such models often fail quickly when asked to accomodate kinematic properties, such as acceleratory extrema, which are not tightly linked to the optimized variables. Other failures include their inability to explain such phenomena as the rate-dependent asymmetry of velocity profiles (Section 6) and the anomalous amplification of peak velocity during target switching (Section 5).

Yet another difficulty derives from the observed variability in V_{max}/V_{avg}. As Ostry *et al.* (1987) noted, the V_{max}/V_{avg} values found for continuous movement were close to those predicted by a minimum-energy model (Nelson, 1983) whereas the values found for discrete movement were close to those predicted by a minimum-jerk model (Hogan, 1984). Rather than attempting to rationalize these outcomes, which are contradictory within any single optimization framework, we interpret such multiple failures as strong evidence that optimization models do not incorporate the rate-limiting designs that generate the parametric structure of arm movement data.

The conceptual problem (see also Grossberg, 1978b, 1982; Jacob, 1977; Partridge, 1982) with most optimization models is their lack of constituents that bear any resemblance to the constituents of real biological systems. We believe it is this conceptual problem that leads to the empirical problem: If the constituents are not correct, then the many emergent properties of the ensemble other than the one explicitly designed in (the one optimized for) will also typically be wrong.

In contrast to these optimization models, the VITE model of trajectory formation assumes that the modern neural system for limb movements includes the minimal anatomy that affords coordinated use of an array of muscles to smoothly cancel the distance between a current and desired motor state, and to maintain synchrony of activity across the array despite large differences in overall movement rate and even larger changes in the contraction rates of individual synergists. To emphasize this basic design is not to deny that evolutionary selection pressures implicitly identified in various optimization models may also have been operative. For example, selection favoring smoother movements may have tuned parameters such as the size and rate of change of the GO-signal during the movement interval.

9. CONCLUSION

The VITE model represents a simple but robust solution to the problem of generating flexible and synchronous goal-oriented movement commands capable of compensating for variations in initial position, desired target position, desired movement speed, and onset-asynchrony. Computer simulation studies of the circuit's emergent properties demonstrate that it provides a compact quantitative explanation of a large behavioral database. Brain analogues of two key circuit constituents have been identified, each on the basis of multiple functional criteria.

The VITE model is but one of several circuits (witness the PUP circuit of Section 3) that are being developed to explain how humans and related mammals achieve dexterous control of goal-oriented limbs. These circuits provide a quantitative instantiation of the functional problems and design principles that have been identified through analysis of the large interdisciplinary data base concerning eye, arm, and speech articulator movements. The manner in which these different circuits globally parse this problem space is no less important than the properties of the individual circuits themselves, since if one tries to solve the wrong set of problems one can never understand in a principled way the dynamics of a complex biological system.

REFERENCES

Abbs, J.H., Gracco, V.L., and Cole, K.J., Control of multi-movement coordination: Sensorimotor mechanisms in speech motor control. *Journal of Motor Behavior*, **16**, 195–231 (1984).

Beggs, W.D.A. and Howarth, C.I., The movement of the hand towards a target. *Quarterly Journal of Experimental Psychology*, **24**, 448–453 (1972).

Bizzi, E., Accornero, N., Chapple, W., and Hogan, N., Posture control and trajectory formation during arm movement. *Journal of Neuroscience*, **4**(11), 2738–2744 (1984).

Bullock, D. and Grossberg, S., Neural dynamics of planned arm movements: Synergies, invariants, and trajectory formation. Paper presented at the symposium on Neural Models of Sensory-Motor Control at the annual meeting of the Society for Mathematical Psychology, Cambridge, MA, August 20, 1986.

Bullock, D. and Grossberg, S., Neural dynamics of planned arm movements: Emergent invariants and speed-accuracy properties during trajectory formation. *Psychological Review*, **95**(1) (1988a).

Bullock, D. and Grossberg, S., Neuromuscular realization of planned trajectories: Adaptive and automatic mechanisms. In preparation, (1988b).

Cohen, M.A., Grossberg, S., and Stork, D.G., Speech perception and production by a self-organizing neural network. In Y.C. Lee (Ed.), **Evolution, Learning, Cognition, and Advanced Architectures**. Hong Kong: World Scientific Publishing, (1987).

Cooke, J.D., The organization of simple, skilled movements. In G.E. Stelmach and J. Requin (Eds.), **Tutorials in Motor Behavior**. Amsterdam: North-Holland, 199–212 (1980).

Evarts, E.V., Hierarchies and emergent features in motor control. In G.M. Edelman, W.E. Gall, and W.M. Cowan (Eds.), **Dynamic Aspects of Neocortical Function**. New York: Wiley, 557–579 (1984).

Fitts, P.M., The information capacity of the human motor system in controlling the amplitude of movement. *Journal of Experimental Psychology*, **47**(6), 381–391 (1954).

Freund, H.-J. and Büdingen, H.J., The relationship between speed and amplitude of the fastest voluntary contractions of human arm muscles. *Experimental Brain*

Research, **31**, 1–12 (1978).

Georgopoulos, A.P., On reaching. *Annual Review of Neuroscience*, **9**, 147–170 (1986).

Georgopoulos, A.P., Kalaska, J.F., and Massey, J.T., Spatial trajectories and reaction times of aimed movements: Effects of practice, uncertainty, and change in target location. *Journal of Neurophysiology*, **46**(4), 725–743 (1981).

Georgopoulos, A.P., Kalaska, J.F., Caminiti, R., and Massey, J.T., The representation of movement direction in the motor cortex: Single cell and population studies. In G.M. Edelman, W.E. Gall, and W.M. Cowan (Eds.), **Dynamic Aspects of Neocortical Function**. New York: Wiley, 501–524 (1984).

Gordon, J. and Ghez, C., Trajectory control in targeted force impulses, III: Compensatory adjustments for initial errors. *Experimental Brain Research*, **67**, 253–269 (1987).

Grossberg, S., Neural pattern discrimination. *Journal of Theoretical Biology*, **27**, 291–337 (1970).

Grossberg, S., A theory of human memory: Self-organization and performance of sensory-motor codes, maps, and plans. In R. Rosen and F. Snell (Eds.), **Progress in Theoretical Biology, Vol. 5**. New York: Academic Press, 233–374 (1978a).

Grossberg, S., Behavioral contrast in short-term memory: Serial binary memory models or parallel continuous memory models? *Journal of Mathematical Psychology*, **17**, 199–219 (1978b).

Grossberg, S., **Studies of Mind and Brain: Neural Principles of Learning, Perception, Development, Cognition, and Motor Control**. Boston: Reidel Press, (1982).

Grossberg, S. and Kuperstein, M., **Neural Dynamics of Adaptive Sensory-Motor Control: Ballistic Eye Movements**. Amsterdam: Elsevier/North-Holland, (1986).

Hasan, Z., Optimized movement trajectories and joint stiffness in unperturbed, inertially loaded movements. *Biological Cybernetics*, **53**, 373–382 (1986).

Horak, F.B. and Anderson, M.E., Influence of globus pallidus on arm movements in monkeys, I. Effects of kainic acid-induced lesions. *Journal of Neurophysiology*, **52**, 290–304 (1984a).

Horak, F.B. and Anderson, M.E., Influence of globus pallidus on arm movements in monkeys, II. Effects of stimulation. *Journal of Neurophysiology*, **52**, 305–322 (1984b).

Hogan, N., An organizing principle for a class of voluntary movements. *Journal of Neuroscience*, **4**(11), 2745–2754 (1984).

Hollerbach, J.M., Moore, S.P., and Atkeson, C.G., Workspace effect in arm movement kinematics derived by joint interpolation. In G. Gantchev, B. Dimitrov, and P. Gatev (Eds.), **Motor Control**. Plenum Press, (1986).

Jacob, F., Evolution and tinkering. *Science*, **196**, 1161–1166 (1977).

Lestienne, F., Effects of inertial load and velocity on the braking process of voluntary limb movements. *Experimental Brain Research*, **35**, 407–418 (1979).

Moore, S.P., and Marteniuk, R.G., Kinematic and electromyographic changes that occur as a function of learning a time-constrained aiming task. *Journal of Motor Behavior*, **18**, 397–426 (1986).

Munhall, K.G., Ostry, D.J., and Parush, A. (1985). Characteristics of velocity profiles of speech movements. *Journal of Experimental Psychology: Human Perception and Performance*, **11**, 457–474 (1985).

Nelson, W.L., Physical principles for economies of skilled movements. *Biological Cybernetics*, **46**, 135–147 (1985).

Ostry, D.J. and Cooke, J.D., Kinematic patterns in speech and limb movements. In E. Keller and M. Gopnik (Eds.), **Motor and Sensory Processes of Language**. Hillsdale, NJ: Erlbaum, 223–235 (1987).

Ostry, D.J., Cooke, J.D., and Munhall, K.G., Velocity curves of human arm and speech movements. *Experimental Brain Research*, **68**, 37-46 (1987).

Partridge, L.D., The good enough calculi of evolving control systems: evolution is not engineering. *American Journal of Physiology*, **242**, R173–R177 (1982).

Piaget, J., **The Origins of Intelligence in Children**. New York: Norton, (1963).

Soechting, J.F., Effect of target size on spatial and temporal characteristics of a pointing movement in man. *Experimental Brain Research*, **54**, 121–132 (1984).

Woodworth, R.S., The accuracy of voluntary movement. *Psychological Review*, **3**, 1-114 (1899).

Zelaznik, H.N., Schmidt, R.A., and Gielen, S.C.A.M., Kinematic properties of rapid aimed hand movements. *Journal of Motor Behavior*, **18**, 353–372 (1986).

Simplicity from Complexity: Archetypal Action Regimes and Smart Perceptual Instruments as Execution-driven Phenomena

M. T. TURVEY

Center for the Ecological Study of Perception and Action, University of Connecticut, Storrs, CT.

and

Haskins Laboratories, New Haven, CT.

ABSTRACT

The coordinate perception and action capabilities characterizing animals at all phyletic levels pose significant challenges for a truly general theory of complex systems. It is suggested that the following are the most notable: identifying the lawful, macroscopic regularities that underwrite perceiving-acting capabilities; conceptualizing information in a way that bears directly on the perceptual guidance of action and connects closely with dynamics; and comprehending the nature and role of intentionality in terms continuous with physical principles and laws. These three challenges converge on the general issue of how the compositional complexity of a living system and its surround can give rise to simple behavior and function. They are discussed in the context of experiments on execution-driven phenomena as exemplified by "archetypal action regimes" and "smart perceptual instruments."

1.0 *Complexity*

In modern usage, the term "complex" as a qualifier of "physical system" refers to the system's composition, or to the system's behavior, or to both. Compositionally, a system is said to be complex when it has properties of the following kind: (a) the numbers of its parts, internal degrees of freedom, and interactions are large; (b) there are nonlinearities; (c) the constraints are nonintegrable (nonholonomic); (d) the parameters or transition matrices change in time (nonstationarity); (e) there is noise, possibly with inconstant statistical properties; (f) there are multiple levels; (g) all three transport modes (diffusion, wave propagation, convection) are present and intertwined; and (h) there is competition over various scales between influences in the equations of motion (e.g., incoherent diffusive transport vs. coherent convective transport) [1]. Behaviorally or functionally, a system is said to be complex when what it does can be characterized as posing extraordinarily difficult problems of predictability, rich patterning, and/or apparent randomness, even though the system might be represented by a simple equation, expressing a simple determinate dynamic.

2.0 *Compositional complexity: Challenges and perspectives*

Living systems and the environmental circumstances supporting them are representative of compositionally complex systems. The perceiving-acting capabilities that characterize these systems pose three major challenges for a general theory of complexity: identifying the lawful, macroscopic regularities that underwrite perceiving-acting capabilities; conceptualizing information in a way that bears directly on the perceptual guidance of action and connects closely with dynamics; and comprehending the nature and role of intentionality in terms continuous with physical principles and laws.

The dynamic of living systems (as compositionally complex systems) warrant two, contrasting kinds of understandings. On the one hand, there is the kind of understanding typified by the question of how a simple change in a single variable can result in very complicated changes in the system's multiple degrees of freedom. (For example, a perturbation in the degree of flexion at the knee in quadrupedal running is a simple local change that results *in* detailed coherent changes in multiple neural, metabolic, muscular and skeletal components.) On the other hand, there is the kind of understanding typified by the question of how complicated changes in the system's multiple degrees of freedom can result repeatedly in a simple change in a single variable. (For example, a change in the speed or in the direction of quadrupedal running is a simple collective function that results *from* detailed coherent changes in multiple neural, metabolic, muscular, and skeletal components.) The first kind of understanding is associated with the general phenomenon that complex consequences can have simple beginnings. The second kind of understanding is associated with the general phenomenon that simple consequences can have complex beginnings. The study of chaotic dynamics has heightened our awareness of both general phenomena. The damped pendulum and the logistic equation are compositionally simple but both exhibit chaos; many dissipative systems that are compositionally complex exhibit asymptotic behavior describable in very few dimensions (they possess an attractor).

3.0 *The notion of "execution driven"*

The functional order of an animal's movements could follow from crude (sparsely detailed, fuzzy) initial conditions if attractor portraits—singular states (attractors) and surrounding gradients (basins)—evolve in the course of execution to constrain the development of the act [2]. Within such an execution driven scheme, the initial conditions need only put the system into the basins of these attractors. The act can (self) organize by converging onto the attractors via the gradients of states associated with the basins. In an execution driven act, with its evolving dynamic, upcoming states depend more upon current states than upon prior (for example, stored, programmed) states. "Execution driven," therefore, contrasts with "program driven" [3, 4], and expresses simultaneously the two themes of complexity from simplicity (though the act is detailed at its many process scales, its beginnings are sparse) and simplicity from complexity (a few macroscopic observables may suffice to describe the result).

Under certain conditions one movement organization gives way suddenly and spontaneously to another. This "sudden reorganization" provides a useful, concrete example of an execution driven process; the basis for the reorganization evolves in the execution of the activity, and it may be conceptualized as a changing attractor portrait. The phenomenon in question can be observed in a paradigm in which a person is required to oscillate the two index fingers (or two hands) at the same frequency (e.g., [5, 6]). Frequency is the control parameter varied by a metronome that the person tracks; the phase relation is interpreted as an order parameter. Results show that there are only two steady states: in phase (homologous muscle groups contracting simultaneously) and antiphase (homologous muscle groups contracting in an alternating fashion). With increasing frequency, antiphase switches rapidly to in phase. In phase, however, does not switch to antiphase, and the antiphase-to-in phase transition is not reversed by a reduction in frequency. Further, the order parameter exhibits critical slowing down and, under certain conditions, the order parameter exhibits critical fluctuations. These conditions are that the fast time scale associated with the relaxation back to antiphase is notably less than the time scale of change in the control parameter, which is notably less, in turn, than the slow time scale associated with the passage from antiphase to in phase. Order parameter fluctuations do not anticipate the sudden behavioral change when the slow time scale is less than the time scale of the control parameter. The evidence suggests that a comparison of behavioral discontinuities in movement systems with phase transitions in physical systems may be more than mere analogy.

The preceding example of a sudden movement reorganization is from the study of the coordination of two limbs connected neuroanatomically within a person. The same phenomenon, exhibiting many of the same criterial properties, has been observed when two limbs are connected optically between two people [7, 8]. In the experiments in question, two seated people each oscillated a leg with the goal of coordinating the two legs out of phase (180 deg) or in phase (0 deg) as the frequency of the movement was increased. To satisfy the goal, the two people watched each other's lower leg. As far as can be discerned, the between-person case matches the within-person case; a transition occurs, it occurs in only one direction (180 deg to 0 deg), and it is accompanied, under certain conditions, by critical fluctuations. The between- and within-person cases involve the same relations among the same observable quantities despite the obvious differences between the two cases (e.g., the nervous systems of two people vs. the nervous system of one person; using vision in between-person coordination vs. not using vision in within-person coordination).

In both cases, the differential stability of the two phase relations and the transition between them can be accommodated under the assumption that the system of coordinated limbs is governed by a smooth potential that is a function of phase (order parameter) and frequency (control parameter). The function can be so defined that the potential minima (attractors) are located at 180 deg and 0 deg, the function's global minimum is at 0 deg, and the180 deg local minimum is annihilated at some critical control parameter value [9]. The equivalence of the between-person and within-person cases points to a perceptual interpretation of the assumed potential, for example, distance from a threshold that expresses a perceptual limit on the ability to resolve phasings of events [3].

If "execution driven" is a fair way to conceptualize the ordinary activities of animals in their ordinary surroundings more generally, then questions must be raised about how to characterize (a) the initial conditions (their nature and their assembly), (b) the evolving attractor portraits, and (c) the resultant macroscopic organizations. The issue of "intentionality" is related to (a), the issue of "information" is related to (b), and the notions of archetypal action regimes and smart perceptual instruments are related to (c). The first order of business is (c).

4.0 *Systems, subsystems, and simplicity*

Simplifying systems is a recognizable strategy in science. A first step in analyzing any system is to generate an abstraction of it, that is, to replace it by a subsystem that retains some, but ignores most, of its features. The replacement is limited to those contexts in which the interactive capabilities evident in the behavior of the system are the capabilities retained in the subsystem [10]. The transformation of a system to a subsystem leaves, therefore, a particular behavioral or functional capability invariant over some range of conditions. It is, in short, a kind of symmetry operation. Classical mechanics provides celebrated examples. A material body is transformed into a mass point, ignoring the body's shape, volume, molecular construction, etc., and retaining only the observables of mass and position. A rigid body (compound) pendulum is transformed into a mathematical (simple) pendulum. The subsystem retains the rigid body's observables of mass and distance from axis and is equivalent to the rigid body with respect to period of oscillation. Biology's best known example is the replacement of the neuron by a binary switch, retaining only the observables corresponding to the states of active and inactive.

Compositionally complex systems can be thought of as capable of doing something similar to what the scientist does in transforming system into subsystem. The symmetry operation, however, is not S —> S* (where system S is *replaced* by subsystem S*, leaving a certain behavior of S invariant) but S* —> S (where S* is *realized* in S, leaving the behavior of S* invariant). With respect to a scientist generating a subsystem, the simplified description (most observables are deleted, few are retained) could be looked at as an effect. It is, perhaps, no more than the result of the scientist's effort to make the detailed complexity of the system manageable. In contrast, when a compositionally complex system makes a subsystem of itself—for example, a person making him- or herself into a "thrower" or, as in the examples below, a "pendulum clock" or an instrument for measuring products of inertia—the simplified description has to be looked at as a cause; the result is a highly particular dynamic.

The symmetry operation S*—> S is a way to conceptualize the assembling of special-purpose, simple movement regimes and perceptual instruments. In performing a given act, a living system (a compositionally complex system) imposes a particular dynamic on its environment. To do so it must realize (in itself) effectively a subsystem S* that retains just those observables capable of inducing the said dynamic. If the observables are not so restricted, then the dynamic imposed by S on the environment will be other than intended. Similarly, in achieving a given perception, a living system has a particular dynamic imposed on it by its environment or, more exactly, *obtains* a particular dynamic [11]. Any measuring device involves a physical system S that has many properties, but which is required to interact with its environment through a single observable *o*. That is, only *o* ought to induce a dynamic on S, ought to change the state of S. If S* is the subsystem that interacts with other systems only through *o*, then the challenge in assembling a meter is to realize S* in S, leaving the behavior of S* invariant.

5.0 *Smart perceptual instruments*

Experimental results and theoretical analyses suggest that the haptic system (the perceptual system by which one knows about the body and surfaces adjacent to the body by means of the body) realizes one or another kind of meter depending on the property to be perceived. These meters or subsystems are "smart perceptual instruments" [12, 13]. As interpreted here, a smart perceptual instrument is a determinate dynamical machine, assembled over the tissues of the body, that executes a single-valued function. It is smart because it capitilizes on peculiarities or invariants of the measurement circumstance. Given the task of perceiving property y it detects property i, a property that maps uniquely to property y and is a special law-based feature of the situation. In the haptic experiments to be described, the law-based property is the inertia tensor.

A person, asked to wield a uniform, visually occluded rod in order to determine where its tip could reach, tends to perceive the distance reachable to a high degree of accuracy [13]. Perceived reachable distance relates linearly to actual reachable distance, with a near zero intercept and a slope that does not differ significantly from unity. This linear relation holds for rods of different densities, for rods wielded in, and perpendicular to, the body's sagittal plane, and for rods wielded at different rates. Manipulating the position of an attached weight on an otherwise uniformly dense rod and manipulating where the rod is grasped, reveals that the moment of inertia of the rod about its axis of rotation is the kinetic basis for this haptic distance perception [13]. Perceived reaching distance relates to moment of inertia by a power formula in which the exponent is approximately 1/3.

The motions of a rod during wielding about axes in the wrist are the three dimensional motions of a rigid body about a fixed center of rotation. Consequently, the moment of inertia in the rod's motion equations is in the form of the inertia tensor. The six off-diagonal terms in this symmetric tensor are the products of inertia; they determine reactive forces at the center of rotation, causing it to wobble. The three diagonal terms are the moments of inertia with respect to the three coordinate axes. A transformation to the principal or symmetry axes of the rigid body eliminates the products of inertia and leaves only the principal moments of inertia on the diagonal. The experimental data suggest that the haptic system behaves in a manner that is functionally equivalent to the principal axis transformation. It separates off the principal moments of inertia from the products of inertia. Moreover, given that the rigid-body motions of a hand-held rod bring about deformable body motions in the tissues comprising the haptic system, the implied functional diagonalization must include the deformation (stress/strain) tensor that characterizes deformable body motions. In sum, in satisfying the intention to perceive the distance reachable with an unseen hand-held rod, the haptic system S realizes a subsystem S* that detects a property of tissue deformation causally determined by the rod's principal moment(s) of inertia.

The execution driven nature of a smart perceptual instrument is illustrated nicely by the fact that subjects can perceive *fractional* lengths when rods are held at positions intermediate between their ends ("how far could you reach with the portion of the rod extending beyond your grasp?"). This additional perceptual achievement depends similarly on principal moments of inertia—but now they are the principal moments of the rod fraction [13, 14]. It would appear that the haptic system, when needed, can decompose the inertia tensor for a given axis of wielding into distinct components corresponding to the mass distributions above and below the given axis. For simplicity, let I be the moment of inertia of the entire rod about a given axis and let I_a and I_b be the moments of inertia of the rod above the axis and the rod below the axis, respectively. When the task is to perceive the whole length of a rod (regardless of hand position), the haptic system behaves in such a way as to realize, in the course of executing the exploratory wielding, a subsystem that registers the dynamics of the whole hand-held object through the principle of superposition or parallel composition ($I = I_a + I_b$). In contrast, when the task is to perceive fractional length, the haptic system behaves in such a way as to realize, in the course of executing the exploratory wielding, another kind of subsystem, one that filters the dynamic of the requisite portion of the rod through subtraction ($I - I_b = I_a$)[14].

The off-diagonal terms of the inertia tensor are associated, as noted above, with bearing forces acting back on the axis of rotation. They cause the axis to "wobble". These variable components of the three-space motion of a wielded object are nullified when the rod's motions are about a symmetry axis of the object. Consequently, a person wielding a hand-held, visually occluded L-shaped rod should be able to identify the direction in which the rod is pointing if he or she can register the orientation at which the forces on the axis of rotation in the wrist are at balance. Recent experiments show that people can perceive the direction with reasonable accuracy [15], suggesting that the haptic system can realize a subsystem that registers the inertia tensor's off-diagonal quantities.

It is helpful to look at the foregoing experimental results in the general terms of an interaction between two systems, the rigid body system of the hand-and-rod (S_r), and the haptic perceptual system (S_h). The three-space motion of S_r about a fixed point, that is produced by the performatory aspect of S_h, induces a dynamic on S_h. Three questions can be posed. First, *what* is the observable o through which the interaction is taking place? Second, *how* is S_h assembled as a subsystem, a meter S^*, that registers o (that is, only takes into account the interactions with states of S_r through the observable o). Third, *why* are o and the property to be perceived linked. (Why, for example, are the principal moment of inertia and reachable distance linked? why are products of inertia and direction linked?) Importantly, the subjects in the experiments were unpracticed. The subjects were merely instructed as to the property that was to be discerned by wielding. The third question, therefore, raises the issue of how to understand the design of a compositionally complex system S such that, given the challenge to perceive a given property, say *reaching distance*, it assembles a part of itself spontaneously into a simple meter S* that registers *principal moments of inertia*. Does a symmetry principle fix the linkages among the propertied intentional states of S, the propertied informational states of a structured energy distribution (see 9.0), and the propertied states of environmental objects [16]? The idea that symmetry dictates design has figured prominently in the physics of inanimate matter [17]. Perhaps it should also figure prominently in an understanding of perceiving-acting capabilities [18].

6.0 *An archetypal dynamical regime*

Attention is now turned to the S* —> S transformation in movement, with S* considered as a simple dynamical regime. At the macroscopic scale of acts, it may be conjectured that there is a finite dictionary of archetypal movement regimes [19, 20]. One such archetypal regime will be described, together with its most significant biological properties. The regime is the pendular, harmonic motion of the individual limbs that characterizes the locomotion of bipeds and quadrupeds, birds and insects.

A preferred period of locomotion can be understood as the resonance period of a hybrid simple pendulum/mass-spring system [21]. That is, with respect to the periodic timing of locomotion, any given biped or quadruped as system S realizes a simple pendulum/mass-spring as subsystem S*. The constituents of S* are a point mass m at a distance l from a fixed axis doing work against two conservative forces. These forces are in the form of gravity acting on the point mass, and a spring of stiffness k acting at a distance b from the axis. The subsystem's characteristic period can be expressed as $\tau_0 = 2\pi[ml^2/(mlg + kb^2)]^{1/2}$ with muscular and other tissues functioning macroscopically as the spring that elastically stores and releases mechanical energy. Regular oscillations are brought about and sustained by muscular forces that are usually close to resonance. The resultant dynamical regime—basically, raising and lowering a mass at regular intervals with respect to gravity—will be called the pendular clocking mode (PCM) of movement organization [2] and the preceding equation will be called the PCM equation.

If the set of limbs of a quadruped is regarded as a compound pendulum, convertible into an equivalent simple pendulum equivalent of length l, then quadruped walking, trotting, and cantering approximate closely the cycle times of a freely oscillating pendulum of length $l/2$, $l/7$ and $l/10$, respectively [2, 21]. Let the torsional stiffness due to gravity, namely mlg, equal G , and let the torsional stiffness due to the body's tissues, namely kb^2, equal K. Values of τ_0 can be generated by the PCM equation for different relations of K to G, and for varying pairs of m, l values constrained by $l \propto m^{1/3}$ (biological lengths scale as biological masses most generally to the 1/3 power, according to geometric similarity [22]). Given K/G = any constant, multiple regression of log τ_0 on log m and log l reveals that the mass exponent vanishes and the length exponent is 1/2. With simple regression the mass exponent is 1/6 and the length exponent is 1/2. The foregoing exponents from multiple and simple regressions accommodate the scaling facts of quadruped locomotion in all three gaits [2, 21, 23]. Further, and perhaps more significant, if $K/G = 1$, then the PCM equation reduces to $\tau_0 = 2\pi(l/2g)^{1/2}$ which characterizes walking; if $K/G = 6$, and $K/G = 9$ then the equation reduces to $\tau_0 = 2\pi(l/7g)^{1/2}$ and $\tau_0 = 2\pi(l/10g)^{1/2}$, which characterize trotting and cantering, respectively. In sum, it can be argued that the elastic restoring torque K assembled by each animal in a given gait is the same constant multiple of the gravitational restoring torque G associated with the magnitude of each animal, despite wide differences in animal sizes and animal shapes. The K/G dimensionless scalars may be looked at as the signatures of attractive states that organize the locomotory systems of (perhaps) all quadrupeds.

Does the archetypal PCM apply even more generally? In addition to expressing the mass of the pendulum as proportional to the cube of its length, the equation $\tau_0 = 2\pi[ml^2/(mlg + kb^2)]^{1/2}$ can be confined to biology more fully by expressing the restoring torque due to the spring as directly proportional to the pendulum's mass. The elastic system—the spring—is constructed in different ways across the roster of animals bounded by the smallest insect and the largest quadruped. In insects, it is built from at least three different materials or structures: (a) the solid skeletal cuticle of the thoracic box; (b) a typical elastomere in the form of the protein resilin; and (c) an elastic component in the myofibrils [24]. Vertebrates cannot duplicate the structure of the insect elastic system [25], but they do approximate its function. Vertebrate muscular tissue stores and uses mechanical energy when the mechanics of an activity demand that contracted muscles be stretched forcibly immediately before shortening, as would be the case in the oscillatory activity of wings and limbs relative to the body [26]. This recovery of mechanical work is made possible by the muscle's elastic elements (the tendons and the cross-bridges between actin and myosin) and the disposition of the muscle's activated contractile machinery to resist lengthening forcibly enough to stretch the elastic elements [26]. Increasing the muscular force resisting lengthening, increases the stretch of the elastic structures and the amount of mechanical energy that can be stored. According to a major interpretation of mammalian muscle control, the tonic recruitment of fibers begins from a specific threshold length, with the recruitment and the resultant muscle force (to a first approximation) increasing as a function of the difference between the current muscle length and the threshold value [27]. Some vertebrates, therefore, possess a mechanism for setting K for a given oscillatory movement. It resides in the neural components and processes that adjust the aforementioned threshold.

Patently, species differ in the details of the system that provides elastic energy, the spring term in the PCM equation. The differences are noteworthy because they result in different behavioral consequences. Thus, insects are restricted to moving their wings at an almost constant frequency, irrespective of flight maneuvers and speed of travel [28, 29]. The same appears to be true of birds [30, 31]. In contrast, quadrupeds can walk, trot, and canter at a number of different frequencies although each gait does exhibit a preferred frequency that appears to be metabolically least costly [32]. Various analyses suggest, however, that despite the wide differences in the construction of the elastic machinery, there is a constant relation across these animals between elastic recoil at the characteristic frequency and body mass, namely, K increases as the first power of body mass [21]. The PCM equation can be constrained in a very general biological way, therefore, as identified at the beginning of the preceding paragraph—m and l are linked quantities ($m \propto l^3$), and so are K and m ($K \propto m$).

The relations derived from the biologically constrained PCM equation between log period and log length, and between log period and log mass, exhibit a number of interesting properties for the range of lengths covering flying insects and quadrupeds. The relations are significantly quadratic and linear. For lengths < 0.1m, period scales to length to the 0.913 power and to mass to the 0.305 power; for lengths > 0.1 m, period scales to length to the 0.543 power and to mass to the 0.183 power. These time allometries are characteristically those found for insects (< 0.1m) and quadrupeds (> 0.1m), respectively. The region of transition between the two slopes, that is, the region of length scales in the vicinity of 0.1m, is the region occupied by large birds, small birds (passerines), and hummingbirds. Where the periods of large birds scale to length close to the 0.5 power, those of small birds scale close to the 1.0 power. Moreover, the relation in double logarithmic coordinates between period and length is largely quadratic for hummingbirds, yielding what has been, traditionally, an odd scaling with an exponent of 1.2 [31]. Collectively, this pattern of results for large birds, small birds, and hummingbirds is rationalized by the PCM equation constrained in the manner of biological systems. The pattern of time allometries does not point to shifts in biological design, but to the physical consequences of PCM operating at particular length magnitudes. Finally, it should be noted that the most general scaling reported in the literature between biological times and biological masses is $time \propto mass^{1/4}$ [33]. This scale is referred to as the fundamental physiological time scale [34], because of its ubiquity. It has not been easy, however, to reconcile the invariant physiological time scale to the fact that biological length is proportional to the cube root of biological mass. The reconciliation, however, may require no special explanations. For the *complete* range of lengths covering flying insects and quadrupeds, the PCM equation constrained by $m \propto l^3$ and $K \propto m$, yields $time \propto mass^{0.24}$ [21].

What happens when PCM is assembled freely and intentionally, as might occur with a human subject in a laboratory? To examine PCM with a single limb, a person holds a weighted pendulum in one hand and oscillates it about the wrist joint. The person is instructed to swing at the most comfortable tempo (to encourage minimal muscular energy and a rhythmic movement at or close to the charcteristic period). For different pendulums the person settles at different tempos. To examine PCM with two limbs, a person holds two pendulums, one in each hand, and swings both simultaneously at the wrists. Whether the two pendulums are swung in phase or out of phase, the person is instructed to swing them at the same tempo, and to do so comfortably. For different pairs of weighted pendulums, the person settles at different, common tempos. By conceptualizing two, coupled wrist-pendulum systems as rigidly connected (a compound pendulum), it is possible to replace them by a single, equivalent simple pendulum with all of its mass m concentrated at a distance l from a single virtual axis of rotation [2]. In this manner, single wrist-pendulum systems and double wrist-pendulum systems can be treated equivalently.

Data were collected from four subjects on the preferred periodic times of 4 left single wrist-pendulum systems, 4 right single wrist-pendulum systems, and 16 coupled wrist-pendulum systems (those produced by combining the left and right systems). The m and l values across these systems (both single and "virtual" single) were purposely unlinked and, therefore, quite different from the m and l values of limbs and wings that are linked by a power formula. A multiple regression conducted on the data of all four subjects revealed that periodic times were proportional to $m^{0.056} l^{0.469}$. The 95% confidence limits were .024 (lower) and .094 (higher) for mass, and .429 (lower) and .540 (higher) for length; both exponents were significant at the .001 level [2]. Where does this time scale come from? Originally the supposition was that it followed from fundamental physiological time scales adjusted to the mechanical constraints of the task. Subsequent investigation showed, however, that it was a direct consequence of assembling PCM over (a) a restricted range of K values (in the vicinity of 1 Nm/rad), and (b) a set of wrist-pendulum systems for which m did not scale systematically to l [21]. Given the circumstances defined by points (a) and (b), $\tau_0 \propto m^{0.052} l^{0.548}$ is the time scale generated by the PCM equation.

Apparently, a person in executing rhythmic pendular movements with one or two limbs, and with the intention to do so comfortably and at a common period, effects the transformation S* —> S, where S* is PCM. That is, the person realizes the functionally simple, abstract subsystem S* characterized by the PCM equation with the compositionally complex tissues and processes of his or her body S.

7.0 *Intentional content*

The preceding two subsections give an inkling of how intentionality can be thought about in ways that are close to the language of physical laws and principles. Consider two simple kinds of intentional contents: "I intend that I perform Y," and "I intend that I perceive Z." Though written, by way of illustration, as spoken sentences, intentional contents are overwhelmingly ineffable. It will be suggested that intentional contents, as aspects of living systems, define a class of soft (nonholonomic) constraints harnessing dynamics [2, 35]. This harnessing is in the form of the S* —> S transformation. That is, the constraints in this class realize abstract subsystems, those that impose selective dynamics on the surround ("... to perform Y"), and those that obtain selective dynamics from the surround ("... to perceive Z"). At the heart of this formulation of intentionality is the notion that perceiving-acting systems are ordinary physical systems that have discovered, and can discover further, extraordinary means of using natural laws and principles. The most prominent sources of regularity for clever exploitation are those that penetrate the scale of perceiving-acting, the ecological scale.

8.0 *Lawful basis of perception and action*

It is not customary to think about perceiving-acting capabilities in general physical terms. Ordinarily, these capabilities are investigated by psychologists, biologists, neurophysiologists, and (more recently) computer scientists. Within the respective disciplines of these investigators, attempts to explain how the individual animal perceives its surroundings and coordinates its actions are guided usually by the metaphor of neural substrate as "efficient cause". According to this metaphor, the nervous system—conceptualized in either neural terms (as in neurophysiology), quasi-neural terms (as in connectionist/neuromorphic machines), or symbolic terms (as in von Neuman machines)—is causally responsible for perceiving-acting phenomena.

It requires the adoption of a different metaphor, that of neural substrate as "supportive medium" [36], to direct thinking toward a more general style of explanation. The perceptually guided actions of animals are macroscopic phenomena at the ecological scale of nature—the scale at which the physical entities of animal and environment are defined. Perceiving and acting are reciprocal functions and comprise one of several kinds of functional order germane to the ecological scale. It is proposed that a theory of this functional order ought to be stated in terms of the suitable observables of the ecological scale, and the relationships that exist among them. There is no reason to expect a priori, however, that the observables actually involved in animal-environment interactions at the ecological scale correspond to those that can be measured directly with the instruments devised thus far to satisfy the various goals of physical inquiry [10]. As expressed originally by Gibson [11, 37, 38], the perceiving-acting capabilities characterizing the ecological scale are underwritten by lawful relations (to be determined) among macroscopic observables (to be discovered and measured). If such lawfulness exists, then living things might be considered as constituting, along with the modulatable energy distributions that envelop them, a rich medium for supporting the linkages among these macroscopic observables. According to this less orthodox metaphor, if we are to understand the causal responsibility for the order and the regularity, the novelty and the adaptability, of perceiving and acting, then we must come to terms with the lawful relations characterizing nature's ecological scale [2, 39]. As the next section details, there is a close connection between these lawful relations and a conception of information suited to perceiving-acting capabilities.

9.0 *Information*

What might be termed an "ecological conception" of information is emerging to meet the challenge identified in 2.0. It links the notion of information with the notion of lawful regularity, particularly, the lawful regularity in evidence at the ecological scale [37]. In some quarters the ecological conception of information is expressed broadly as the idea that any flow morphology that can be defined reliably on a low energy field (that is, a field whose kinetic terms are very small relative to those of the systems it envelops or connects) is potentially a source of information about the dynamic that gave rise to it [2, 40].

The perceiving-acting interactions between animal and environment are said to occur through a class of macroscopic, nonkinetic (kinematic, geometric, spectral) observables. The observables in this class are alternative descriptions of ambient energy distributions (optical, acoustical, mechanical, chemical, etc.). Specifically, they are not those aspects of an energy distribution that are most conveniently and easily measured by contemporary scientific instruments, as noted above. Rather, they are mass-independent, primarily qualitative and large scale, properties of the distribution. Further, for each observable in this class, there is an invariant (that is, lawful) relation with either an observable of surface layout (the basis of exteroception), or an observable of self-movement (the basis of proprioception), or an observable of the surface layout-moving animal relation (the basis of exproprioception, [41]). Consequently, the observables in this class are said to be *information about, in the sense of specific to,* properties of the animal-environment system.

Efforts thus far to identify these "information" observables have used the techniques of the vector calculus and the insights of hydrodynamics and deformation analysis (e.g.,[42]). Imagine a "point of observation" (that could be occupied by an eye of an animal) in a transparent medium (air, water). The light multiply reflected from the surrounding surfaces, that is ambient to (surrounds completely) the point of observation, has structure. It has structure at multiple nested scales, and does so because of the simple ecological fact that the surrounding surfaces have structure at multiple nested scales [37]. If the point of observation is in motion, then each component of structure can be described by a vector with the totality of vectors comprising a velocity vector field. If the field is described (for convenience) with the aid of a sphere surrounding the observation point, then the changes in velocity in a local neighborhood (on the tangent plane) of any component can be defined by taking the derivative of the velocity vector function with respect to distance (on the tangent plane). The totality of such changes is a tensor quantity. The optical flow field so described can be decomposed, in preliminary fashion and in standard ways, into three major components identifying aspects of local structure, namely, curl (amount of rotation), divergence (amount of dilation and contraction), and deformation (amount of angular strain).

The preceding rough sketch permits, by way of example, the identification of two observables of optical flow that bear significantly on action's control. First, the inverse of the relative rate of change in any locally defined dilation, referred to as *Tau* [43], is specific to the *immediacy* of contact with the environmental surface generating this particular local region of optical structure. That is, there is a phenomenological law of ecological optics of the following kind: Rectilinear motion toward substantial surface *E* generates the optical property *Tau* that is specific to the time that will elapse before contact is made with the surface under the current conditions of motion [42, 43, 44, 45]. Second, the rate of change in any locally defined *Tau* is specific to the *harshness* of contact with the environmental surface generating this particular local region of optical structure. That is, there is a phenomenological law of ecological optics of the following kind: Rectilinear motion toward substantial surface *E* generates the optical property $d(Tau)/dt$ that specifies a violent contact (large momentum transport) with the surface under the current conditions of motion if $d(Tau)/dt < -0.5$, and a soft contact (no momentum transport) if $d(Tau)/dt \geq -0.5$ [44].

These two examples merit closer examination. There is a system *A-E* (animal-environment system). At a specific instant of time t_p there is (a) a particular layout of substantial surfaces, (b) a direction of motion in one component, namely *A*, that is toward a substantial surface E_i and (c) a pattern of forces producing this motion. Collectively, they comprise a state of the system *A-E*. This state can unfold, as it were, into a state at time t_r in which *A* (the component previously in motion) and E_i (a component previously not in motion) come into contact. The optical structure generated by the state of *A-E* at t_p contains properties that are *specific to this unfolding*. To repeat, *Tau* specifies when contact will be made, d(*Tau*)/d*t* specifies the type of contact that will be made. These optical properties have been referred to as observables on the *A-E* state at t_p. They are, however, observables of an odd kind.

Ordinarily, the concept of *observable* refers to some characteristic of a system that is measurable directly, in principle, and logically coextensive with the concept of state [10]. One forces a distinction, however, between observable and state by appending to the concept of observable the idea of a concrete procedure for evaluating an observable on a state, that is, a measuring instrument or meter [10]. In the ecological system under inquiry here, the animal's perceptual apparatus comprises the meters in question, but they are not required for distinguishing observable from state. This is because an "information" observable on a state, and on a state's pending evolution, is not logically or conceptually identical with or similar to the state—d(*Tau*)/d*t* < -0.5, for example, is *not* a hard collision. In very general terms, information about a state is *not* the state (although more conventional approaches to perception often seem to claim that it is). A kinetic state of affairs (call it *x*) structures the optical distribution in such a way that a nonkinetic macroscopic optical property (call it *o*) can be defined that is specific to what the state of affairs will evolve into (call it *y*) if the conditions continue. The structure of this piece of perceiving-acting is theoretically challenging, and representative. (We encountered it before in the discussion of smart perceptual instruments.) It can be summarized in the following manner: There is an observable *o* lawfully linked to, but conceptually distinct from, a state *y* of *A-E*; any system with a meter that detects *o*, is a system that perceives *y*.

It is interesting to speculate how one might try to explain the changes in the forces an animal is producing currently, to achieve a given kind of contact imminently, in the absence of knowledge about d(*Tau*)/d*t*. Clearly, the explanation could become quite involved. There is a moral here. An animal-environment system is, most obviously, a compositionally complex system. But we must be prepared for the striking possibility that the *macroscopic* observables that characterize this system, and which are responsible for the detailed interactions exhibited between its components, permit an enormously simplified description of behavior and function.

Another aspect of the preceding information discussion deserves comment. "Information" observables, such as *Tau* and d(*Tau*)/d*t*, are impermanent. They are generated by activity and desist when activity stops. During their existence, however, they are specific to the conditions of an activity and provide macroscopic boundary conditions shaping the manner of its execution. The preceding suggests how we might comprehend the evolving attractor portraits in the generalized notion of execution driven action (Section 2.0). They are the morphologies of nonkinetic flow fields [2, 46].

10.0 ***Advancing the analysis of optical structure***

The stationary and transforming optic array is an object of study in its own right. It defines the context within which mobile visual creatures evolved and within which they conduct their daily affairs. Serious treatment of optical flow structure in search of the observables supporting vision and its control of action is a comparatively new enterprise (see [47], for an overview). Much of what has been achieved to date rests largely on the application of well tried mathematical methods and physical ideas (as represented in the sketch of optical analysis given above). The optic array's detailed structure at many scales and certain peculiarities suggest, however, the need for novel approaches to its analysis.

One especially noteworthy peculiarity follows from the simple ecological facts that the ambient optic array is completely filled with structure (there are no gaps) and that the ground is usually cluttered with objects [37]. Consequently, the optical structure that completely and continuously fills the ambient optic flow at a moving point of observation is being deleted and accreted continuously (at locally different rates and by locally different amounts) as environmental surfaces become hidden and unhidden. This peculiarity makes the use of projective geometry as a general method of optical analysis suspect, and directs attention to new methods (mathematical and physical) that could address explicitly the relations between rich optical patternings of structural accretion and deletion and the animal-environment circumstances that generate them. In this regard, the indefinite nesting of optical structure, and the fact that it is generated by the faces and facets of surfaces that are variably sized, variably oriented, irregular, cracked, rough, pitted, pocked, textured at multiple scales, and so on, suggests that it may have to be interpreted in fractal terms [48]. The traditional Euclidean description of what constitutes a layout of surfaces, the geometric portrayal of the environment, fits the ecological facts poorly; a fractal description fits them better.

11.0 ***Autonomy of the dynamic at the ecological scale***

To conclude, a central assumption can be made regarding the perceiving-acting capabilities of living systems, and the general issue of how functional simplicity is distilled from compositional complexity: The macrodynamic of the ecological scale is autonomous and sovereign, and plays back as boundary conditions on the dynamic of a more microscopic nature [46]. Appreciating this assumption requires an attitude like that commonly taken toward the cell by students of brain function, namely, that the boundaries of cells (the cells' membranes) border the domain of the relevant interactions and delimit, thereby, the most suitable observable quantities. Processes contained within the membrane, processes interior to the cell, are part of a separate dynamic—one for which the intercellular interactions serve as dynamic boundary conditions. The point is similarly made with respect to perception and action: The *animal* (not its receptor units, not its motor units, etc.) sets the scale on the relevant interactions and the observables through which they occur [11]. Understanding the perceiving-acting capabilities of animals at all phyletic levels—how they distill functional simplicity from compositional complexity—ought to be sought in terms of these observables and their dynamic.

Acknowledgements. Preparation of this paper was supported in part by a James McKeen Cattell Fellowship. The research reported was made possible, in part, by ONR Grant NOOO14-83-C-0083 and BRS Grant RR-05596, awarded to the Haskins Laboratories, and a grant from the University of Connecticut Research Foundation awarded to the author.

References

[1] Yates, F. E., "Outline of a physical theory of physiological systems," *Journal of Physiology and Pharmacology*, **60**, 217-248 (1982).

[2] Kugler, P.N., & Turvey, M.T., *Information, natural law, and the self-assembly of rhythmic movement*. Hillsdale, NJ: Lawrence Erlbaum Associates (1987).

[3] Kugler, P. N., "A morphological perspective on the origin and evolution of movement patterns," In M. G. Wade & J. Whiting (Eds.), *Motor skill acquisition in chidren: Aspects of coordination and control*. The Hague, Netherlands: Martinus Nighoff (1986).

[4] Yates, F. E., "Quantumstuff and biostuff: A view of patterns of convergence in contemporary science," In F. E. Yates (Ed), *Self-organizing systems: The emergence of order*. New York: Plenum Press (1987).

[5] Kelso, J.A.S., Scholz, J.P., & Schöner, G., "Nonequilibrium phase transitions in coordinated biological motion: Critical fluctuations," *Physics Letters*, **118**, 279-284 (1986).

[6] Scholz, J. P., Kelso, J. A. S., &Schöner, G., "Nonequilibrium phase transitions in coordinated biological motion: Critical slowing down and swithching time," *Physics Letters*, **123**, 390-394

(1987).

[7] Schmidt, R. C., Carello. C., & Turvey, M. T., "Visual coupling of biological oscillators," *Perceiving-Acting Workshop Review (Technical Report of the Center for the Ecological Study of Perception and Action, University of Connecticut)*, **2**, 22-24 (1987).

[8] Schmidt, R. C., Carello, C., & Turvey, M. T., "Phase transitions and critical fluctuations in visually coupled oscillators," (in preparation).

[9] Haken, H., Kelso, J. A. S., & Bunz, H., "A theoretical model of phase transitions in human movements," *Biological Cybernetics*, **51**, 347-356 (1985).

[10] Rosen, R., *Fundamentals of measurement and representation of natural systems*. New York: North Holland (1978).

[11] Gibson, J.J., *The senses considered as perceptual systems*. Boston: Houghton Mifflin (1966).

[12] Runeson, S., "On the possibility of 'smart' perceptual mechanisms," *Scandinavian Journal of Psychology*, **18**, 172-179 (1977).

[13] Solomon, Y., & Turvey, M.T., "Haptically perceiving the distances reachable with hand-held objects," *Journal of Experimental Psychology: Human Perception and Performance* (in press).

[14] Solomon, Y., Turvey, M.T., & Burton, G., "Perceiving the partial and total lengths of rods by wielding: Haptic diagonalization and decomposition of the inertia tensor," (submitted).

[15] Solomon, Y., Turvey, M. T., Burton, G., & Runeson, S., "Haptically perceiving the direction in which a hand-held object is pointing," (in preparation).

[16] Turvey, M. T., "The challenge of a physical account of action: A personal view," In H. T. A. Whiting & O. G. Meijer (Eds.), *A natural-physical approach to movement*. Amsterdam: North-Holland Press (in press).

[17] Zee, A., *Fearful symmetry*. New York: MacMillan (1986).

[18] Shaw, R. E., & M. McIntyre., "Algoristic foundations to cognitive psychology," In W. Weimer

& D. Palermo (Eds.), *Cognition and the symbolic processes.* Hillsdale, NJ: Lawrence Erlbaum Associates (1974).

[19] Gelfand I. M., Gurfinkel, V. S., Tsetlin, M.L., & Shik, M.L., "Some problems in the analysis of movements," In I. M. Gelfand, V. S. Gurfinkel, S. V. Fomin, & M. L. Tsetlin (Eds.), *Models of the structural-functional organization of certain biological systems* . Cambridge, MA: MIT Press (1971).

[20] Iberall, A. S. personal communication.

[21] Turvey, M. T., Schmidt, R. C., Rosenblum, L. D., & Kugler, P. N., "On the time allometry of coordinated rhythmic movements,"*Journal of Theoretical Biology* (in press).

[22] Peters, H. P., *The ecological implications of body size.* Cambridge, MA: Cambridge University Press (1983).

[23] Pennycuick, C., "On the running of the gnu (*Connochaetes taurinus*) and other animals," *Journal of Experimental Biology*, **63**, 775-799 (1975).

[24] Buchtal, F., & Weis-Fogh, T., "Contribution of the sarcolemma to the force exerted by resting muscle of insects," *Acta Physiologica Scandinavia*, **35**, 345-364 (1956).

[25] Weis-Fogh, T., "Quick estimates of flight fitness in hovering animals, including novel mechanisms of lift and production," *Journal of Experimental Biology*, **59**, 169-230 (1973).

[26] Cavagna, G.,"Elastic energy in muscle," *Exercise and Sports Science Reviews*, **5**, 89-129 (1977).

[27] Feldman, A. G., "Once more on the equilibrium-point hypothesis (lambda model) for motor control," *Journal of Motor Behavior*, **18**, 17-54 (1986).

[28] Sotavalta, O., "The essential factor regulating the wing stroke frequency of insects in mutilation and loading experiments and in experiments in subatomospheric pressure," *Annals of the Zoological Society 'Vanamo'*, **15**, 1-66 (1952).

[29] Sotavalta, O., "The effect of wing inertia on the wingstroke frequency of moths, dragonflies, and

cockroach," *Annales Entomologici Fennici*, *20*, 93-101 (1954).

[30] Greenewalt, C. H., "The wings of insects and birds as mechanical oscillators,"*Proceedings of the American Philosophical Society*, **104**, 605-611 (1960).

[31] Greenewalt, C. H., "The flight of birds," *Transactions of the American Philosophical Society*, **65**, 1-67 (1975).

[32] Hoyt, D. F., & Taylor, C. R., "Gait and the energetics of locomotion in horses," *Nature*, **292**, 239-240 (1981).

[33] Calder, W. A. III., *Size, function and life history.* Cambrige, MA: Harvard University Press (1984).

[34] Lindstedt, S. L., & Calder, W. A., "Body size, physiological time and longevity of homeothermic animals," *Qarterly Review of Biology*, **56**, 1-16 (1981).

[35] Turvey, M. T., "Intentionality: A problem of multiple reference frames, specificational information, and extraordinary boundary conditions on natural law," *Behavioral and Brain Sciences*, **9**, 153-155 (1986).

[36] Ben-Zeev, A., "The Kantian revolution in perception." *Journal for Theory of Social Behavior*, **14**, 69-84 (1985).

[37] Gibson, J.J., *The ecological approach to visual perception.* Boston: Houghton Mifflin (1979).

[38] Reed E.S., & Jones, R. (Eds.), *Reasons for realism: Essays of J.J. Gibson* (pp. 61-75). Hillsdale, NJ: Lawrence Erlbaum Associates (1982).

[39] Turvey, M. T., Shaw, R. E., Reed, E., & Mace, W., "Ecological laws of perceiving and acting: In reply to Fodor and Pylyshyn (1981)," *Cognition*, **9**, 237-304 (1981)

[40] Kugler, P. N., Turvey, M. T., Carello, C., & Shaw, R. E., "The physics of controlled collisions: A reverie about locomotion," In W. H. Warren & R. E. Shaw (Eds.), *Persistence and change.* Hillsdale, NJ: Lawrence Erlbaum and Associates (1985).

[41] Lee D. N., & Lishman, R., "Visual control of locomotion," *Scandinavian Journal of Psychology*, **18**, 224-230.

[42] Koenderink, J., "Optic flow," *Vision Research*, **26**, 161-180.

[43] Lee, D. N., "A theory of visual control of nraking based on information about time-to-collision," *Perception*, **5**, 437-459.

[44] Lee, D.N., "Visuo-motor coordination in space-time," In G. E. Stelmach & J. Requin (Eds.), *Tutorials in motor behavior* (pp. 281-293). Amsterdam: North Holland (1980).

[45] Todd, J. T., "Visual information about moving objects," *Journal of Experimental Psychology: Human Perception and Performance*, **7**, 795-810 (1981).

[46] Kugler, P. N., & Turvey, M. T., "Self organization, flow fields, and information," *Journal of Human Movement Studies (Special issue: Self organization in human movement systems)* (in press).

[47] Turvey, M. T., & Carello, C., "The ecological approach to perceiving-acting: A pictorial essay." *Acta Psychologica*, **63**, 133-155 (1986).

[48] Mandelbrot, B. *The fractal geometry of nature*. New York: Freeman (1977).

Dynamical Approaches to the Development of Behavior

ESTHER THELEN

Department of Psychology
Indiana University
Bloomington, Indiana 47405
USA

ABSTRACT

Principles of dynamical systems can be extended to phenomena in ontogenetic time. Here I characterize the well-known transitions in the development of upright stepping in human infants from a dynamical perspective. I suggest that developmental change may result from three processes: phase shifts, stabilizing attractors, and acquisition of new control parameters. A dynamical perspective offers new empirical strategies for developmental research.

1. FROM REAL TO DEVELOPMENTAL TIME

Principles of dynamical systems offer powerful concepts and methods for understanding how complex biological systems produce patterned behavior. The theorists and experimentalists at this conference have applied these principles to behavior emerging in "real time"--- the time of neural networks, motor systems, physiological processes, and the psychological domains of perception, cognition, and action. Here I explore a dynamical approach to order and complexity that emerges over a much longer time scale, that of ontogeny. I focus particularly on the development of behavior, although the approach may well generalize to other morphogenetic processes.

To enter the developmental domain, we must expand our conventional time metric by many orders of magnitude. The fundamental questions, however, are independent of time scale. How do new behavioral forms and functions appear from antecedent conditions which do not contain those forms and functions? Can order emerge without autonomous central executives or pre-existing codes? What gives developing behavior both stability of form and flexibility of function? How can continuous processes lead to discontinuous outcomes? What is the relation between the organism and its environment?

In their search for theoretical principles that address these difficult core issues, developmentalists, like other biological scientists, have evolved two very different world views [1,2]. One view uses the machine as the root metaphor. The organism is composed of parts to be analyzed: genetic instructions, responses to stimuli, neural or anatomical units, or information-processing elements. Development is then these orderly relationships within units writ large; the process is no more than can be accounted for by knowledge of the units. The second view, in the tradition of von Bertalanffy, Weiss, and others, is the systems or organismic model. Here development is seen more as process than product, and the emphasis is on emergent properties, multicausality, autonomy, and indeterminancy. I think it is fair to say that over the years, the organismic view has held all the theoretical cards, but that the mechanistic view is winning the empirical game. Developmental studies are most commonly designed from a mechanistic model, usually with strong linear predictions of outcome. After the data are analyzed, however, they are often interpreted with an organismic bias. Despite the theoretical appeal of systems formulations, most developmentalists use them more as post hoc explanations for noise, variability, and prediction failures than as principles to operationalize the study of developmental process.

Here I suggest that contemporary synergetic principles may provide both a theoretically elegant and operationally useful approach to systems changing over developmental time, just as we have seen their elegance and power in real-time biological systems. Behavioral development is notoriously noisy and nonstationary, and we often have limited experimental options, especially with immature humans. Thus, we may never approach the fine mapping of data onto theory accomplished by Kelso and his colleagues, for example, in the motor domain. What we may expect, however, is a strategy to bring the promise of an organismic ontogeny into an operational reality.

In complex systems, interesting things happen at transitions. The plan of this paper is to describe several well-known transitions in early human motor development and to present studies which explore the processes which lead to these transitions. I then offer a dynamical or synergetic account at a descriptive and qualitative level of these phenomena and suggest how such principles may generalize to other developmental domains (See also [3,4]).

2. TRANSITIONS IN THE DEVELOPMENT OF UPRIGHT STEPPING

Figure 1 is an illustration from Myrtle McGraw's classic 1945 monograph, The Neuromuscular Maturation of the Human Infant, depicting seven stages in the development of upright locomotion [5]. The developmental course of upright stepping is marked by three rather dramatic discontinuities or phase shifts. In stage A, the newborn period up to about 2 months, infants will perform characteristic step-like movements when held upright, as shown. In stage B, usually from 2 to 8 months, the same position elicits no or very few

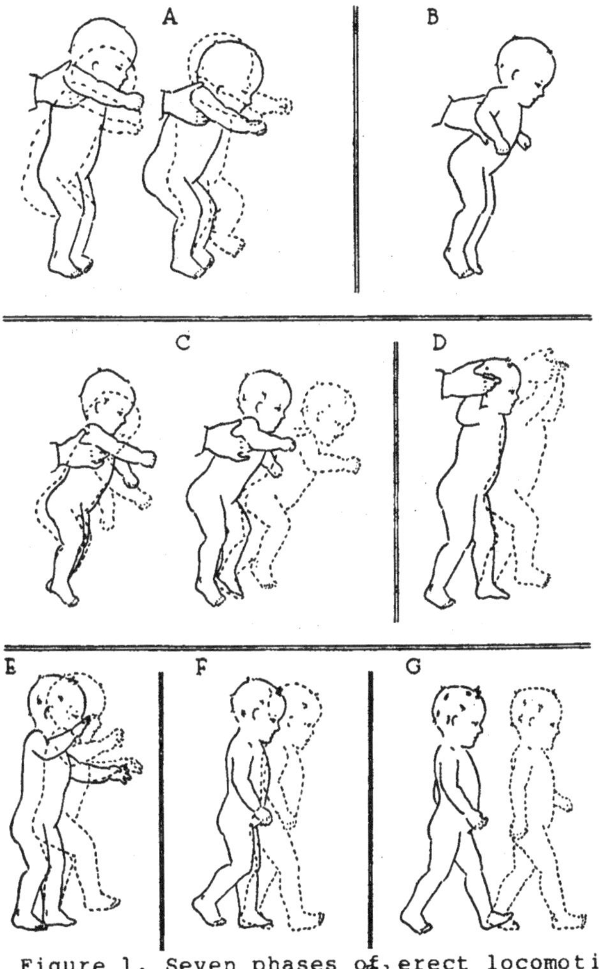

Figure 1. Seven phases of erect locomotion in human infants. From McGraw[5], reprinted with permission.

stepping movements. This ability, which seemed to have magically disappeared, reappears again in stage C, around 8-9 months, after which stepping performance gradually improves. The third shift occurs in stage E, when infants walk independently. The prevailing interpretation of this sequence has been that the maturing cortex first inhibits lower centers and then regains "voluntary" control over the stepping generator [5]. That is, that development of a brain-based executive function was the primary cause of the new forms of behavior.

A number of observations convinced me that the process was more complex and dynamic than the simple brain-to-behavior model would suggest. Let us look at the first transition from newborn stepping to no stepping. When stage A babies were given a few minutes of practice of the stepping response each day, they never entered stage B at all, but went directly to stage C [6]. In contrast, when we added small weights to the legs of stage A babies, we could shift them to stage B prematurely: they stopped stepping [7]. Alternating step-like movements could be restored to stage B infants by several simple manipulations. One manipulation was just to lay them down. When supine, infants performed leg kicks identical in kinematic structure to the steps of stage A infants [8]. Likewise, stage B infants stepped when we held them chest-deep in a tank of water [7]. Some stage B infants would not step when they were calm, but stepped when excited or crying [9].

The later transitions appeared equally mobile. By supporting stage B or C infants over a motorized treadmill, I elicited stepping responses that were kinematically similar to stage D stepping or even better. That is, infants who stepped not at all or in a poorly coordinated manner when the treadmill was turned off, immediately produced movements that approximated much more mature behavior in the patterns of both intra- and interlimb phase relations when the treadmill was running [10]. The stepping of stage E infants, in turn, was accelerated to the coordinative level of stage F infants by the simple action of holding their hands in support [11].

Such observations call into question, if not the reality of McGraw's stages of development, the assumption that a central executive function acts like a switch, moving the infant from one behavioral stage to another. It is unlikely that a brain switch could be activated or deactivated when the baby is placed supine or immersed in warm water! I offer instead a characterization of this developmental trajectory based on the same principles of dynamic pattern formation as in real-time systems.

3. A DYNAMICAL APPROACH TO TRANSITIONS IN UPRIGHT STEPPING

3.1. Self-organization in the Developing Motor System

Human infants are, of course, high-dimensional systems. Potentially, the anatomical elements and physiological processes are free to assemble in a very large number of ways. Behavior, however, is highly constrained. Movements of the legs, for example, can be described by relatively few kinematic parameters of joint excursions and phase relations among the articulated segments --collective variables or order parameters which capture the compression of the degrees of freedom. Movement output can be highly patterned in space and time. Even the actions of very young infants which appear disorganized, and are certainly involuntary, can have a great deal of underlying structure. For example, Figure 2, which shows phase plane representations of the knee excursions during spontaneous kicking of a premature infant at two ages, assumes the topography typical of limit cycle attractors [12]. Over developmental time, degrees of freedom are further compressed so that only discrete behavioral constellations are associated with particular age ranges.

While there is simply no question that networks in the central nervous system are capable of elaborate pattern generation, the ontogenetic course of infant movements cannot be understood on the basis of CNS pattern generators alone.

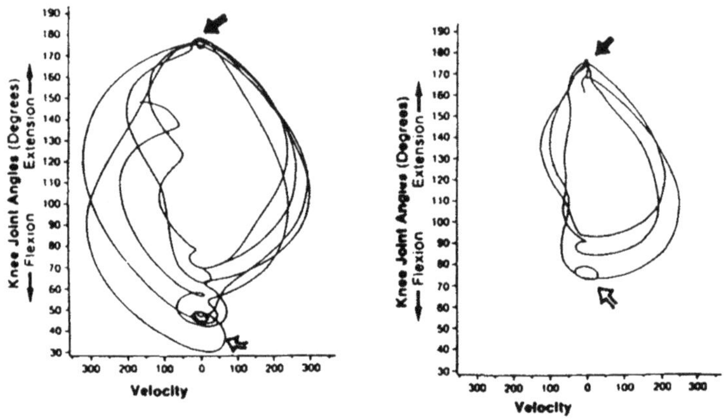

Figure 2. Phase plane trajectories of angular rotations of the knee joint of 10 seconds of spontaneous rhythmical kicking movements of a premature infant at 34 weeks gestational age (left panel) and the same infant at 40 weeks. Reprinted from Heriza[12], with permission.

Rather, the performance-- the actual assembly of the low-dimensional behavior-- is a systems product of these neural substrates in their entire anatomical and physiologcal matrix, and most importantly, only within the task context of the infant. Thus, the output behavior is truly self-organized by the cooperativity of the elements within the task constraints and not driven by CNS commands alone. Organism and context are inseparable. The relations of the elements are different in the supine infant than they are in the upright infant, and submerging the torso in water compresses the degrees of freedom differently again. The treadmill in particular provides a facilitative condition for a high degree of symmetrical, complex patterned movement to emerge. Presumably, the dynamic and biomechanical effects of moving the legs backwards act to organize the parts into a coherent and active whole.

I showed previously the dramatic effect of the treadmill on seven-month-old, non-stepping infants, but we can demonstrate treadmill stepping in much younger infants. In a longitudinal study, Beverly Ulrich and I found well-coordinated, alternating treadmill stepping movements in infants as young as 1 month of age, and nearly all of the infants were stepping well on the treadmill by 3-4 months, although performance gradually improved. At all ages, treadmill stepping was remarkably flexible and adaptive. When we increased the treadmill speed, infants increased their step rate in a nearly linear fashion and with the asymmetrical step cycle phase compensations seen in adult locomotion. Infants maintained regular alternation of the legs even when we perturbed their symmetry quite drastically by driving one leg on a belt moving twice as fast as the belt supporting the opposite leg [13]. These self-adjusting properties of the system suggest that treadmill stepping is neither a traditional reflex, nor the stereotyped output of an intrinsic motor program.

Stepping performance is thus dependent more on context-- the presence of the treadmill-- than on age or phase of development. The central nervous system has potential locomotor-like pattern generating capabilities perhaps as long as a year in advance of actual locomotion, but this ability remains cryptic without the critical component, the stretching of the legs backward. Presumably, when infants assume the erect posture and have sufficient strength and balance control to shift the weight over a single-limb base of support, they provide their own dynamic and biomechanical conditions to step. By giving unstable infants our hands or a table to hold onto, we again substitute for a component essential for the assembly of the behavior, and infants perform at a level more advanced than without the support.

3.2 Developmental Stages as Dynamic Attractors

How then do we account for the apparent reality of McGraw's stages and the equally apparent mobility of the system to assemble and re-assemble at different ages and under different task constraints? In dynamic terminology, developing organisms can be envisioned as occupying preferred, but not obligatory regions of their abstract state space. That is, given the particular developmental status of the components and the particular real-time context of the organism, certain cooperative interactions are more stable than others, and the system tends to seek those stable solutions. When supported upright, infants during their first year show discrete and stable topographies of leg movements, as McGraw and others have indeed documented. When they are placed supine or on a treadmill or their arousal level increases, the components are stable in other regions of the space. As components change continuously over time, the system is itself continuous. As the compression of the components shifts to different regions of the state space, the outcome behavior is discontinuous.

4. PROCESSES OF DEVELOPMENTAL CHANGE

This dynamic formulation allows us to suggest several explicit processes underlying ontogenetic change.

4.1 Phase Shifts or Bifurcations

The appearance of new forms during development may result from phase shifts or bifurcations that arise when the scaling of a component disrupts the stability of the system. Developing systems, like many real-time systems, are nonlinear. Small perturbations do not disrupt their dynamic stability, but at a critical point of some control parameter, internal noise is so amplified as to drive the system to new stable solutions. Complexity may increase by such repeated bifurcations.

As in other complex systems, the control parameter involved in a developmental phase shift does not itself contain a prescription for change. Control parameters may be essential components of the ensemble, but nonspecific to the behavior of interest. For example, human infants may be shifted into new forms of expressive behavior, not by changes in the neurological bases of affect or improvements in some abstract communicative ability, but by increases in muscle strength and tone that allow the infant to assemble specific and sustained gestures [4]. Muscle strength and tone are not expression-specific, but when a threshold level is reached, they provide a facultative matrix for the emergence of other behavior forms.

The first transition in the development of stepping-- the disappearance of an apparently well-coordinated behavior-- illustrates how such a nonspecific control parameter may serve to reorganize the system. As mentioned above, the pattern generating capabilities of the CNS are not lost at this time, but changes in other components of the system must mask their expression. My colleagues and I have suggested that the control parameter for this developmental shift is the very rapid deposition of body fat in infants within the first 2 or 3 months of life [8]. This disproportional acquisition of subcutaneous fat in relation to muscle mass makes it difficult for infants to lift their legs when held in the biomechanically demanding upright position. Contextual manipulations which again shift this critical biodynamic assembly- adding weights, changing posture, submerging, providing the treadmill-- allow the system to seek another dynamic stability or attractor state, that of cyclical stepping or kicking. The body build changes that shift the system are nonspecific to that system, and indeed may have profound reverberations for other behavioral outcomes as well.

It is not difficult to find examples in young infants where neural and anatomical growth may serve as control parameters. However, variables external to the organism but essential for the system assembly, or experience with the structured environment, may also effect developmental shifts. These may include behavior of social partners, where caregivers provide elements seemingly "missing" for the performance of a behavior[4]. Social behavior itself only emerges within the context of the dyad; even young infants can engage in elaborate and patterned interactions with a sensitive partner. Experience with the nonsocial environment is equally important for organizing the system into coherent and increasingly adaptive behavior. For example, developmentalists have shown that the onset of self-propelled locomotion (crawling) and the subsequent experience of active engagement with the environment from different perspectives leads to new, more adaptive levels of spatial cognition and social relations[14].

Developing systems may shift into new modes by the same underlying process as other physical and biological systems: the amplification of naturally-occuring fluctuations in the components to the point where dynamical stability is disrupted[15]. If so, we should be able to detect evidence of instability at the point of transition by increased behavioral variability and increased vulnerability to perturbations, as measured by the time to return to a dynamically stable state.

In fact, there is abundant evidence in the developmental literature of both humans and other animals of such instabilities at the points of transitions[16]. It is common to see infants and young children at the cusp of a new developmental stage to perform at one level of complexity at one testing and seemingly revert to a more simple level the very next day. Such reverberations are not seen when the stage is well-consolidated. At these transitions, the children are especially sensitive to contextual and procedural differences. Infants may complete a task in the home, but fail at the very same task when in a laboratory. A simple demonstration may accelerate performance and a small distraction may retard it. The behavior is assembled not only as a function of abstract structures or representations, but it coalesces with the task context. When the ensemble is fragile, it is easy to move it into different configurations, as we have also shown with the development of stepping.

4.2 Attractor States Stabilize

New forms arise in ontogenetic time by the shattering of stable forms and the subsequent formation of new ensembles. Another process that may be important for understanding developmental phenomena is the consolidation of those new forms into reliable and reproduceable actions. Over time, therefore, the attractor state becomes more stable; the walls of the hypothetical potential well become steeper and more perturbation is required to disrupt the preferred dynamic state. Simply put: youngsters often get better at what they do.

Practice of an action, for example, may result in a linear increase in performance without a qualitative shift into a new level. We often know little about the nature of such practice; that is, which of the components contributing to the ensemble is responsible for making actions more rapid, accurate, or adaptable, or in reducing nonfunctional variability. This is especially problematic in developing systems because practice and maturation are confounded. Infants, for example, show significant improvements in hand-eye coordination between 3 and 6 months. Are these the result of maturational changes in the central nervous system or of repeated practice in the task?

The developmental course of the treadmill task illustrates well the reduction of variability characterizing increased stability and suggests something of the underlying dynamics of this skill. We can see in the individual profiles that all infants show a period of gradual improvement in the task, as indexed in the number of steps they performed on the treadmill. Even infants such as SL and JF, who were relatively late in their shift into reliable stepping, improved their performance in a linear fashion for several months (Figure 3).

Let us look in more detail at the developmental profile of one infant, CH, who showed a typical gradual increase in number of steps on the treadmill after month 2. In this longitudinal study, we gradually increased the speed of the treadmill in equal increments over seven trials as shown in Figure 4, which reports the average cycle durations of CH's right leg as a function of age and treadmill speed. If the infant is adjusting the steps to the speed of the treadmill, we should see decreasing cycle durations with increasing speed.

Infant CH's steps were marginally responsive to speed even the first few months (top panel), but the variability was significantly reduced in months 6 through 9 (bottom panel). Stepping become a more stable response both with age and with treadmill speed, as the higher speeds elicited a more consistent response.

Treadmill stepping is intriguing because infants were improving their performance without practice of the specific skill. Two processes may explain this. First, the neuromotor system may "settle in" to less variable output independent of experience. Or, and perhaps more likely, essential components of this motor skill may change as a result of participating in other behavior, and may bring those changes

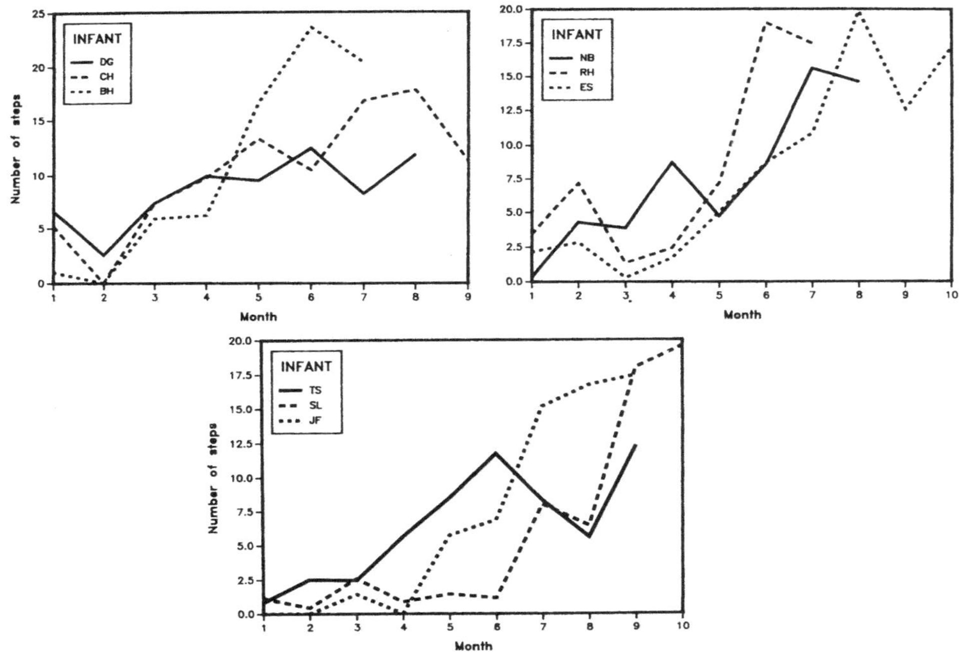

Figure 3. Individual developmental profiles of treadmill-elicited, alternating stepping in nine infants. Performance on their "better" day is plotted. Note gradual improvement in performance without specific practice of the skill.

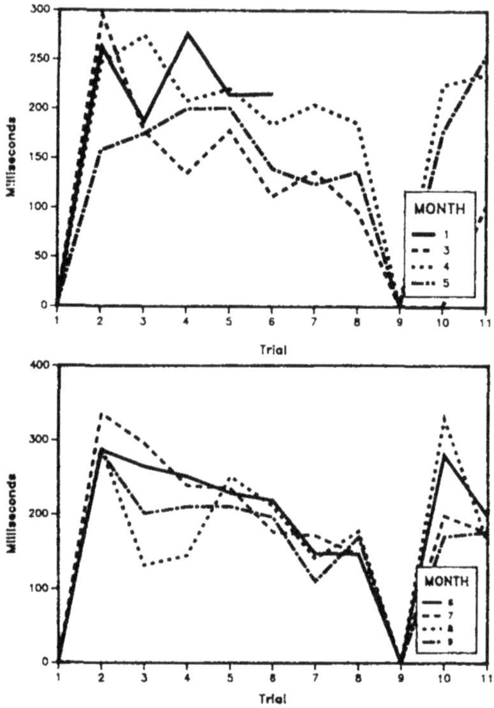

Figure 4. Cycle durations of a single infant, CH, as a function of age and treadmill speed. In trials 1 and 9 the treadmill was turned off. In trials 2-9, we increased the speed of the treadmill in equal increments from .11 to .29 m/s. Trials 10 and 11 were "split-belt" trials. Top panel, ages 1-5 months; CH did not step at 2 months. Bottom panel, ages 6-9 months.

to the particular assembly elicited by the treadmill. For example, increases in muscle strength, especially in the dorsal extensor muscles, may assist in maintaining upright posture and stretching the leg backward necessary for treadmill stepping. Muscle strength, however, may increase as a result of both organic changes and other movement and postural activities. As muscle strength improves within these nonspecific contexts, it contributes to the treadmill task when the treadmill is presented. Such a formulation eliminates the need for a "treadmill stepping generator" or other prescriptive codes for the behavior.

4.3 Control Parameters Change

If we imagine that the ontogenetic state space of an organism contains all the actual and potential states of the essential variables, we might envision that at any point in developmental time, the organism occupies only a circumscribed range of positions in that state space. These positions are constrained by the developmental status of the components and the range of tasks to which the organism can respond. As the developmental status of the components change through the processes of incremental growth and phase reorganization, and as that development opens up new task domains, the range of positions migrate on the state space creating a "tunnel" which itself has trajectory through the space.

This means that the control parameters driving phase shifts in behavior change during development. The new dynamic assembly of components may move the system into a region of stability where it is no longer sensitive to the scalar changes of the old control parameters. These new attractors may be so stable that, for practical purposes, the system is irreversible; it can never regress back to critical sensitivity to the control parameters of earlier stages.

In addition, it is well-known that the component structures and processes of the developing organism themselves change in asynchronous and asymmetric ways. The organism does not march forward as a piece; some processes show periods of rapid growth, then little change, others are more linear. Development also includes periods of regression and loss. This adds even more nonlinearity to the ontogenetic course.

These points may again be illustrated with the transitions in infant stepping. I suggested above that the control parameter for the first transition was the rapid deposition of body fat which masked stepping performance in certain contexts. What, then, underlies the transition from stage B (no stepping) to subsequent phases where the behavior reappears? Although body composition and proportions are always important contributors to movement trajectories, these variables are changing much more slowly in the second half of the first year. At this later age, it appears that the critical factors may be more those contributing to upright stability for the difficult task of independent bipedalism. Thus, the problem may not be ability to lift the leg, but to stabilize the base and shift the weight from leg to leg. Control parameters may then be the muscle strength needed to provide a single-leg pillar of support and the neuromotor ability to rapidly correct for deviations of the center of gravity over the base of support. We can improve stepping performance in 10 or 11-month-old infants by enhancing balance through additional support. The same control parameter is not effective in shifting the system at earlier ages, or when locomotion is well-established.

Because each developmental stage has a unique configuration (as well as elements of continuity), the nature of the control parameters cannot be assumed to be the same as those of earlier or later stages. They can only be derived empirically. Even in a single domain, developmental shifts may have different underlying dynamics.

5. HOW DYNAMICS CAN CHANGE THE STUDY OF BEHAVIORAL DEVELOPMENT

If the dynamical principles of complex systems in real time can be extended to systems in developmental time, there are implications for both the conceptualization of development and for empirical studies. On the theoretical level, this means that development must be viewed as a series of more-or-less stable attractors that are task assembled, and where complexity arises as successive bifurcations that may be deterministic, but are not predetermined. Development is deterministic because at the point of transition, the phase shift is inevitable; the system can never assemble itself in exactly the same configuration as the precursor state. Development is not predetermined, however, because stochastic processes underly the initiation of the phase shift.

Such a perspective means abandoning single-cause models of development as elaborated representations or schemas, or as a linear read-out of the genetic code, or as the direct mapping of the environment onto the organism. The evidence from every domain of concern to psychologists--social, emotional, physical, cognitive, perceptual, motor, abnormal--supports, I believe, such a more fluid perspective.

Our empirical task is even more daunting. The first step toward a dynamical model of any developmental phenomenon is understanding the behavior of the collective variables over time [17]. Here we face the nontrivial (and maybe even monumental!) task of choosing the appropriate collective variable: how best to capture the compression of the degrees of freedom. As we extend the time scale, it is not clear that the psychological meaning of a collective variable remains stationary. The same behavior may mean different things at different ages. Thus, the trajectory of the collective variable must be plotted on an appropriate time scale.

Because it at the transition between new forms that we can explore the dynamics of developmental change, a dynamical understanding almost always requires intensive longitudinal studies. If behavior samples are too widely spaced, interesting transitions may well be missed. If infant stepping were sampled only in newborns and one-year olds, the whole rich and complex trajectory would be lost. Longitudinal studies are expensive and frustrating because one is tied to the variables originally chosen. On the other hand, these studies nearly always tell us something worthwhile.

Transitions should be characterized by increased variability and vulnerability to perturbations. It is at these points where variables can be manipulated to discover the control parameters. Within our experimental and ethical constraints, we can withhold or enrich components to see if we can accelerate or retard the behavior in question. For example, at the transition to independent walking, additional support accelerates coordination, while giving the infant inappropriate visual flow field information causes stumbling and falling[18]. How these potential control parameters interact is the goal of a process-oriented approach.

As I have argued with examples from the development of upright stepping, the principles of dynamical systems have explanatory power currently missing in other contemporary accounts of the diverse and often paradoxical phenomena of early life. Adopting such a perspective means that developmentalists must broaden their traditional domain-specific concerns to ask how change can arise from the cooperation of many subsystems and processes assembled in a fluid, and task-specific manner. While we have not yet attained the goal of precise quantitative modeling, we gain in understanding of the processes by which new behavioral forms arise.

ACKNOWLDGEMENTS

I thank Jody Jensen, John Merrill, and Beverly Ulrich for their helpful comments, Beverly Ulrich and David Niles for their collaboration on their research, and Scott Kelso and Alan Fogel for many stimulating discussions. Supported by NSF BNS 85-09793, PHS RO1 HD 22830, and an NIH Career Development Award.

REFERENCES

1. Reese, H.W., & Overton, W. F. (1970). Models of development and theories of development. In L. R. Goulet & P. B. Baltes (Eds.), Life-span development psychology: Research and theory (pp. 115-145). New York: Academic Press.

2. Sameroff, A. J. (1983). Developmental systems: Contexts and evolution. In P. H. Mussen (Ed.), Handbook of child psychology. 4th ed., Volume I. History, theory, and methods (pp. 237-294). New York: Wiley.

3. Thelen, E., Kelso, J. A. S., & Fogel, A. (1987). Self-organizing systems and infant motor development. Developmental Review, 7, 39-65.

4. Fogel, A., & Thelen, E. (in press). The development of expressive and communicative action in the first year: Reinterpreting the evidence from a dynamic systems perspective. Developmental Psychology.

5. McGraw, M. B. (1945). The neuromuscular maturation of the human infant. New York: Columbia University Press.

6. Zelazo, P. R., Zelazo, N. A., & Kolb, S. (1972). "Walking" in the newborn. Science, 177, 1058-1059.

7. Thelen, E., Fisher, D. M., & Ridley-Johnson, R. (1984). The relationship between physical growth and a newborn reflex. Infant Behavior and Development, 7, 479-493.

8. Thelen, E., & Fisher, D. M. (1983). Newborn stepping: An explanation for a "disappearing reflex." Developmental Psychology, 18, 760-775.

9. Thelen, E., Fisher, D. M., Ridley-Johnson, R., & Griffin, N. (1982). The effects of body build and arousal on newborn infant stepping. Developmental Psychobiology, 15, 447-453.

10. Thelen, E. (1986). Treadmill-elicited stepping in seven-month-old infants. Child Development, 57, 1498-1506.

11. Clark, J. E., Whitall, J., & Phillips, S.J. (in press). Human interlimb coordination: The first 6 months of independent walking. Developmental Psychobiology.

12. Heriza, C. (1986). The organization of spontaneous movements in premature infants. Unpublished Ph.D. dissertation, Southern Illinois University.

13. Thelen, E., Ulrich, B. D., & Niles, D. (1987). Bilateral coordination in human infants: Stepping on a split-belt treadmill. Journal of Experimental Psychology: Human Perception and Performance, 13, 405-410.

14. Bertenthal, B. I., Campos, J. J., & Barrett, K. C. (1984). Self-produced locomotion: An organizer of emotional, cognitive, and social development in infancy. In R. Emde & R. Harmon (Eds.)., Continuities and discontinuities in development (pp. 175-210). New York: Plenum.

15. Kelso, J. A. S. Scholz, J. P., & Schoner, G. (1986). Non-equilibrium phase transitions in coordinated biological motion: Critical fluctuations. Physics Letters A, 118, 279-284.

16. Thelen, E. (in press). Self-organization in developmental processes: Can systems approaches work? In M. Gunnar (Ed.), Systems and Development: 22nd Annual Minnesota Symposium on Child Developmentn. Hillsdale, H. J.: Lawrence Erlbaum.

17. Kelso, J. A. S., & Schoner, G. (in press). Toward a physical (synergetic) theory of biological coordination. In R. Graham (Ed.), Lasers and synergetics, Heidelberg, Berlin: Springer.

18. Stoffregen, T. A., Schmuckler, M. A., & Gibson, E. J. (1987). Use of central and peripheral optical flow in stance and locomotion in young walkers. Perception, 16, 113-119.

Critical Behavior in Perception-Action Systems

WILLIAM H. WARREN, JR.

Departments of Psychology and
Cognitive and Linguistic Sciences
Brown University
Providence, RI 02912, USA

ABSTRACT

The perceptually guided behavior of organisms appears to be characterized by goal-directed action modes punctuated by abrupt transitions between them. Such critical behavior typically occurs when some continuous change in external or internal parameters yields a qualitative change in the organization of action. Traditional theories in psychology and cognitive science have explained these qualitative transitions in terms of *a priori* cognitive or neural structures imposed upon behavior, such as switching between preexisting "motor programs." In contrast, following recent work I suggest that they may be understood as a consequence of the underlying dynamics of the larger organism-environment system, with behavior being guided by information rather than driven by the dynamics deterministically. Examples of behavioral transitions in locomotor gaits, stair climbing, and walking through apertures are discussed qualitatively.

It is apparent that the global behavior of humans and other organisms is strikingly nonlinear. Under some circumstances, highly stable patterns of behavior persist in the face of large perturbations, while under others a small change in external or internal conditions can yield a qualitative reorganization in behavior. Consider, as a prosaic example that I will develop at some length here, the case of gait transitions in legged locomotion. Over a range of speeds, human bipeds adopt a stable walking gait in which the body behaves much like an inverted pendulum, conservatively exchanging kinetic and potential energy as it pivots up and over the ankle on each step, and maintaining foot contact with the ground at all times. But a slight increase in speed above a critical value prompts an abrupt transition to a running gait in which the body behaves more like a bouncing ball, storing elastic energy when it lands and projecting itself back into the air, yielding a flight period with no foot contact. How is it that, in the service of forward progression, the musculo-skeletal system is suddenly reorganized from a pendulum into a bouncing ball?

This sort of case is actually quite general. It appears that the behavior of organisms is structured into goal-directed *action modes,* characteristic activities such as walking, running, grasping, throwing, chewing, swallowing, and so on, with qualitative transitions between them [1-3]. Action modes are directed toward the functional properties or "affordances" of the environment and act to preserve the system's integrity in the face of environmental fluctuations. Each involves an organization of the musculature into task-specific groupings, called "synergies" or "coordinative structures," that establish temporary couplings between muscles. Such an organization reduces the high dimensionality of the action system (whether one counts the 100 degrees of freedom of the joints or the 792 degrees of freedom of the muscles) down to a few free parameters that are regulated to adapt the stable pattern to local conditions. The dynamic self-regulation of behavior then consists of switching between (or simultaneously nesting) these action modes as required to maintain the system.

The phenomenology of action modes raises a number of questions. Primary among them is the source of these stable patterns -- where the motor organization comes from. A related problem is how the system shifts appropriately from one stable organization to another. Two sorts of answers to these questions are currently about in the motor control literature. The dominant view ascribes the appearance of a regular pattern of movement to a preexisting cognitive structure such as a motor program [4] or a preexisting neurological structure such as a central pattern generator [5,6]. Thus, the organization is assumed to exist *a priori* in some independent form ready to be imposed upon the musculature, regressing the question of its origin onto the phylogenetic or ontogenetic history of the organism. Transitions between action modes are correspondingly conceived as switching from one motor program to another [7]. In effect, the boundaries between action modes are thought to be inserted into the continuum of possible behavior by a representational structure or prewired neurological mechanism.

The theme of this meeting, in contrast, is that patterns in nature and the transitions between them emerge naturally from the dynamics of sufficiently complex systems [8,9]. This view has inspired recent efforts to establish a basis for motor control and coordination in similar principles of self-organization applied at the scale of behaving organisms [10-13]. Rather than imposing order on the action system, the role of the central nervous system may rather be, as McMahon has remarked only half-jokingly, to prevent interference with the natural mechanical resonances of the system. In this paper, I want to pursue this view for the problem of transitions in behavior by elaborating several purely qualitative examples. A major problem, of course, lies in developing an appropriate description and identifying the variables that will reveal the order and simplicity that resides in the apparent complexity of behavior.

The first point I'd like to make is that, in applying such principles to the macroscopic behavior of organisms, what constitutes the "system" is not just the nervous system, nor the physical body and its biomechanics, but must be enlarged to include the environmental context of constraint -- what I will call the animal-environment system [14]. In the case of locomotion, these environmental constraints obviously include a gravitational field, an irregular ground surface cluttered with obstacles, and the resulting envelope of reaction and inertial forces [15]. For an action such as grasping, they include the properties of objects such as size, shape, mass, and friction that, when taken in relation to the properties of the hand, yield transitions between characteristic grips. For example, an increase in object diameter yields transitions from an opposable pincer grip, to a full power grip, to a two-handed grip [16]. Thus, as the relationship between properties of the environment and properties of the organism is continuously varied, *critical points* may be reached at which a transition to a new mode of action occurs and a qualitative reorganization of the musculature is required [17]. Such critical points emerge not only from the internal dynamics of the action system, but also from the dynamics of the organism-environment system.

My second point is that the adoption of action modes and the transitions between them are not deterministically driven by the dynamics of the system, but rather guided by *information* (see [13] for a systematic development of this idea). The behavior of organisms is not determined by physical laws, and yet it exhibits regularities that can often be related to them. An organism that is sensitive to its own dynamics and those of its interactions with the environment can exploit the system's stabilities and instabilities to control its behavior. This requires that there be information, in Gibson's [18] sense, that is *specific* to behaviorally relevant aspects of the system's dynamics -- haptic patterns defined over skin, joint, and muscle receptors, optical patterns available at a moving point of observation, and so forth. In the high-dimensional space of possible postures and movements, only small regions are stable and functionally useful [19]; an actor sensitive to those regions, their boundaries, and the transitions between them can control action stably and efficiently. Thus, the nervous system must not simply prevent interference with these resonances, it must actually sense and ride them.

GAIT TRANSITIONS AS A HIERARCHY OF INSTABILITIES

I'd like to begin with a thought experiment about critical behavior in perception-action systems by examining the case of gait transitions in more detail. As a quadruped increases its speed of locomotion, the animal passes through a sequence of stable action modes, from standing, to walking, trotting, and galloping. Each gait may be characterized by a *duty factor,* the proportion of the step cycle that each foot is on the ground, and the *relative phase* of the four limbs, the proportion of the step cycle that each foot lags behind the left front foot [20]. Each of these gait patterns is qualitatively stable over a range of speeds, but then appears to bifurcate at a critical speed. The sequence of quadruped gait transitions looks rather like what Haken [21] has called a "hierarchy of instabilities," in which continuous change in a control parameter yields a succession of qualitative changes in behavior. Moreover, the gait patterns and transitions are fundamentally similar across a wide variety of quadrupeds, suggesting that there may be only a few stable solutions for progression on four legs which have been discovered independently by diverse species [22]. Can we characterize these transitions qualitatively, and infer the role of information therein?

For the purposes of such an analysis, a speed-related variable seems to be a reasonable candidate as the control parameter. It is a functionally significant dimension of locomotion for animals in catching prey, escaping predators, and reaching other goals -- pragmatically, what a gait change enables one to do is go faster. Exploiting the theory of dynamical similarity, Alexander & Jayes [23] proposed a general speed parameter that scales speed (v) to leg length (l), yielding a dimensionless Froude number for all quadrupeds regardless of size: $\boldsymbol{F} = v^2/gl$. Let's adopt this scale-independent number as a control parameter. As for an index of the qualitative organization of gait, the phase difference *phi* between limbs appears to be a good candidate, reflecting the stability of the pattern [24].

Consider, initially, the task of standing up in the case of a human biped, with $\boldsymbol{F} \approx 0$ and the legs locked in phase at *phi* = 0. Making the simplifying assumption that the body is essentially rigid during stance and simply pivots about the ankles, a standing posture can be modelled as an inverted pendulum upon a small base, thereby having a small stable region. To maintain this static equilibrium, the basic requirement is to keep the body's center of gravity (c.g.) over the quadralateral base of support provided by the feet. This is something that people do quite well even with eyes closed [25] as well as under novel conditions, such as standing in line with a suitcase at the Ft. Lauderdale airport and shifting it from hand to hand. We can also approach the edge of the base of support, that critical boundary between standing and falling, in a controlled manner by shifting our weight around.

Apparently, a standing human senses the region of stability and the critical boundary by means of propriospecific information, a complex spatiotemporal pattern defined over the feet, joints, and muscles. Particularly important seems to be the morphology of the distribution of pressure on the soles, whose maximum one can feel creeping toward the edges of the feet as one approaches the boundary. This information is used to control the torques generated at the ankles to maintain stance, which reciprocally influence the pattern of pressure, establishing a perception-action control cycle. Presumably, an invariant morphology defined over these spatiotemporal transformations specifies the critical margins of the stable region.

However, during the initiation of gait, the system must move off the stable region and cross the critical boundary, making a transition from stationary stance to a new periodic regime [26]. In order to start walking, one first falls forward and then catches oneself with the leading foot, initiating a repeating cycle. The transition from standing to walking is achieved within one half-cycle and over a constant time interval, for by the time the trailing foot leaves the ground the walker has reached the steady-state walking velocity and frequency, analogous to an inverted pendulum [27]. This involves a smooth shift from a static equilibrium with *phi* = 0 in which the c.g. is positioned above a constant base of support, to a dynamic equilibrium with *phi* = 0.5 in which the moving c.g. is seldom above the base of support. The system thus appears to bifurcate from one stable equilibrium point into a periodic mode that is stable and a transitory equilibrium point that is unstable, as the c.g. passes briefly over the supporting foot.

If we represent this in terms of what happens at a single joint such as the hip, the phase portrait that plots hip angle *theta* against its angular velocity shifts from a single point to a periodic orbit with amplitude *a* (Figure 1). This bears some resemblance to the Hopf bifurcation, in which an equilibrium point is transformed into a limit cycle oscillation [28]. The bifurcation diagram in Figure 2 represents the amplitude of the hip angle oscillation as a function of the control parameter ***F***, illustrating the stable periodic branch associated with walking and the unstable branch associated with the transitory equilibrium point. Notice, however, that the system's bifurcation behavior is not determined -- rather than shifting to a walking gait upon passing through the critical region, people can, and occasionally do, fall down. Nevertheless, there is typically a systematic transition from one action mode to another. This suggests that the actor is perceiving the critical point, exploiting it to regulate behavior and yield an appropriate shift from standing to walking. It thus appears that such qualitative transtions in macroscopic behavior are not just produced physically, but can be induced by haptic information specific to the system's own dynamics.

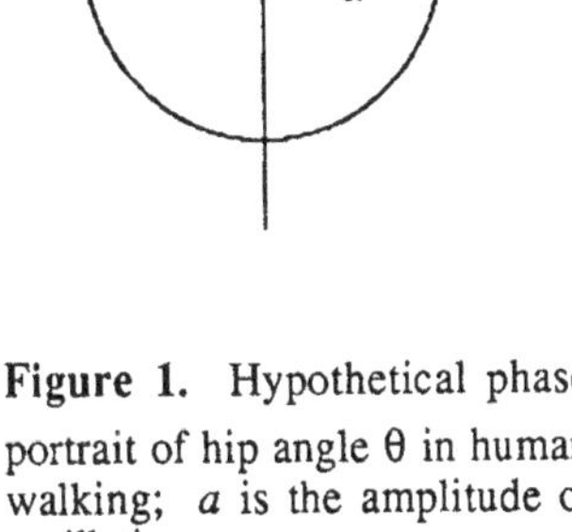

Figure 1. Hypothetical phase portrait of hip angle θ in human walking; *a* is the amplitude of oscillation.

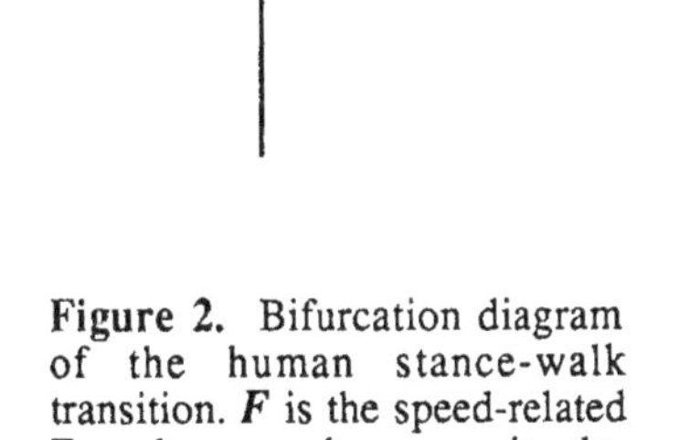

Figure 2. Bifurcation diagram of the human stance-walk transition. ***F*** is the speed-related Froude number, *a* is the amplitude of hip angle oscillation.

Moving up the hierarchy of instabilities, let's consider the gait transition between trotting and galloping in the quadruped. The trot is an "alternate" gait in which two limbs on the same girdle, such as the forelimbs, move out of phase with *phi* = 0.5. The gallop, on the other hand, is a so-called "in-phase" gait in which limbs on the same girdle move nearly in phase, with *phi* = 0.2. In observations of free overground gaits in nine quadrupedal mammals, Alexander & Jayes [23] found that as the Froude number increased, relative phase dropped sharply at a critical value of around $F = 2.5$ (Figure 3). Similarly sudden shifts have been reported by Pennycuick [29] and Heglund, Taylor, & McMahon [30]. There is an obvious similarity between this abrupt shift in limb phase and the analogous shift from out-of-phase to in-phase oscillatory movements of two fingers studied by Kelso and his collegues [24,31,32].

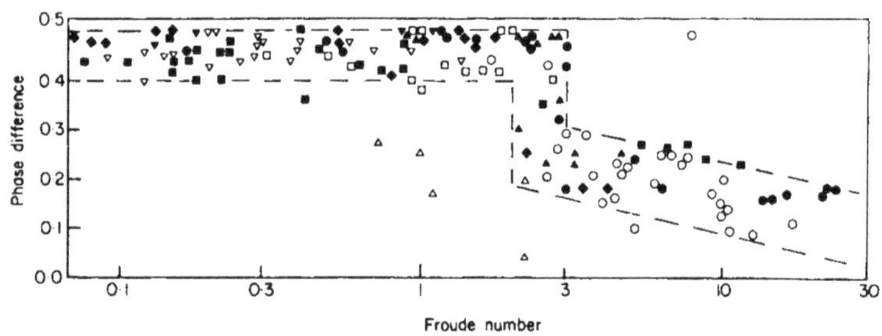

Figure 3. The trot-gallop transition. Forelimb phase difference ϕ as a function of Froude number F for nine quadrupedal mammals. (From Alexander & Jayes [23]).

Figure 4. Hypothetical diagram of trot-gallop transition, showing hysteresis.

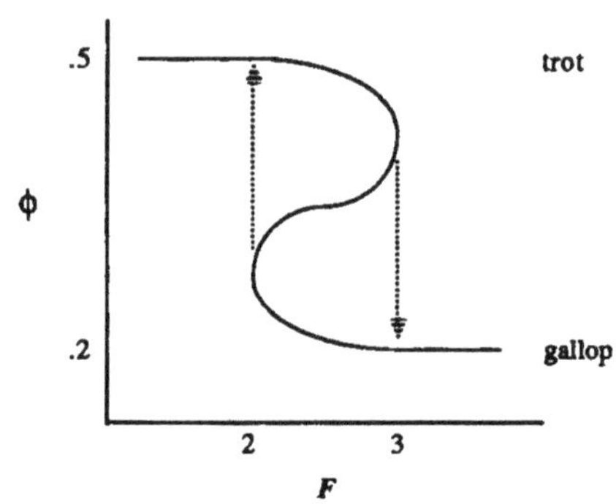

The form of the trot-gallop transition is suggestive of a pair of cyclic fold bifurcations represented in Figure 4, which includes a hysteresis loop. On this model, if we start the animal in a trot and gradually increase the Froude parameter, the system reaches a bifurcation point at, say, $\boldsymbol{F} = 3$ and jumps abruptly to a new periodic orbit that corresponds to galloping. On the other hand, if we start the system in a gallop and decrease the Froude number, it will jump back to a trot at a different bifurcation point, say $\boldsymbol{F} = 2$. Humans seem to exhibit just this sort of hysteresis when we speed up into a run and slow back down into a walk. This might be expected from the inertial differences between accelerating and decelerating the body mass and the consequent demands on each gait, such a maintaining fore-aft balance when decelerating from a run to a walk. However, there are no empirical data available that compare an approach to the trot-gallop transition from each side of the bifurcation. Heglund, et. al. [30] appear to have elicited the transition in the mouse, rat, and dog by increasing the speed of the treadmill belt, thus identifying the first bifurcation point. Alexander & Jayes [23] analyzed films of steady-state locomotion at various speeds, rather than driving individual animals through the gait transition. It is conceivable, therefore, that their broad band of transitional Froude numbers, between 2 and 3, reflects a hysteresis loop as in Figure 4 rather than individual differences between animals. This is a clear case in which thinking in terms of bifurcation analysis would lead us in new experimental directions.

The transition between walking and trotting, in which the phase shift occurs between the front and rear girdle, is similarly abrupt in some species, such as the dog and the horse, and thus I suspect that its characteristics would be qualitatively the same. However, in other species such as the sheep the shift is more gradual [23] and may be a canard rather than a strict bifurcation. This suggests that a different mechanism may underly the walk-trot transition in these species.

A study by Hoyt & Taylor [33] takes a step toward undertanding the underlying dynamics of gait transitions by examining the energetics of locomotion in horses. They measured energy expenditure per meter of travel during forced walking, trotting, and galloping on a treadmill at speeds that extended beyond the normal range for each gait. Three overlapping U-shaped functions were obtained, with energy minima at a characteristic speed for each gait (Figure 5). These minimum energy values were approximately equivalent, indicating that the three gaits were equally efficient in terms of the distance travelled per calorie of energy expended. Comparing these results to free overground locomotion in one animal, they observed that the horse only adopted speeds that were close to the minimum in each gait, except when shifting gaits. The same phenomenon has been found for preferred walking speeds in humans [34], and for preferred stairway dimensions in human stair climbing [17]. These observations suggest that stable action modes may have a basis in the qualitative features of an underlying energy manifold.

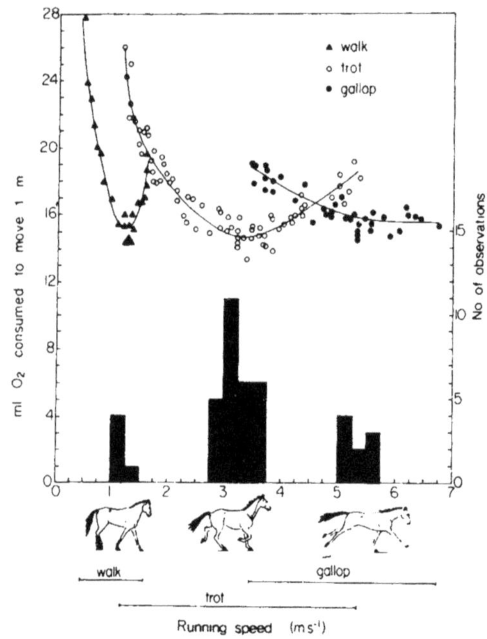

Figure 5. Oxygen consumption per meter of travel as a function of speed for three gaits in three horses on a treadmill. Histogram shows preferred speed in each gait for one horse during overground locomotion. (From Hoyt & Taylor [33]).

These results may also shed light on the gait transitions. Hoyt & Taylor's [33] energy curves for trotting and galloping intersect at a speed of 4.6 m/s, or about F = 2.1. Thus, this is the point at which galloping becomes energetically more efficient than trotting. Using Heglund, et. al.'s [30] allometric equation that scales the transition speed (in m/s) to body mass (kg) ($v = 1.5m^{0.24}$), we can estimate the transition speed for the horses in Hoyt & Taylor's study, yielding a value of 4.9 m/s or F = 2.4. This agrees very closely with the value at which the energy curves intersect, suggesting that horses change gaits to move toward an energetically preferred or stable state. Similar observations have been made for the walk-run transition in humans. Alexander [35] has shown formally that the biomechanical upper limit on normal human walking falls at about 3 m/s or $\boldsymbol{F} = 1.0$, but the walk-run transition occurs at 2.4 m/s or $\boldsymbol{F} = 0.6$, the point at which it becomes more efficient to run than to walk [36]. In sum, the gait transitions can be related to the underlying energetics of legged locomotion, but the physical origin of the energy landscape remains to be accounted for.

Once again, it appears that these gait transitions are perceptually induced rather than dynamically determined. For example, the horses in Hoyt & Taylor's treadmill study were physically capable of maintaining extended gaits, such as trotting at a high speed of 5.25 m/s, although it was energetically costly to do so. They were not compelled to make the trot-gallop transition at the critical point of 4.7 m/s. This suggests that the animals can haptically sense and, under natural conditions, will ride the underlying energetic manifold as a means of ordering their behavior.

VISUALLY INDUCED TRANSITIONS

The significance of environmental constraints is highlighted by actions that encounter more complex objects than the idealized flat, open surface of support. Gibson's [14] paradigmatic example was guiding locomotion through an environment cluttered with obstacles, openings, and irregularities in the ground surface. In this case, the locomotor movements of the observer must be adjusted to local contingencies, often in advance of actual contact with them, and thus visual information becomes increasingly important. Do we find analogous qualitative transitions in such complex behavior, and can they be induced by visual information as well as haptic information?

One type of locomotion over an irregular surface that I have studied empirically is the case of human stair climbing, an action mode closely related to level walking but laboring under tight environmental constraints [17]. As the fit between the riser height of a stair (R) relative to the leg length of the climber (L) is varied, a critical point is reached that entails a shift from bipedal climbing to a quadrupedal "clambering" gait. To analyze this behavioral transition, one must of necessity consider dimensions of the environmental structure in relation to those of the action system. By scaling relevant environmental variables to complementary action system variables, we can obtain scale-independent, dimensionless *pi* numbers, analogous to the Froude number, that capture these relations. Essentially, this employs dimensions of the organism as a natural standard for measuring reciprocal dimensions of the environment, helping to reduce the dimensionality of the system. As a simple example, scaling riser height (*R*)to leg length (*L*) yields a dimensionless ratio, $pi = R/L$. A specific value of this *pi* number expresses the intrinsic fit between stairway and climber, independent of the absolute scale of the system, and is an obvious candidate for a control parameter in our analysis of behavioral transitions.

The critical point between bipedal climbing and quadrupedal clambering modes should occur when riser height becomes too high to permit the climber to get his or her center of gravity over the top foot, which provides the base of support when raising the body mass up onto the next step. This critical point can be estimated by a simple biomechanical model of maximum leg flexion [17], which yields a value of $pi = .88$. It follows from similarity theory that critical values of *pi*, if properly derived, should be constant over scale changes in the system [37] -- that is, critical riser height should be a constant proportion of leg length regardless of the absolute size of the climber. Thus, we might expect a bifurcation at this critical point similar to the one in Figure 4, in which the gait pattern changes qualitatively with a continuous change in the control parameter $pi = R/L$.

To determine whether this transition can be induced visually, we presented short and tall observers with slides of stairs varying in riser height and simply asked them to judge whether the steps would be climbable or unclimbable using a normal bipedal gait. The results appear in Figure 6. When the judgments are plotted as a function of absolute riser height *R*, there is a statistically significant difference in between short and tall observers. However, when the same data are replotted as a function of *R/L*, the curves are nearly congruent, which indicates that the critical point is indeed scale-independent. The critical value for both groups falls at precisely $pi = .88$, as predicted by the biomechanical model. Thus, it appears that a qualitative transition would occur with a continuous change in *pi*, and further, that such a transition could be visually induced.

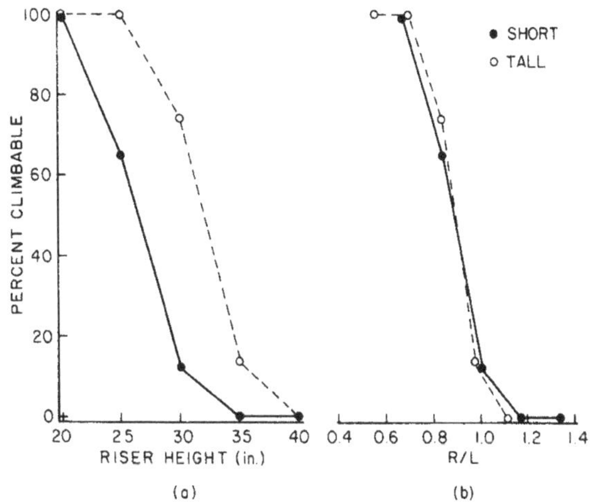

Figure 6. Stair climbing: Percentage of "climbable" judgments as a function of (a) absolute riser height *R*, and (b) intrinsic ratio of riser height to leg length *R/L*, for short and tall male subjects.

We have also measured a transition between action modes in the case of postural adjustments that are made when walking through narrow openings such as doorways [38]. The relevant *pi* number here is the ratio between the width of the aperture (*A*) and the walker's widest body dimension, which in both men and women is shoulder width (*S*): $pi = A/S$. As aperture width decreases toward shoulder width and this dimensionless ratio approaches $pi = 1$, the walker must introduce shoulder rotation to pass through the opening. This represents another behavioral transition, from a mode of frontal walking to walking with shoulder rotation, which we would expect to observe at a scale-invariant critical value of the control parameter $pi = A/S$.

We videotaped large and small subjects walking through apertures of different widths, using a method of limits in which the aperture was progressively decreased in a series of "descending" trials, and then progressively increased in a series of "ascending" trials. Each trial consisted of walking once through an aperture of a given width. The data for shoulder rotation (in degrees from a frontal position) as a function of absolute aperture width *A* appear in Figure 7, yielding a significant difference between the small and large walkers. However, when the data are replotted as a function of *A/S*, the curves collapse and the critical point falls at $pi = 1.3$ for both groups, once again invariant over scale

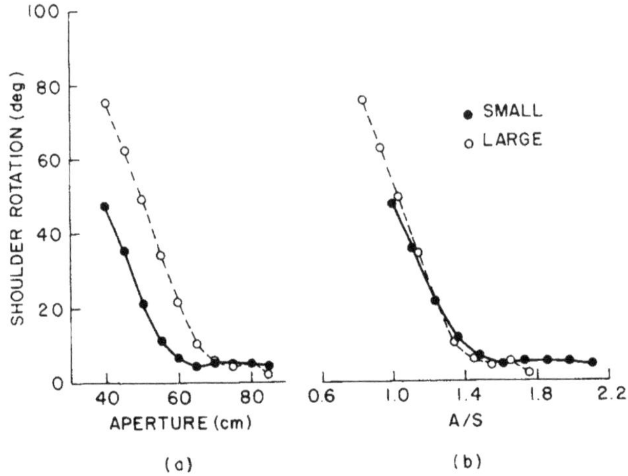

Figure 7. Walking through apertures: Angle of shoulder rotation as a function of (a) absolute aperture width A, and (b) intrinsic ratio of aperture width to shoulder width A/S, for small and large male subjects.

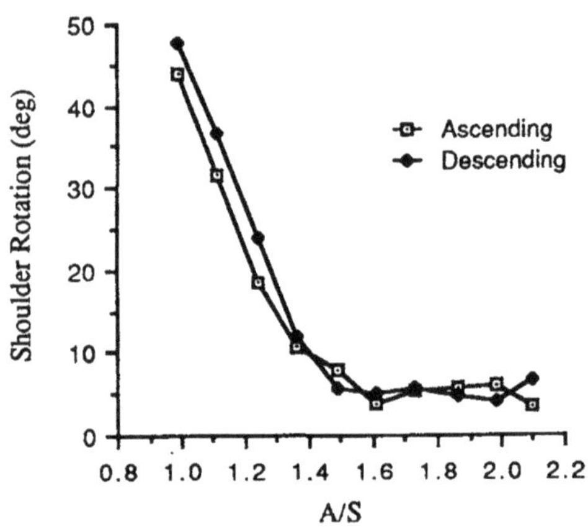

Figure 8. Shoulder rotation as a function of the aperture to shoulder width ratio A/S in ascending and descending series of trials, for small subjects. No evidence of hysteresis.

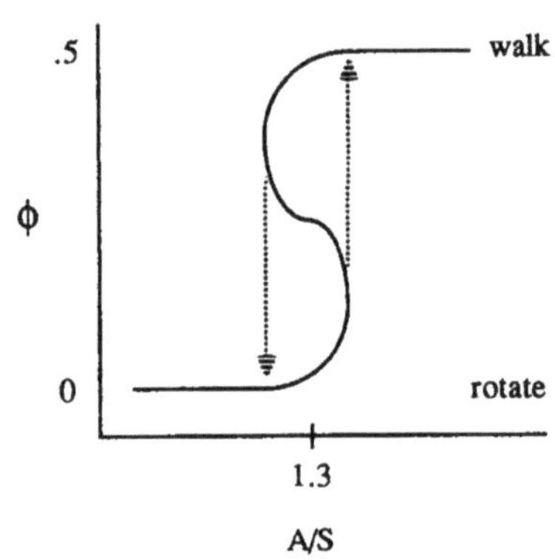

Figure 9. Hypothetical diagram of walking through apertures, showing hysteresis. Phase difference ϕ between hip and shoulder oscillation as a function of the aperture to shoulder width ratio A/S.

changes. Thus, there appears to be a bifurcation point at which a transition to shoulder rotation mode occurs, about which we might expect a hysteresis loop. We can look for this by comparing the data from ascending and descending trials (Figure 8). As these two curves are essentially congruent near the critical point, there is no evidence here for hysteresis in the perception-action system. However, this could be an artifact due to the coarseness of our manipulation of the order parameter, incrementing aperture width in relatively large 5 cm steps, or because the variation occurred over separate trials. Finer steps or variation of the order parameter within a single trial, such as widening or narrowing the aperture as the subject approached, could provide a more sensitive measure of perceptual-motor hysteresis.

A better index of the reorganization of action in this task might be the phase relation between the shoulders and hips. In the normal walking mode, the shoulders tend to counter-rotate slightly against the hips, such that they are out of phase at *phi* = 0.5. Thus, the leading shoulder is contralateral to the leading leg, and the arm on the leading shoulder and the contralateral leg swing forward together. This counter-rotation acts to balance the inertia of the upper and lower body about the longitudinal axis of the body, preventing the build-up of torque at the ankles -- as becomes apparent to anyone trying to walk with shoulders and hips locked in phase.

Now, rotating the shoulders to pass through an aperture could be done in either of two ways. One could simply modulate the shoulder amplitude of the established anti-phase regime, requiring no reorganization of the musculature. However, the body gets rather twisted up this way, threatening postural imbalance. We have noticed informally that rotation of the shoulders to pass through an aperture generally occurs *in* phase with the hips, with *phi* = 0. Thus, there appears to be a change in both rotation amplitude and relative phase, analogous to the phase shifts that occur in gait transitions and finger oscillation, which requires a qualitative reorganization of the musculature. This transition can be represented as a bifurcation in Figure 9, in which the phase difference between shoulder and hip rotation is plotted against the control parameter *pi* = *A/S*. The anticipated hysteresis could be tested by measurements of this phase relation.

In both of these examples, it is clear that behavioral transitions between action modes can be induced visually, and do not occur simply as a consequence of deterministic processes. In the case of walking through apertures, the qualitative shift in action is initiated prior to arrival at the opening and is carried through with no physical contact between the actor and the inducing surfaces, which are linked only by a low-energy optical field. This nondeterministic perceptual coupling between physical conditions and the resultant behavior implies the existence of optical (as well as haptic, acoustic, etc.) information that is specific to the critical points and other dynamics of the animal-environment system [13,38,39].

In sum, qualitative transitions in the behavior of organisms can be approached as instances of critical behavior in complex systems, leading us to explore them in new ways. The goal is to explain the stable patterns and transitions as emerging from the underlying dynamics of the organism-environment system, rather than assuming their existence *a priori*. In contrast to physical systems, however, it does not seem as though perception-action systems are deterministically driven by their dynamics. Rather, the informational coupling permits organisms to be sensitive to the dynamical constraints and to exploit them, selecting over the regions of stability and instability to organize and guide behavior.

REFERENCES

1. Iberall, A.S., "A field and circuit thermodynamics for integrative physiology. I. Introduction to general notions", *Am. J. of Physiology* **233**, R171-R180 (1977).

2. Iberall, A.S. & McCulloch, W., "The organizing principle of complex living systems", *Trans. ASME J. of Basic Engineering* **19**, 290-294 (1969).

3. Warren, W.H., "Action modes and laws of control for the visual guidance of action", in O. Meijer & K. Roth (Eds.), *Movement behavior: The motor-action controversy*. Amsterdam: North Holland (in press).

4. Schmidt, R.A., *Motor control and learning*. Champaign, IL: Human Kinetics Publishers (1982).

5. Grillner, S., "Locomotion in vertebrates: Central mechanisms and reflex interaction", *Physiological Rev.* **55**, 247-304.

6. Selverston, A., "Are central pattern generators understandable?" *Behavioral and Brain Sciences* **3**, 535-571 (1980).

7. Shapiro, D.C., Zernicke, R.F., Gregor, R.J., & Diestel, J.D., "Evidence for generalized motor programs using gait pattern analysis", *J. of Motor Behavior* **13**, 33-47 (1981).

8. Haken, H., *Synergetics: An introduction* (3rd Ed.) Berlin: Springer-Verlag (1983).

9. Prigogine, I., *From being to becoming*. San Francisco: Freeman (1980).

10. Kelso, J.A.S., "Contrasting perspectives on order and regulation in movement", in J. Long & A. Baddely (Eds.), *Attention and performance IX*. Hillsdale, NJ: Erlbaum (1981).

11. Kugler, P.N., Kelso, J.A.S., & Turvey, M.T., "On the concept of coordinative structures as dissipative structures: I. Theoretical lines of convergence", in G.E. Stelmach & J. Requin (Eds.), *Tutorials in motor behavior*. Amsterdam: North Holland (1980).

12. Kugler, P.N., Kelso, J.A.S., & Turvey, M.T., "On the control and coordination of naturally developing systems", in J.A.S. Kelso & J.E. Clark (Eds.), *The development of movement control and coordination*. New York: John Wiley (1982).

13. Kugler, P.N. & Turvey, M.T., *Information, natural law, and the self-assembly of rhythmic movement*. Hillsdale, NJ: Erlbaum (1987).

14. Gibson, J.J., *The ecological approach to visual perception*. Boston: Houghton-Mifflin (1979).

15. Bernstein, N., *The coordination and regulation of movements*. London: Pergamon (1967).

16. Newell, K.M. & Scully, D.M., "The development of prehension: Constraints on grip patterns", unpublished manuscript, Univ. of Illinois at Urbana-Champaign (1987).

17. Warren, W.H., "Perceiving affordances: Visual guidance of stair climbing", *J. of Experimental Psychology: Human Perception and Performance* **10**, 683-703 (1984).

18. Gibson, J.J., *The senses considered as perceptual systems*. Boston: Houghton-Mifflin (1966).

19. Thom, R., "Less cybernetics, more geometry...", *Behavioral and Brain Sciences* **8**, 166-167 (1985).

20. McMahon, T.A., *Muscles, reflexes, and locomotion*. Princeton, NJ: Princeton Univ. Press (1984).

21. Haken, H., "Synopsis and introduction", in E. Basar, H. Flohr, H. Haken, & A.J. Mandell (Eds.), Synergetics of the brain. Berlin: Springer-Verlag (1983) (b).

23. Alexander, R.M. & Jayes, A.S., "A dynamic similarity hypothesis for the gaits of quadrupedal mammals", *J. of Zoology* (London) **201**, 135-152 (1983).

24. Haken, H., Kelso, J.A.S., & Bunz, H., "A theoretical model of phase transitions in human hand movements", *Biological Cybernetics* **51**, 347-356.

25. Dornan, J., Fernie, G.R., & Holliday, P.J., "Visual input: Its importance in the control of postural sway", *Arch. of Physical Medicine and Rehabilitation* **59**, 586-591 (1978).

26. Cook, T. & Cozzens, B., "Human solutions for locomotion: III. The initiation of gait", in R.M. Herman, S. Grillner, P.S.G. Stein, & D.G. Stuart (Eds.), *Neural control of locomotion*. New York: Plenum (1976).

27. Breniere, Y. & Do, M.C., "When and how does steady state gait movement induced from upright posture begin?" *J. of Biomechanics* **19**, 1035-1040 (1986).

28. Thompson, J.M.T. & Stewart, H.B., *Nonlinear dynamics and chaos*. New York: John Wiley (1986).

29. Pennycuick, C.J., "On the running of the gnu *(Connochaetes taurinus)* and other animals", *J. of Experimental Biology* **63**, 775-799 (1975).

30. Heglund, N.C., Taylor, C.R., & McMahon, T.A., "Scaling stride frequency and gait to animal size: Mice to horses", *Science* **186**, 1112-1113 (1974).

31. Kelso, J.A.S., "Phase transitions and critical behavior in human bimanual coordination", *Am. J. of Physiology* **246**, R1000-R1004 (1984).

32. Schoner, G., Haken, H., & Kelso, J.A.S., "A stochastic theory of phase transitions in human hand movement", *Biological Cybernetics* **53**, 247-257 (1986).

33. Hoyt, D.F. & Taylor, C.R., "Gait and the energetics of locomotion in horses", *Nature* **292**, 239-240 (1981).

34. Ralston, H.J., "Energy-speed relations and optimal speed during level walking", *Int. S. Angew. Physiol. einschl Arbeitsphysiol.* **17**, 277-283 (1958).

35. Alexander, R.M., "Walking and running", *Am.Scientist* **72**, 348-354 (1984).

36. Margaria, R., "Energy metabolism of walking and running at various speeds and slopes", *Accademia nazionale dei Lincei, Rome, Classe di scienze Fisiche, mathematiche e naturali, Series 6*, **7**, 299-368 (1938).

37. Rosen, R., *Fundamentals of measurement and representation of natural systems*. New York: North Holland (1978).

38. Warren, W.H. & Whang. S., "Visual guidance of walking through apertures: Body-scaled information for affordances", *J. of Experimental Psychology: Human Perception and Performance* **13**, 371-383 (1987).

39. Mark, L.S., "Eyeheight-scaled information about affordances: A study of sitting and stair climbing", *J. of Experimental Psychology: Human Perception and Performance* **13**, 361-370 (1987).

POOR SIGNAL TO NOISE RATIO IN SCIENCE[1]

Rolf Landauer
IBM T. J. Watson Research Center
P. O. Box 218, Yorktown Heights, NY 10598, USA

The progress of science depends on our exchange of information with colleagues. We need to know what others have done, and we need the satisfaction that others are paying attention to the results of our investigations. Despite the critical nature of our communications, it is my contention that we do it very badly. Unfortunately, problems cannot be corrected, unless they are understood and articulated, and this particular problem seems to be the family secret that no one wants to admit. We can find a good many published discussions about government funding of science, about job opportunities, about the treatment of minorities, about disarmament, about overhead costs; but rarely do we admit how poorly we talk to each other. At most, we have persistent letters to the Editor, expressing frustration with the details of the manuscript selection process.

The most central aspect of the problem is the endless amount of information offered to us. There has been a tremendous proliferation of literature. Some of our most creative scientists, e.g. Onsager or Feynman, published a few dozen papers in a career. Now we expect to see that many for an Assistant Professor. A recent conference paper contained thirty-eight citations, twenty-eight were to the work of that group. Did they really need a twenty-ninth? Journals have become as thick as telephone books, with indices so long that they no longer tempt anyone to scan them. New journals, often ultra-specialized, keep appearing. The proliferation makes any single paper invisible; as a result, we have to repeat ourselves. If we say it only once, it will be assumed that we have changed our minds. Thus, we have positive feedback: The larger the literature becomes, the more we are under pressure to repeat our publications. We do not, of course, repeat them literally; that is against the rules. Instead we change them modestly, with a letter, a paper in an archival journal, a contribution to a conference, all closely related and following each other. Then, eventually, we write a "review," which is hardly a survey of a broad field, but a compilation of several papers written by the author and a few like-minded col-

---

1 This essay is adapted from an after-dinner talk presented at the conference on Dynamic Patterns in Complex Systems.

leagues. The proliferation of literature allows the appearance of journals, by private publishers, costing two-thousand dollars per year.

The proliferation of published information, and the resulting relative invisibility of any one item, is the principal reason why we all have become public relations specialists. It is that same environment which also permits an occasional highly successful P.R. campaign, which creates something out of nothing. The proliferation also puts a great burden on us in the production of many journals, through refereeing. That is a cost to our society which does not show up as part of the subscription cost, nor as a line item in the budgets of our institutions.

The refereeing process, with all its arguments, consumes a lot of energy. Does it accomplish something? From the viewpoint of the critical referee it seems ineffective. The rejected paper eventually finds a more cordial (or lazy?) referee, perhaps at another journal, and appears anyway. The harassed author is equally dissatisfied. After all, the referee wants to do his chore in a few hours, whereas we all know that reading a paper, particularly one with novel conceptual content, requires a long persistent effort; sometimes it takes years. We all know what would happen if an unknown Swiss patent examiner were to express a sweeping and provocative new viewpoint in a paper sent to Physical Review Letters. Even a generous referee would say: "The author may have something, but hasn't really made it clear. Furthermore, the author has not explained the relationship to Shinglehofer's work." Shinglehofer, of course, is the referee. Manuscript selection, through referees, is a slightly biased lottery, based on a pretense that we can quickly tell whether something is right or wrong, important, or unimportant. This is part of the widespread pretense that we read a great many papers. But reading a paper is hard and time consuming. While we all *scan* hundreds of papers a year, we only *read* a small handful. But no one ever admits that. The pretense that it is easy to recognize the truth, unfortunately, also shows up in review papers. Authors of review papers need to be encouraged to admit: "I really can't figure out whether this paper is right or wrong. It's important, and I hope somebody will do it."

I used to believe that refereeing could be made more responsible by identifying the referee to the author and the reader. Eventually, however, I had occasion to study a good many book reviews, over a period of a few weeks, in connection with an "Invited Opinion" column, in the American Journal of Physiology. I discovered that the reviewers had, in a surprising number of cases, not read their book. Only a small percentage of the reviews gave real evidence of conscientious attention to the volume in question. Now, if clearly identified book reviewers do not do a good job, we cannot expect public identification of the referee to do much for the manuscript selection process.

Referees need to take themselves less seriously as the ultimate judges of truth. The negative comment that needs to be made most often, unfortunately, rarely gets stated. Referees for the Journal of Relativistic Metallurgy, or the Reviews of Subterannean Integrated Circuitry, are unlikely to say, "Enough petty little contributions in this narrow field. Stop it."

The young scientist often assumes that it is lack of membership in the *establishment* which causes the scientist's work to be ignored. But that's wrong, the work of established scientists tends to get only a little more attention, at best. Indeed, "establishments" are little corners in a many-dimensional space, and no really novel line of thought is protected by membership in an establishment. After all, it is the existence of a set of shared and well-accepted concepts which form the basis for a typical micro-establishment.

In fact, the work of world-famous scientists can get overlooked as easily as a Ph.D. thesis, if the work is not addressed to the right community at the right time. Einstein, in 1917, published a paper on the old Sommerfeld-Wilson quantization schemes which preceded the final advent of quantum mechanics in 1925 and 1926. There are semi-classical approximations to the Schrödinger equation which were first developed in the one-dimensional case. That is called the WKB approximation, and was extended to the many-dimensional case much more slowly. These approximations picture a quantum mechanical wave propagating along classical orbits, and are intimately related to the older Sommerfeld-Wilson schemes. The many-dimensional versions of the WKB approximation are sometimes called EBK approximations. The "E" in this stands for Einstein and the central nature of his 1917 paper in this field. But that represents a modern awareness. Einstein's paper, in fact, remained obscure for decades. My own Ph.D. thesis, completed in 1950, dealt with this subject. I did not know about Einstein's paper. Eight years later I saw a citation to it in a paper by J. B. Keller, but in the usual fashion ignored that. Many years after that, Martin Gutzwiller finally told me in some detail about Einstein's paper.

An even more extreme example comes from von Neumann, whose contribution, in fact, received widespread initial attention. Nevertheless, it then went into obscurity and later waves of closely related activity ignored his invention completely. In 1954 von Neumann filed a patent application, issued in 1957, as U.S. Patent #2,815,488. The same concept was discovered independently by E. Goto, in Japan. These proposals utilize parametric excitation, i.e. the excitation of subharmonics in nonlinear circuits, to carry out the logic functions in a computer. These inventions stem from a time when junction transistors were still very slow, and the inventions represented an attempt to move toward faster computing.

The von Neumann and Goto inventions generated great interest, originally. Intensive efforts to develop these inventions came into existence in the United States and Japan. In Tokyo, working "parametron" computers were built. Despite that, the work and concepts went into complete obscurity, quickly. The work was not remembered in the early 1960's, when interest in optical parametric schemes arose. In recent years, when work in chaos and strange attractors have again brought attention to parametric excitation, von Neumann and Goto remain forgotten. How can von Neumann, one of the most renowned scientists of the twentieth century, suffer this fate? There are a number of parts to the answer, but a key point relates to the many-dimensional nature of our "science establishment." von Neumann was not a card-holding member of the quantum electronics union; he never attended their conferences. His health started to fail in 1955, and he died in 1957. Even if he had lived longer, however, it is unlikely that a man with his vast range of interests and activities would have participated in early quantum electronics meetings. The very word "establishment" is misleading. It is participation in informal collegial mutual interest groupings that counts. Designation as a world famous scientist may even be a handicap.

The pressure on authors, for excessive publication, takes its toll in many ways. Sloppy citation practices are an obvious symptom. There is no dearth of citations, they are thrown in liberally to appease colleagues. But they are thrown in hastily at the end, with a citation dispenser, and with little regard to the real content of the citations. I've often been tempted to invent a citation, and see if it would be copied and propagate. Eventually, however, I learned that the experiment was not needed, it already existed. Science Citation Index for 1976 carries 36 citations to two papers by a non-existent J. C. Maxwell-Garnett. Enough to qualify him for tenure at many universities! The same Science Citation Index volume also carries five citations to two papers which, remarkably enough, occur on the same journal pages; and these are by the once very real J. C. M. Garnett. Presumably the five citations come from those who actually looked at the Garnett papers, and did not just copy the author's name from earlier papers. Garnett's two papers, from 1904 and 1906, dealt with the optical properties of glasses with small metallic precipitates. At a 1977 conference, I insulted many of the offending authors by displaying, via transparencies, the relevant pages from Science Citation Index. In the published proceedings I included some biographical material about the man with this mysterious name. It had a modest effect. A decade later, in the 1986 Science Citation Index, correct and incorrect entries are about equally abundant.

We have discussed the published literature in great detail; we will touch upon conferences much more briefly. The range of problems, however, is very similar. Conferences fall into three categories. First of all, there are the huge annual gatherings of the major professional societies. Then there are the perennial social alumni reunions of slightly more informal groupings: The 23rd Annual International Workshop on Relativistic Metallurgy, or the 11th biennial conference on Subterranean Integrated Circuitry. One meeting of that sort, held last year, had 500 poster papers and 250 short talks. Most of the 500 poster papers were displayed simultaneously! Now these meetings do have a purpose and utility. But they reinforce the existing fads and mini-establishments. The money and effort that goes into them would be far better spent on more exploratory sessions, like the meeting represented in this volume, which permit new connections to be formed, and allow genuine cross-fertilization. Funding agencies should have a self-imposed rule: No support for a session after five iterations. And, incidentally, no support for meetings run by the same perennial organizers. We do not need perennial centers for conference organization, or at least should support them only in the very few cases where an organization has demonstrated an ability and dedication to go beyond narrow self-serving goals. Finally there are the small workshops. Typically these consist of a mixture. Half the papers repeat the last related workshop. After all, it's safer to invite the same old reliable and highly regarded cast again, than to take a chance on new faces. And, in any case, it would be a good deal of work for the organizing committee to investigate suggestions for less well known speakers. Then there is the other half of the small workshop. Those are papers by the organizing committee, their associates, their close buddies, and by past and future organizers of similar conferences. Frank Moss, in a recent book review of a conference publication told us *...over half the invited papers were authored or coauthored by members of the International Programme Committee.* Such an honest description is remarkable; we need more of that.

The world isn't all bad, there are a few examples of remarkably effective and well thought out meetings. Hermann Haken, in 1972, had the first interdisciplinary meeting which somewhat resembled the session which gave rise to this volume. Interdisciplinary meetings existed before then, but they typically had a high component of flaky stuff. Probably not everything presented at Haken's session in 1972 has stood up, but all of it was serious and represented real intellectual depth and effort. Participation was a breath-taking experience for me; for the first time I found myself among people with comparable interests and a comparable sense of values. I was no longer an orphan! And the meeting had another earmark of a good conference: The conference had broad representation in its selection of speakers; it was not dominated by the organizer and his close associates.

Science is heavily dominated by fads and fashions. That's not totally bad. Some clustering of interest is needed to permit effective information exchange and stimulation. But the positive feedback in the system does seem excessive, and there are strong pressures to submit. If there are not enough people to understand and appreciate your work, can you get a tenure appointment or funding? If you don't have at least two or three competitors, no one will organize a session of invited papers into which your work will fit. Much of the tendency for fads represents a tool orientation; we do that which we can easily do. Nonlinear dynamics and fractals are two subjects in which we have gained grand new insights, which were slow in coming.

They were slow in coming, in part, because the people who were perceptive and understood the nature of chaos, and understood fractals, did not fit the fashions. But now these fields have, in turn, become dominant fashions, and in part because it is easy to go to the computer and *do something* in these areas. It doesn't take years of mental struggle or apparatus building, to generate one more paper.

The real pioneers, like Shannon, Onsager, Mandelbrot, generate fashions, they do not follow them. They recognize the possibility of intellectual order, where others do not notice it. Then we go and structure our communication tools so as to make that particularly unrewarding! In recent weeks (in the Fall of 1987), I am often asked about the remarkable succession of Nobel prizes, for two years in a row, in the tiny Swiss branch of IBM Research. Everybody is allowed to arrive at their own post-facto analysis of such a rare event, just like in the case of the almost simultaneous stock market setback. But there is one obvious common denominator to the scanning tunneling microscope, and to high temperature superconductivity. In both cases, the investigators were following their own private high risk convictions, far from the existing fashions. Would they have gotten NSF grants if they had been dependent on that?

Earlier, we mentioned J. C. M. Garnett. The next major step in the evolution of that subject, the electrical properties of heterogeneous media, was taken by D. A. G. Bruggeman, in the mid-thirties. Bruggeman presented the first modern effective medium theories, giving explicit recognition to the statistical nature of the problem. His work in turn, after a few citations by his Dutch fellow citizens, became obscure for several decades, until the field became seriously active. Why was Bruggeman perceptive, why did he appreciate a question which did not receive widespread attention until much later? Probably because he was a high school teacher. If he had been at a major center of learning he would have been caught in the tides of his time and done something that was fashionable then, e.g. molecular spectroscopy.

The basically oriented scientist might assume that the problems we have discussed are not present in technology. After all an invention works or doesn't work. In that case there should be no room or need for public relations activities. But it's not that simple. Yes, when it finally comes to a product, to be sold, it has to work and meet competition. But in the advanced technology stages it is very different. There, too, we have much the same organization as in science, with funding agencies, and with fads. There are many proposals which became major thrusts, only to be abandoned again subsequently. An adventurous technological climate *must reward risk-taking*, and *must allow failures*. But there are too many cases where everyone eventually sheepishly realizes that the flaws or disadvantages were really understood at an early stage, but the dissident voices were not heeded. The system is organized to let that happen. The proponents, in need of funding, put forth the positive view, write the papers and organize the workshops. Even the funding agencies, after a first hard and sincere decision, are then emotionally locked into it. The contract supervisor wants to demonstrate to his management that the original decision was sound. By contrast, the skeptic doesn't have equal motivation to speak up. He won't, in any case, be cited in the literature, or be invited to the conferences, dominated by the proponents. We give no awards for whistle-blowing, except perhaps in connection with public safety matters. Then why bother? Furthermore, once the thrust is abandoned, there is no incentive for a technical obituary. There will be no record from which others can learn.

We have alluded to a number of problems, and have suggested little in the way of solutions. But, as already stated, problems must first be recognized and articulated. The system needs some negative feedback and needs more honesty. The protective self-serving influence of our mini-establishments must be diluted. But how one can do that is not obvious. As with all social problems, theory is a poor guide and predictor; experimentation is needed. Semi-popular scientific journals have a key role in all this; they are free to comment on the scene. The funding agencies, also, have a key role. They can be leaders, rather than followers tied rigidly to a peer consensus.

The scientific community needs more sensitivity to conflict of interest questions. To most of my academic colleagues that is something that arises when a representative of a pharmaceutical company serves on a committee evaluating drugs made by the committee member's employer. Or, perhaps, when the principal investigator on a contract employs a relative. Clearly, however, conflict of interest questions are broader than that. The conference that is supported by public funds is executing a public trust and has the obligation to use the money accordingly, and not for self-serving ends. A major journal reviewed a book authored by the chairman of the board of the journal's parent organization, and selected as reviewer a lifetime associate of the author from the author's institution. Undoubtedly, the staff members of that journal are proud of their ethical conduct; they don't place personal phone calls at their journal's expense.

Finally, I must ask a more disturbing question. All of us plead, almost continuously, for increased funding for science. If, however, we generate more information than we can absorb, perhaps we should put our house in order, first. Additional scientists will only cause the signal to noise ratio to deteriorate even more.

Poster Presentations

J. J. Buchanan, J.A.S. Kelso & G.C. de Guzman Florida Atlantic University	Further phase transitions in biological coordination: Rhythmic movement in 3-dimensions
D. R. Chialvo, D. C. Michaels & J. Jalife SUNY Health Science Center	Experimental observations and difference-equation model of nonlinear dynamics and chaos in non-Oscillatory cardiac tissues
V. D. Deshmukh & L.S. Russo, Jr. University of Florida	The Pd 265 potential: A sign of dynamic patterning of language-related cerebral cortex
M. Feingold, D.L. Gonzalez, O. Piro & H. Viturro James Franck Institute University of Chicago & Universidad Nacional de la Plata	Driven excitable systems vs. driven oscillators: Global dynamics
A. Fogel Purdue University	Dynamic systems in human development: The inhibition of attention to mother in early infancy
I. Giaever & C. R. Keese General Electric Corporate Research & Development	Fractal mammalian cell behavior in culture
J. J. Jeka & J.A.S. Kelso Florida Atlantic University	Dynamic patterns of multi-limb coordination
R. M. Kiehn University of Houston	Aging as a topological perturbation process
E. Meron & P. Pelce James Franck Institute, University of Chicago & Universite de Provence	Dynamics of spiral wave formation in excitable media
J.W.L. Merrill & R. F. Port Indiana University	A genetic learning algorithm for dynamic neural networks
D. C. Michaels, R. Chialvo & J. Jalife SUNY Health Science Center	Phase-locking and irregular dynamics in a mathematical model of repetitive vagal stimulation of the sinus node
R.F. Port Indiana University	Language and control of speech production
G. Schöner & J.A.S. Kelso Florida Atlantic University	A theory of learning and recall in biological coordination
H. Shimizu & Y. Yamaguchi University of Tokyo	The context-dependent self-organization of semantic information in a synergetic network
J. Y. Suh Indiana University	Applying distributed genetic algorithms to evolving dynamic systems
B. Tuller & J.A.S. Kelso Florida Atlantic University	Environmentally-specified patterns of movement coordination in normal and split-brain subjects

FURTHER PHASE TRANSITIONS IN BIOLOGICAL COORDINATION: RHYTHMIC MOVEMENT IN 3-DIMENSIONS

J.J. Buchanan, J.A.S. Kelso, G.C. deGuzman
Center for Complex Systems
Florida Atlantic University
Boca Raton, Fla. 33431

We explore a serendipitous, pattern formation phenomenon first noticed by Kelso and Scholz (in H. Haken, ed., Complex Systems, Springer Verlag, Berlin, 1985). When subjects, prepared initially in an anti-phase pattern of rhythmic hand movement in the transverse plane (i.e. as one finger flexes, the other extends), increased the frequency of oscillation, a spontaneous transition to an in-phase pattern was initially observed -- followed, at a higher frequency, by a transition to an in-phase motion but now in the vertical plane. Kelso and Scholz noted that the motion frequently became rotary in nature before this transition to a simultaneous flexion-extension pattern occurred. This higher-order phase transition in a system which, because of additional degrees of freedom, has multiple patterns available, is systematically explored in the present study. Data on six subjects were collected using the WATSMART 3-D motion analysis system.

Of particular interest is the problem of identifying collective variables (or order parameters) and their dynamics in a biological system with multiple (more than 2) patterns. Is there a generalizable measure that describes coordination similar to the relative phase? What additional attractors are present and observed? Are the same phenomena as seen in earlier work (enhanced order parameter fluctuations, critical slowing down) also observed in this experimental system? Does switching among multiple patterns follow a particular path, i.e. is it history dependent? It is an important step of generalization to determine whether the balance between a pattern's stability and fluctuation governs the evolution of additional temporal patterns.

[Work supported by NIMH, ONR, and AFOSR]

EXPERIMENTAL OBSERVATIONS AND DIFFERENCE-EQUATION MODEL OF NONLINEAR DYNAMICS AND CHAOS IN NON-OSCILLATORY CARDIAC TISSUES.

Dante R. Chialvo, Donald C. Michaels and Jose Jalife. Dept. of Pharmacology
SUNY Health Science Center at Syracuse, NY 13210

Chaotic patterns of excitation and impulse propagation can be demonstrated in cardiac tissues that do not oscillate spontaneously. We investigated the dynamics of such phenomena in isolated sheep Purkinje fibers superfused with Tyrode solution and driven by brief depolarizing current pulses applied through a suction pipette. Microelectrode recordings were obtained 3 to 5 mm from the stimulus site. By using various regimes of stimulus strength, duration and frequency, we defined the behavioral structure in the parameter space. In all experiments, stable locking patterns occurred at relatively low frequencies and levels of current strength. The hierarchy always followed Farey's series and, for a given period-adding sequence, the relation between period of locking and pulse strength was similar to that described for the rotation number in the circle map. Chaotic-like patterns were observed at higher frequencies and current strengths. In a second group of fibers, the dynamics were studied during action potential propagation along a non-homogeneous cell strand. The bifurcation parameters in this case were driving rate and changes in the effective conduction time (latency) induced by segmental uncoupling between cells. At the normal levels of coupling, stimulation at increasing rates during 1:1 led to progressive increase in latency, until 2:1 locking occurred. When the degree of conduction impairment was high, changing the driving rate yielded various locking patterns that resembled those in the first group. Yet, at intermediate levels of coupling, period doubling and irregular dynamics were found as the driving rate was changed between the 1:1 and 2:1 domains. All these observations were reproduced semi-quantitatively in a simple analytical model. The dynamics in this model are determined by two empirically derived functions: rate-latency of activation and rate-action potential duration. As the driving rate is increased, ordered locking or period doubling bifurcations may occur, depending on the precise shape of the functions. In conclusion, locking phenomena and chaos can be demonstrated in cardiac Purkinje fibers that do not undergo self-sustaining oscillations. Moreover, the model results suggest that the mechanism of such dynamics may be related to the non-monotonicity of the rate-latency function.

The Pd 265 potential: a sign of dynamic patterning of language-related cerebral cortex

Vinod D. Deshmukh and Louis S. Russo, Jr.

Topographic mapping of brain electrical activity was performed during linguistic tasks in 9 healthy volunteers 21-53 years of age. 7 were right-handed and 2 left-handed as determined by the Edinburough Handedness Inventory. The linguistic tasks included listening to target tones, counting subvocally (verbalizing), counting aloud (vocalizing) and responding manually by lifting left index finger. Standard and modified P-300 paradigm were used. The event-related potentials (ERP) and EEG were recorded from 16 electrode positions.

The mean ($\bar{x}$) and standard deviations (sd) of the latencies in milliseconds of the ERP components were:

Auditory process: $N_2=228\pm10.9$, $P_3=343\pm37.4$, $N_4=449\pm26$, N_2-N_4 interval$=220\pm36$.

Audioverbal process: $N_2=223\pm17.8$, $P_3=341\pm25.8$, $N_4=462\pm24$, N_2-N_4 interval$=241\pm29$.

Audiovocal process: $N_2=228\pm34.2$, $P_3=346\pm35.3$, $N_4=474\pm31.8$, N_2-N_4 interval$=250\pm31.9$.

Audiomanual process: $N_2=204\pm9$, $P_3=351\pm15.4$.

During the audioverbal process, when the right hemispheric activity was subtracted from the corresponding left hemispheric activity, a new speech-related potential was seen in the N_2-P_3 interval with regional positivity over the primary language area of the dominant hemisphere. Since its mean latency in 9 subjects was 265 ± 10.9 ms, it was labeled as the Pd265 potential (an interhemispheric difference in electrocortical positivity) related to linguistic processing.

Driven excitable systems vs. driven oscillators: Global dynamics

Mario Feingold,[1] Diego L. Gonzalez,[2] Oreste Piro[1] and Hector Viturro[2]

[1]*The James Franck Institute, The University of Chicago, Chicago IL 60637.*
[2]*Departamento de Física, Universidad Nacional de La Plata, C.C. 67, (1900)La Plata, República Argentina.*

The distinctive attribute of excitable systems is their response to external stimulation. Any stimulus whose amplitude is below a certain threshold only produces small changes in the system leaving it always close to the quiescent state. In contrast, an upper-threshold stimulation triggers transitions to a state which is qualitatively and quantitatively different from the initial one. Moreover, this excited state relaxes to the quiescent one in a characteristic time which is insensitive to the stimulus amplitude. The simplest model which captures those essential features is the van der Pol-Fitzugh-Nagumo (vdPFN) equation. While a great deal of work has been done on the self-oscillating regime of this model, the dynamical behavior of the non-self-excited version in presence of periodic excitations is much less understood. In this work, we investigate the dynamics of driven non-self-oscillating excitable systems by means of the vdPFN model. We find that the global bifurcation structure is similar to the phase locking diagram of driven oscillators in spite of the absence of an internal frequency. This similitude appears both in the topology and in the scaling properties of the diagrams. The upper-critical behavior is also of the same type as in forced oscillators and circle maps including the presence of period doubling and chaotic trajectories. An analytical argument is developed to connect the two classes of systems. The most important implication of our results is that non trivial dynamics like phase locking, quasi-periodicity and chaotic responses can be generated by excitable media regardless of whether they are oscillatory or not in the autonomous case.

Dynamic systems in human development: The inhibition of attention to mother in early infancy

Alan Fogel
Department of Child Development and Family Studies
Purdue University
West Lafayette, IN 47907

The reigning metaphor for human development is that new abilities emerge as a result of centrally organized maturational processes; advances in different skill domains are thought to occur in synchrony. Dynamic systems theory offers an alternative in which skill domains develop at different rates and in different contexts, and new abilities emerge as the result of the dynamic interaction of components across domains. In many cases all but one of the requisite components of a skill are present, but the child will not manifest the skill until the missing component is acquired.

This report examines the developmental shift in attention to people vs. objects occurring between the ages of 4 and 6 months. Before the shift infants are oriented almost entirely to adults, after the shift to objects with only occasional glances to adults. We predicted that a variety of components--all changing at about the same age--might play a role in inducing the shift: increased skill in grasping objects, the transition from lying to sitting unsupported, decline of fussiness, and general maturation.

The results show that a stable emergent state of gazing away from mother can be created in pre-sitting infants who can reach simply by experimentally providing the missing control parameter of sitting upright. This stable state cannot be created by the postural manipulation in pre-reaching infants. The supine behavior of pre-sitting infants who can reach for objects shows both within and between individual variability characteristic of critical fluctuations of a dynamic system in a transition region between two stable states.

Fractal Mammalian Cell Behavior in Culture*

Ivar Giaever and Charles R. Keese
General Electric
Corporate Research and Development Center
P.O. Box 8, Schenectady, NY 12301

A sensor to detect the behavior of cells growing in tissue culture has been developed. Mammalian fibroblasts, cultured on small electrodes, are subjected to weak AC fields. Although these fields have no apparent effect on the cells, the attached fibroblasts cause a marked change in the measured impedance of the electrode and also cause its value to fluctuate within time.

The changes in impedance of the electrode can be correlated with several aspects of cell behavior including attachment, spreading and locomotion. Based on the dependence of the impedance upon the frequency of the applied signal, we have shown that this system can probe, in real time, the narrow space between the ventral surface of a layer of cells and the substrate. The fluctuations in the impedance of an electrode confluent with cells is a measure of the variation with time of this narrow space. The fluctuations have a distinct fractal behavior and have been extensively analyzed by R/S statistics introduced by Hurst. Attempts are presently underway to distinguish between normal and transformed cells using this technique.

* Paper presented at Conference on *Dynamic Patterns in Complex Systems*. Manuscript not available at time of publication.

Dynamic Patterns of Multi-Limb Coordination

J.J. Jeka and J.A.S. Kelso

Center for Complex Systems and Department of Psychology
Florida Atlantic University
Boca Raton, FL 33431

In general, the movements of animals, including people, require the cooperation among many components, and thus can be said to exhibit the property of *coordination*. The present research studies multi-limb tasks involving the combined action of the arms and legs in humans, a situation in which many coordinative patterns are potentially available. Our aim is to empirically identify the order parameters (collective variables) characterizing these patterns and to study their dynamics, including relative stability and loss of stability, as parameters are varied.

An apparatus specifically designed for this purpose, allows us to measure the kinematics of (up to) four rhythmically moving limbs of a seated subject. Frequency of motion acts as a control parameter and is manipulated by an auditory metronome to which the subject must synchronize his/her actions. Patterns are differentiated by qualitative changes in behavior, observed through the collective variable, relative phase. Several different phase transition phenomena are observed in a set of three experiments. When the system is prepared in a pattern in which the phase and frequency-locked $(1:1)$ movements of non-homologous limbs are in opposite directions (e.g., an arm moving up as a leg moves down and vice-versa), a transition occurs at a critical frequency to a pattern in which both components move in the same direction. However, no such transition occurs when the system is prepared with non-homologous limbs moving in the same direction. At a still higher driving frequency, the arms and legs can no longer be characterized by a single pattern, but display much richer dynamics, including a variety of frequency $(2:1, 3:2\ldots)$ and phase relationships, as well as epochs of non-stationary behavior. Nonlinear oscillator modelling and various measurement tools from the field of dynamical systems (e.g., return maps, phase plane trajectories,etc...) are used to explore these observations in greater detail.

Work supported by NIMH, ONR and AFOSR.

AGING AS A TOPOLOGICAL PERTURBATION PROCESS

R. M. Kiehn, Physics Department, University of Houston

The biological system, as it evolves through its life cycle of birth, maturity and aging, and ultimate death, is the classic example of an irreversible system which reaches and maintains a non-equilibrium mature state of low entropy by synergetic processes. Ths slow ageing of the mature state is apparently continous, but irreversible, and is therefore NOT a homeomorphism which by definition preserves topology. Topology must change during the aging process.

A true equilibrium state evolves without change of topology; therefore the slow aging process must involve a topological perturbation from the equilibrium state. The development of a theory to understand the concept of aging must be capable of dealing with questions of topological evolution and topological stability. Almost all current perturbation theories are based on geometrical variations, and presume, apriori, invariant topology.

It is suggested that Cartan's theory of exterior differential forms may be used to produce a continuous, but topologically varying, theory of evolution which is capable of describing irreversible processes, including the production and evolution of defects. It would appear that the evolution of energy (Newton's law of motion) is to be associated with the evolution of 1 dimensional topological properties induced by a 1-form of Action, A. The evolution of entropy is to be associated with the 3-form of Torsion-Integrability, A^dA, which is irreducibly three dimensional or more. If the 3-form A^dA vanishes, the system is in a Caratheodory equilibrium state of constant entropy. A topological pertubation would carry the system from a zero to a non-zero value for the integrability 3-form.

DYNAMICS OF SPIRAL WAVE FORMATION IN EXCITABLE MEDIA

Ehud Meron
The James Franck Institute, University of Chicago, Chicago IL 60637.

and

Pierre Pelce
Laboratoire de Recherche en Combustion, Universite de Provence,
13397 Marseille Cedex 13, France.

ABSTRACT

A model for the velocity of a wavefront, having a free edge and moving in an excitable medium, is proposed. The model propagates arbitrary initial conditions into "fully developed" rotating spiral waves. It is based on the following physical observations. The normal velocity of a convex (concave) front is reduced (enhanced) with respect to the velocity of a planar front. This diffusion induced effect stabilizes the front against short wavelength perturbations. In the vicinity of a free edge a normal velocity profile develops due to tangential diffusion. The velocity profile imposes on the front's free edge a persistent bias to curl. The curvature at the free edge eventually saturates because of a finite width effect: the wavefront is followed by a refractory tail consisting of sites which are not reexcitable. When the tip starts probing these sites curvature saturation begins. The proposed expression for the front velocity is used to derive a partial differential equation for the curvature of the front. This equation is solved numerically, using the Crank-Nicholson scheme, and once the curvature field is known the wavefront can be constructed. Dimensional analysis is used to relate the model parameters to physical ones and to derive scaling relations between experimentally observable quantities.

A genetic learning algorithm for dynamic neural networks

John W. L. Merrill and Robert F. Port *
Departments of Linguistics and Computer Science
Indiana University
Bloomington, Indiana 47405

September 29, 1987

Abstract

We discuss an algorithm to train abstract neural networks (ANNs). Most such algorithms are limited to small networks because of inefficiency, or are limited to feed-forward networks because of limitations in the algorithm itself. The most attractive feature of ANNs is their implicit ability to operate on several time-scales simultaneously, so those algorithms would unduly limit the applicability of ANNs.

Our algorithm is both efficient and applicable to all network structures. In addition, it can be used to train networks which cannot be controlled instant–by–instant or which can only be evaluated in vague terms (*i.e.* "better" or "worse," instead of "Increase x_1 by 0.6, x_2 by $-.2$, ...")

Based on Holland's Genetic Algorithms for function optimization, it differs from standard genetic search algorithms in several ways: the inclusion of simulated annealing, and use of a non-standard binary search operator. Each modification improves performance. We also show that the best way to combine simulated annealing into genetic search lies in modifying the behavior of the blending (cross-over) operator.

We draw three major conclusions. First, the recent trend away from stochastic learning algorithms and towards deterministic learning algorithms among those studying machine learning with neural networks is premature. Second, the performance of both simulated annealing and genetic search can be improved by a judicious fusion of both techniques. Finally, we conjecture that any adequate learning algorithm must contain both stochastic and deterministic elements, since the most attractive features of these styles are complementary.

*Research partially supported by NSF grants DCR–85–18725 and DCR–85–05635

PHASE-LOCKING AND IRREGULAR DYNAMICS IN A MATHEMATICAL MODEL OF REPETITIVE VAGAL STIMULATION OF THE SINUS NODE.

Donald C. Michaels, Dante R. Chialvo and Jose Jalife. Dept. of Pharmacology
SUNY Health Science Center at Syracuse, NY 13210

A model of the cardiac pacemaker was used to study the dynamics of the response to repetitive vagal stimulation. The central region of the sinoatrial node was simulated as a 15 x 15 array of resistively coupled pacemaker cells. Cells were randomly assigned one of ten intrinsic cycle lengths. Under control conditions, the pacemakers mutually entrained to a common frequency and a dominant region became apparent. Single acetylcholine (ACh; "vagal") pulses applied to a randomly selected 60% of the cells resulted in a shift of the dominant pacemaker region to another region of the array, a delay in the next spontaneous firing, and an increase in apparent conduction time. When ACh pulses were applied repetitively over a range of stimulus intensities and frequencies, a variety of stimulus:response patterns were obtained including phase-locking (e.g., 1:1, 2:1, 3:2, etc) and irregular dynamics. A map of the parameter space indicated that a lower stimulus intensities, phase locking followed the typical Farey sequence. At higher intensities, regions of irregular dynamics were found. In another group of simulations, qualitatively similar results were obtained when a single pacemaker was perturbed with repetitive ACh pulses. Irregular dynamics in both types of simulation were investigated using phase-plane (orbital) plots, Poincare mapping, and return mapping. On the route to truly irregular dynamics, period doubling bifurcations (2:2, 4:4 and 8:8) were obtained both spatially and in the parameter plane. The attractor in the return map during chaotic activity resembled the Lorenz tent map. However, the interactions of neighboring cells exhibiting chaotic dynamics resulted in characteristic alterations of the attractor geometry that could not be simulated by the addition of random noise. Our results suggest that irregular dynamics obeying the rules derived from other chaotic systems are present during vagal stimulation of the sinus node. These irregular dynamics may result in clinically relevant disturbances of cardiac rate and rhythm.

Language and the Control of Speech Production.

Robert F. Port,Departments of Linguistics and Computer Science

Languages often exhibit 'phonological rules' that change certain aspects of articulation depending on their position in the syllable. Thus, the English /t/ phoneme is aspirated when it is at the absolute beginning of a syllable (cf. *t*[h]*op* vs. *stop*) and is flapped when intervocalic (as in *cut* vs. *cutter*). Linguistic descriptions of these facts emphasize a *symbolic* model in which a 'rule' changes the values of one or more 'features' of the segments involved. Our recent research, however, (Port & O'Dell, 1986; Fourakis & Port, 1986; Port, Crawford & Rominger, 1987; Port, Reilly & Maki, 1987) suggests that many phonological rules result in speakers actually manipulating a very large number of dynamic variables of speech production. We present evidence from *neutralization rules* (phonological rules that destroy a distinction) in English and German. Thus, *ride* and *write* are distinct words, but when a vowel-initial suffix is added, as in *rider–writer*, the /d/ and /t/ seem to neutralize to a 'flap' sound. But the evidence from detailed phonetic studies is clear that they do *not* neutralize to the same speech sound. In most cases examined, the rule of neutralization does not succeed in fully destroying the difference. Instead, the phonological 'rule' appears to be exerted as continuous, graded dynamic control over the course of the speech gestures, not the discrete all-or-nothing rules traditionally employed. The data show that this control must be exerted by a dynamic system that embodies a very large number of articulatory variables. The existence of the 'phonological rule' apparently reflects an attempt by the language itself to achieve neutralization at the symbolic level (since English *writer–rider* and German *bunt– Bund* really do sound more or less the same to native speakers, indicating that no useful phonetic information is available there). Still the neutralization is incomplete and is retained in subtle temporal phenomena (phenomena that listeners can exploit for perception if forced to, as we have also shown).

Following suggestions of Pattee (1977) it is possible to observe here the discrete neutralization effects of the symbol-like unit at the level of the language itself (a social institution that 'slaves' individual speakers). Yet, at the same time, we can see that actual implementation of the neutralization is achieved by changes in the order parameters of the syllable control system of the speaker. Changes in order parameters typically cause complex dynamic effects. These data demonstrate the appropriateness of dynamic models to account for the linguists' paradox of discrete vs. continuous processes in language.

A theory of learning and recall in biological coordination

G. Schöner and J.A.S. Kelso

Center for Complex Systems
Florida Atlantic University
Boca Raton, FL 33431

In earlier work (e.g. Kelso & Schöner, *Springer Proc. Phys.* **19**, 224, 1987) a synergetic strategy for understanding biological coordination was developed that stressed collective (behavioral) variables and their (intrinsic) dynamics, stability and fluctuations. The influence of environmental information and of learned coordination patterns were understood as perturbations of the vector-field of the intrinsic dynamics, leading to modifications of the stable behavioral patterns (Schöner & Kelso, *Biological Cybernetics* (in press)). Here we ask how the processes of learning a behavioral pattern and activation of a learned pattern (variously termed recall or recognition) can be understood within our approach.

Both learning and recognition are defined here in terms of their influence on the intrinsic dynamics. Memory is viewed as a part of the dynamics of the behavioral variables, i.e. a perturbation of the intrinsic vector-field that attracts the behavioral variables toward the memorized pattern. The memorized patterns and the relative strength of their influence evolve dynamically during learning and recall. We show that new degrees of freedom–memory and recall variables–must be introduced. The synergetic strategy can be extended to determine learning and recall dynamics if the corresponding time scales are taken into account. We derive a general form for these dynamics that resembles models used in micro-evolution theory (Eigen & Schuster, *The Hypercycle*, Springer V., 1979).

Even though the theory possesses generality (e.g. to a variety of behavioral and neural patterns) we corroborate our findings by addressing a concrete example that involves temporal patterns in rhythmic movements (e.g. Kelso, Schöner, Scholz & Haken, *Physica Scripta* **35**,79 (1987)). All relevant dynamics are explicitly determined making use of earlier results. Numerical solution of these dynamics shows that the system learns and recalls with the correct time scale relations. Several distinctive features–not present in more formal theories–are discussed that are open to experimental test, including: (a) the influence of the intrinsic dynamics on the recall process; (b) the presence of qualitative changes (phase transitions) in the dynamics of the behavioral variables during the process of learning a required temporal pattern; and (c) the presence of such phase transitions during the recall of a previously learned pattern.

Work supported by NIMH, ONR and AFOSR. G.S. acknowledges support by the Deutsche Forschungsgemeinschaft, Bonn, W.-Germany.

The Context-Dependent Self-Organization of Semantic Information in a Synergetic Network

H. Shimizu and Y. Yamaguchi

Faculty of Pharmaceutical Sciences,
University of Tokyo,
Hongo, Tokyo 113, Japan

The self-organization of semantic information in a synergetic network system with ability of self-organizing an information carrier will be studied. Each piece of elementary information is encoded in the internal dynamics of a synergetic element, a holon, of the system. The internal dynamics has a selectable set of freedom, a set of internal modes, and the elementary meaning of the encoded piece is specifically determined by the internal mode to be selected. We succeeded in finding such a network system which self-organizes a collective mode of the internal dynamics of the holons by selectively activating only relevant ones of the internal modes while sppressing irrelevant ones. In this self-organization, the collective mode automatically encodes semantic information composed of the elementary information in such a way that the elementary meanings of the encoded pieces are consistent with the holistic meaning of the self-organized information, i.e. context. Thus, our synergetic system has its own semantic information dynamics, which enables us to discuss the creation of new semantic information. In addition, our synergetic network system is not based on a "softwared-controlled intelligence" but spontaneously generates its operational information, the software for the information processing, due to its synergetic ability of the holons, without being controlled by an externally given program. In fact, it would be one of important characteristics of the brain, a kind of a synergetic computer composed of holons.

Applying Distributed Genetic Algorithms to Evolving Dynamic System.

Jung Y. Suh
Department of Computer Science
Indiana University

Genetic Algorithms (GA) have been applied to such problems as the optimization of real-valued functions, learning strategies for games and the combinatorial optimization problems. Its main feature is that it operates on populations of structures instead of one. The GA stochastically applies two operators, called *recombination* and *mutation*, to structures in the population, improving the population's performance toward to the global optimum.

D. Van Gucht and I have applied this algorithm to the Traveling Salesman Problem using heuristic recombination and simulated annealing instead of mutation. Performance improves dramatically. We can consistently get solutions within 5 % of the optimum tour. Now the algorithm runs on a multiprocessor parallel machine in a distributed asynchronous fashion. This version, the Distributed Genetic Algorithm (DGA), employs a "local selection" step that minimizes the sequential overhead. Thus the algorithm is greatly simplified: a collection of identical processes, randomly interacting without central control. Not only is this a better algorithm for parallel implementation but also more natural and simpler. The DGA results in linear speed-up (computing speed proportional to the number of processors) – unusual for non-trivial combinatorial optimization. The quality of solution obtained with DGA is as good as that of GA.

Now we intend to apply DGA to the evolution of dynamic systems. Imagine a population of dynamic systems. Each dynamic system has a set of fixed points, which encode set of some features. We will evolve this population, using local selection, and appropriate recombination and simulated annealing, toward one which responds to the environment in the most efficient manner, i.e., which can represent stimulus input in an optimal way. This task is to evolve the representation, not the configuration of structure. To date, GAs can only find the best configuration given a representation scheme. Ultimately, we envision a DGA which can evolve both configuration and representation.

Environmentally-specified patterns of movement coordination in normal and split-brain subjects

Betty Tuller and J.A.S. Kelso

Center for Complex Systems and Department of Psychology
Florida Atlantic University, Boca Raton, FL 33431

Previous studies have documented that the phasing among components of multidegree of freedom systems (such as the speech apparatus, and the bimanual system) can remain stable across various kinds of parameter changes (e.g., Kelso, Saltzman, & Tuller, *J. Phonetics* **14**, 29-59 (1986)). Yet some phase relationships may be more stable than others. Specifically, Kelso (e.g., *Am. J. Physiol.* **15**, R1000-R1004 (1984)) has shown that subjects, asked to increase the frequency of antisymmetrical finger cycling motions, shift abruptly to an in-phase, symmetrical mode. These two modal patterns correspond to attractors of the collective variable, relative phase, bistable in one parameter regime and monostable beyond a critical frequency. Haken, Kelso, & Bunz (*Biol. Cyber.* **51**, 347-356 (1985)) were able to explain the experimental results using concepts from synergetics and the mathematical tools of nonlinear dynamics. Here we explore the attractor layout further by studying rhythmic movement patterns between the hands in response to environmental signals. First, we examine the mean and variability of actually produced relative phase in unskilled subjects and skilled musicians to determine whether the relative stability of in-phase and antiphase patterns is still apparent when the environmental signals require intermediate phase values. Second, split-brain studies suggest that if lights determining each finger's timing are projected to separate cerebral hemispheres then (because control of finger movements appears to be strongly lateralized) the movement patterns produced by commissurotomy patients should be solely a function of this environmental information. Alternatively, an intact corpus callosum may be necessary for a subject to use environmentally-specified timing information to modify the system's intrinsic dynamics.

The results are unequivocal. Only two phase-locked states–in-phase and antiphase–are shown to be stable for all subjects irrespective of musical skill. Split-brain subjects show an even greater predominance of these patterns, supporting the notion that reduced cortical interaction between the hemispheres actually diminishes the ability to modify behavior on the basis of environmental information. Moreover, differences in trajectories produced by normal individuals and those without an intact corpus callosum are remarkable. The resultant patterns of coordination in normal and split-brain subjects allow us to generalize previous findings of stable phase relationships in bimanual movements to environmentally-specified movement patterns. A recent theory (Schöner & Kelso, *Biol. Cyber.*, in press a,b) explains these results at both the level of the collective variable dynamics and the individual oscillatory components. Key ideas concern whether environmental information cooperates or competes with the intrinsic dynamics (corresponding to the two stable behavioral patterns). Environmentally-specified patterns of coordination are shown to arise in a self-organized fashion from a cooperative coupling of the components and their interaction with the environment.

Work supported by NINCDS, NIMH and ONR.

List of Participants

Frederick Abraham
Blueberry Brain Institute
Gregg Hall
Waterbury Center, VT 05677

John Bendler
General Electric
Corporate Research & Development
Post Office Box 8
Schenectady, NY 12301

Vinod Chandra
Florida Atlantic University
Electrical Engineering
Boca Raton, FL 33431

* Avis H. Cohen
Cornell University
Div. Biological Science
Seeley G. Mudd Hall
Ithaca, NY 14853-2702

Gonzalo C. deGuzman
Florida Atlantic University
Center for Complex Systems
Boca Raton, FL 33431

Max Dresden
SUNY
Institute for Theoretical
Physics
Stony Brook, NY 11794-3840

D. Emmerick
University of Illinois
Freer Hall
906 S. Goodwin Avenue
Urbana, IL 61801

Peter Beek
The Free University
Post Office Box 7161
Amsterdam, The Netherlands

G.O. Cai
Florida Atlantic University
Center for Applied
Stochastic Research
Boca Raton, FL 33431

Dante Chialvo
SUNY
Health Service Center
766 Irving Avenue
Syracuse, NY 13210

James Dammann
Florida Atlantic University
Electrical Engineering
Boca Raton, FL 33431

Vinod D. Deshmukh
University of Florida
University Hospital
655 West 8th Street
Jacksonville, FL 32209

Minh Duong-Van
University of California
L. Livermore National Lab
Livermore, CA 94550

* Doyne Farmer
Los Alamos National Lab.
B 258
Los Alamos, NM 87545

E. B. Fernandez
Florida Atlantic University
Electrical Engineering
Boca Raton, FL 33431

Alan Fogel
Purdue University
Child Development &
Family Studies
West Lafayette, IN 47907

* Ary Goldberger
Harvard Medical School
Beth Israel Hospital
330 Brookline Avenue
Boston, MA 02215

Howard Hock
Florida Atlantic University
Psychology Department
Boca Raton, FL 33431

Wangye Jiang
Florida Atlantic University
Center for Complex Systems
Florida Atlantic University
Boca Raton, FL 33431

R. M. Kiehn
University of Houston
Physics Department
Houston, TX 77004

Stephen Koslow
Neuroscience Branch, NIMH
Room 11-105
5600 Fishers Lane
Rockville, MD 20857

Peter Kugler
University of Illinois
Department of Kinesiology
Urbana, IL 61801

Y. K. Lin
Florida Atlantic University
Center for Applied Stochastic
Research
Boca Raton, FL 33431

Thomas Fine
Medical College of Ohio
CS #10008
Toledo, OH 43699

* Ivar Giaever
General Electric
Research & Development Center
P.O. Box 8, Building K1, 3C334
Schenectady, NY 12301

* Stephen Grossberg
Center for Adaptive Systems
Boston University
111 Cummington St., 2nd Floor
Boston, MA 02215

John Jeka
Florida Atlantic University
Center for Complex Systems
Boca Raton, FL 33431

J.A.S. Kelso
Center for Complex
Florida Atlantic University
Boca Raton, FL 33431

* Nancy Kopell
Boston University
Mathematics Department
111 Cummington St.
Boston, MA 02215

Ron Kostoff
Office of Naval Research
Technical Assessment
Code 10PA, 800 N. Quincy St.
Arlington, VA 22217

* Rolf Landauer
IBM
T.J. Watson Research Center
Post Office Box 218
Yorktown Heights, NY 10598

Donald Lloyd
Loats Associates, Inc.
24 North Court St.
Westminster, MD 21157

Arnold Mandell
University California, San Diego
Biological Dynamics and Theoretical Medicine
M-003
La Jolla, CA 92093

John W. Merrill
Indiana University
Department of Linguistics
Lindley Hall
Bloomington, IN 47405

Donald C. Michaels
SUNY
Health Service Center
766 Irving Avenue
Syracuse, NY 13210

* George J. Mpitsos
Oregon State University
Mark O. Hatfield Marine Science Center
2030 Marine Science Drive
Newport, OR 97365-5296

* Norman H. Packard
University of Illinois Champaign-Urbana
Center for Complex Systems Research
508 South Sixth Street
Champaign, IL 61820

Robert Port
Indiana University
Department of Linguistics
Bloomington, IN 47405

* Paul E. Rapp
Medical College of Pennsylvania
Physiology & Biochemistry
3300 Henry Avenue
Philadelphia, PA 19129

* Benoit B. Mandelbrot
IBM Research Center
Post Office Box 218
Yorktown Heights, NY 10598

Ehud Meron
Columbia University
Astronomy Dept.
538 West 120 Street
New York, NY 10027

Janet Metcalfe
Indiana University
Psychology Department
Bloomington, IN 47405

Roger Minich
L. Livermore National Lab
University of California
Livermore, CA 94550

Allan Nash
Florida Atlantic University
Psychology Department
Boca Raton, FL 33431

Oreste Piro
James Franck Institute
University of Chicago
5640 South Ellis Avenue
Chicago, Il 60637

* Itamar Procaccia
The Weizmann Institute
Chemical Physics Dept.
Rehovot, 76 100 Israel

David Rigney
Harvard Medical School
25 Shattuck Street
Boston, MA 02115

Mary Ann Roberton
University of Wisconsin-Madison
310 Lathrop Hall
Dept. Physical Education
Madison, WI 53706

* Robert Rosen
Dalhousie University
Physiology & Biophysics
Halifax, NS, Canada B3H 4H7

* Otto Rössler
University of Tubingen
Physics & Theoretical Chemistry
D-7400 Tubingen 1
Federal Republic of Germany

Zvi Roth
Florida Atlantic University
College of Engineering
Boca Raton, FL 33431

Elliot L. Saltzman
Haskins Laboratories
270 Crown Street
New Haven, CT 06510

Jeffrey Saven
New University
5700 N. Trail, Box 14
Sarasota, FL 32423

Ron Schoenfeld
Neuroscience Branch, NIMH
Room 11-105
5600 Fisher Lane
Rockville, MD 20857

* Gregor Schoner
Florida Atlantic University
Center for Complex Systems
Boca Raton, FL 33431

* Allen I. Selverston
University of California
Department of Biology
B-022
La Jolla, CA 92093

Ravi Shankar
Florida Atlantic University
Electrical Engineering
Boca Raton, FL 33431

* Hiroshi Shimizu
University of Tokyo
Faculty Pharmaceutical Science
Bunkyo-ku, Tokyo 113
Japan

R. Sudhakar
Florida Atlantic University
Electrical Engineering
Boca Raton, FL 33431

Jung U. Suh
Indiana University
Department of Computer Science
Bloomington, IN 47405

* Harry L. Swinney
University of Texas at Austin
Department of Physics
Center for Nonlinear Dynamics
Austin, TX 78712-9990

* Harold Szu
Office of Naval Research
Code 5709
Naval Research Laboratory
Washington, DC 20375-5000

Betty Tuller
Florida Atlantic University
Center for Complex Systems
Boca Raton, FL 33431

Daniel A. Vasco
University of Texas at Austin
Department of Zoology
Austin, TX 78712

* William Warren
Brown University
Department of Psychology
Providence, RI 02912

* Bruce J. West
University of California-SD
Institute for Nonlinear Science
La Jolla, CA 92093

Yan Yong
Florida Atlantic University
Center for Applied
Stochastic Research
Boca Raton, FL 33431

* Esther Thelen
Indiana University
Department of Psychology
Bloomington, IN 47405

* Michael T. Turvey
Haskins Laboratories
270 Crown Street
New Haven, CT 06510

Stephen Wallace
University of Colorado
Dept. of Kinesiology
Boulder, CO

Rolf Wåsen
Metoflex A.B.
Alpheddan
S-59041 Remforsa, Sweden

Yoko Yamaguchi
University of Tokyo
Faculty of Pharmaceutical Science
Bunkyoku Tokyo 113, Japan

* Invited Speaker

This work relates to Department of Navy Grant N00014-J-1012, issued by the Office of Naval Research. The United States Government has a royalty-free license throughout the world in all copyrightable material contained herein.

www.ingramcontent.com/pod-product-compliance
Lightning Source LLC
LaVergne TN
LVHW071223011026
849280LV00007B/35
9789971504854